Forty-Five Years of Research at the NBRL,

BOSTON, MASSACHUSETTS

Forty-Five Years of Research at the NBRL, BOSTON, MASSACHUSETTS

CHAOTIC OBSERVATIONS, SERENDIPITY, AND PATIENCE

C. Robert Valeri, MD, and Gina Ragno Giorgio, Esq.

ISBN: 978-1-4834-0277-2 (e)
ISBN: 978-1-4834-0276-5 (sc)
ISBN: 978-1-4834-0278-9 (hc)

Library of Congress Control Number: 2013913184

Lulu Publishing Services rev. date: 8/26/2013

TABLE OF CONTENTS

AUTHORS AND ACKNOWLEDGEMENTS

C. Robert Valeri, M.D. attended Tufts College where he was elected to the Phi Beta Kappa\Society and he graduated Summa Cum Laude, earning his Bachelor of Science degree in 1954. He attended Harvard Medical School and received his Medical Degree in 1958. He performed his internship and residency in internal medicine at Boston City Hospital and the New England Medical Center and a fellowship in clinical hematology with Dr. W.C. Moloney and Dr. Jane Desforges at Boston City Hospital. Dr. Valeri is board-certified in internal medicine and a fellow of the American College of Physicians (FACP).

Following his residency training, Dr. Valeri served in the Medical Corps of the United States Navy from 1962 to 1985 when he retired as CAPT, MC, USN. Dr. Valeri has been the Director of the Naval Blood Research Laboratory from 1965 to 2010. The laboratory has been involved in several areas of research primarily designed to optimize the treatment of military and civilian casualties. Dr. Valeri has authored more than 500 scientific articles and 4 books reporting the results of the research performed at the NBRL.

Gina Ragno Giorgio, Esq. is a magna cum laude graduate with a Bachelor of Arts in Philosophy from Boston University; class valedictorian graduate cum laude and law journal executive articles editor of Southern New England School of Law, North Dartmouth, MA; and co-author of 70 peer-reviewed scientific publications.

Acknowledgements: The authors acknowledge the secretarial assistance of Ms. Janet Clegg and the editorial assistance of Ms. Cynthia Ann Valeri. The authors acknowledge the support of the Office of Naval Research (ONR) especially that of Dr. Michael Given to publish this book.

PREFACE

In this volume, Dr. Robert Valeri presents a retrospective of his scientific contributions that were published in 534 peer-reviewed journal articles, 4 books and 6 monographs, describing his 45-year career at the Naval Blood Research Laboratory. He offers readers a personal perspective of how his many contributions to the science of transfusion medicine came about, adding an additional 407 pages and 12 chapters to his scientific *oeuvre.* This brief introductory Preface can hardly capture the permanent importance and creativity of Dr. Robert Valeri's voluminous scientific contributions. History will make that determination, which I know will be positive and glowing. In the interim, I offer the following perspective of these contributions by an innovative scientist and admired colleague whose impact on medicine in the United States will be significant and long lasting.

In Chapter One, Dr. Valeri shares with readers his quest for an appropriate title for this memoir. After considerable deliberation, he settled on the title "The 45 Years of Research at the NBRL, Boston Massachusetts: Chaotic Observations, *Serendipity*, and Patience" (italics added). Several times in the text, Dr. Valeri returns to the theme of serendipity. He attributes to "serendipitous observations" the sparks that resulted in his developing methods for freezing red blood cells in a –80C mechanical freezer, for freezing platelets using 6% DMSO in -80 C mechanical freezers, and for a series of studies that are the basis for contemporary protocols for transfusion support for military and civilian traumatic injuries. He considers it serendipitous that his first assignment after volunteering to join the United States Navy in 1962 was at the Blood Research Laboratory at the Chelsea Naval Hospital where between 1968 to 1974 servicemen returning from the Vietnam War required intensive and meticulous care for combat injuries. His first-hand experiences with these casualties alerted him to the inadequacy of methods for obtaining basic physiological measurements, such as central blood volume in patients who were chronically hypovolemic and anemic. He considers it serendipity that as a result of this chance encounter, he was soon immersed in the basic pathophysiology of acute blood loss and the quest for optimal management of transfusing blood components.

Dr. Valeri's choice of the word "serendipity" for inclusion in the title and for repeated reference in the text is, to a certain degree, appropriate. He quotes a personal hero, Niccolo Machiavelli, whose Prince advocated that it was "better to be lucky than smart." However, even if we accept Dr. Valeri's attribution to serendipity for his opportunities to initiate his research, the conclusion is incomplete. According to an ancient Persian tale, there were three princes of Serendip (Sri Lanka) who had the good fortune of repeatedly discovering treasure by accident. While seeking one treasure, they happened by chance on another treasure of even greater value. In the case of Dr. Robert Valeri, the well-known quotation from Louis Pasteur's 1854 lecture at the University of Lillie is more suitable, "In fields of observation, chance favors only the prepared mind." Dr. Valeri's mind was prepared during his medical education at Harvard Medical School and his post-graduate training at the Boston City Hospital and the New England Medical Center. Among his mentors in clinical medicine were the world-renown clinicians, Drs. Sam Proger and Ted Astwood (New England Medical Center), Francis Moore (Peter Bent Brigham Hospital) and Drs. William C. Moloney and Jane Desforges in hematology (Boston City Hospital). Pasteur's observation can be applied aptly for many advances and discoveries in clinical medicine, in general, but none more appropriately than for those advances and discoveries that emerged during Dr. Robert Valeri's 45-year tenure at the Naval Blood Research Laboratory.

Like the contributions to science by Louis Pasteur, those by Dr. Robert Valeri reflect the creativity of an informed and critical observer and innovator. As I read this memoir, I am constantly reminded of our contemporary emphasis on "translational research," e.g., clinical research that is founded on basic science, but conducted with the goal of translating basic observations into practical applications for improving health and well-being. In my reading of this volume, I did not encounter the words "translational research." Much of Dr. Valeri's work was 45-years before its time. It was – nearly all of it – what we would today call translational research. As readers prepare to engage themselves in this exciting, thoughtful and informative memoir, I encourage them to keep track of the many practical and lasting contributions to American medicine that emerged from Dr. Robert Valeri's creativity and from Boston's Naval Blood Research Laboratory. It is a very impressive record.

S. Gerald Sandler, MD, FACP, FCAP
Professor of Medicine and Pathology
Georgetown University Medical Center
Washington, DC

CHAPTER 1

GENERAL COMMENTS

I have threatened to write this book over the past 5 decades. I've had a wonderful life as a physician and a scientist. I have had the privilege of working in exciting times and to be able to contribute to many different aspects of contemporary medical science. I want to share some of my thoughts and experiences with colleagues and students. The following is the story of my thought process and responses to chaotic and serendipitous events. There are selected references listed in the general commentary whereas the chapters provide numerous references to support these comments. This book describes a personal journey. I hope the reader enjoys reading about it as much as I have enjoyed traveling it.

Various titles for this book have been considered – "Only the Names Change"; "You Cannot Escape Your Upbringing"; "Tachyphylaxis" and "Music and Lyrics – revisited". I have found that the management of people has been the most difficult thing for me to do. In fact, John Kimbell, former CEO of Fenwal Laboratories, said that I was the worst manager that he observed over a 15-year period. I thought I was a good manager who gave collaborators and employees freedom and unlimited opportunities. It became apparent that most people are only interested in financial and self-rewards and only few are interested in learning, thinking, and using information. The behavior of the 400 or more individuals that I was responsible to interact with demonstrated that self-interest and monetary reward were the major motivations of these individuals. The idea that individuals are interested in knowledge for the satisfaction that knowledge provides without financial and self-reward is foreign to 95% of the people that I observed over the past 5 decades. Are you listening? Do you understand? Are you tone deaf from "Music and Lyrics"? Dr. Sidney Burwell stated that 50% of the information reported by the lecturer is not correct – unfortunately the lecturer does not know what 50% is correct! Dean George Packer Berry at the Harvard Medical School reported that you can train a monkey, dog, cat and medical students to perform specific assignments. Unfortunately you cannot train a physician to think!

Dr. Mark David Altschule reported that the published peer-reviewed papers should provide detailed material and method and results sections. The introduction and discussion sections of the paper will change over time when new information is provided to properly interpret the results reported in the paper. As new information is provided, the data reported in peer-reviewed published papers can be properly interpreted.

In the movie Music and Lyrics – Hugh Grant (Alex Fletcher) and Drew Barrymore (Sophie Fischer) report that lyrics are as important as the music. In a similar manner, the words are critical to present the significance of the science that is reported.

Hypothesis driven research demands that the investigators find what they are looking for and many fail to recognize the elephant sitting in the middle of the room or chaotic observations that occur but are not investigated.

Are you listening? Do you understand? Fifty percent of what you hear and 50% of what you read is not correct. Unfortunately, the speakers and the authors do not know what 50% is correct. The listener and the reader must be able to determine the 50% of the spoken information and 50% of the written information which are correct. If the information doesn't sound right, then the portion of the lecture and the written document should be questioned. Eventually the portion of the spoken and written information will be reevaluated and the misinformation will be identified. As David Danon told me in 1965, the word "research" means to search again in order to discover the significance of the observation. My experiences over the past 5 decades have provided me with the following opinion: I am an advocate of the book "The Prince" by Niccolo Machiavilli, who reported that it is better to be lucky than smart; the behavior of an individual is good until for self-interest the person needs to do dishonest things to protect his or her self-interest; and behavior is controlled by the fear that bad behavior may be punished. Fear or threat of potential punishment may control the behavior of the people in a society. The fear of the effect of religious consequence of bad behavior is now replaced by litigation and lawyers. Lawyers and potential litigation now attempt to control the behavior of individuals.

Clinical medicine has been replaced by molecular biology. Molecular biologists measure molecular events but have failed thus far to relate these molecular events to function. Studies of structure and function relationships are needed to obtain insights into the complex biological events.

Usually males, unlike females, do not think about details and outcome with regard responses to specific issues. Females usually respond to events in thoughtful and careful manner, whereas the response of males is passive and usually not carefully thought out with regards the consequences of their actions. The second X chromosome in the female is the controlling chromosome, i.e. female XX and male XY sex chromosome.

What is the question? Who am I and what am I going to do about it? The person is able to do whatever he or she wants to do, the only problem is that most individuals do not know what they want to do. Education is from the Latin word "educere" which means to pull out of you what is within you. The genetic code of the individuals is subjected to transcription, i.e. your genotype is expressed into your phenotype by the environmental events that you are exposed to. The transcription of your genotype to produce your phenotype usually identifies who you are. Are you selected or are you elected to your final endeavor? Selection is due to someone who identifies your potential which you may not be aware of. Very few people know when they are 10 years old that they want to be President of the United States like Bill Clinton.

One over zero is infinity – until recently you were born with unlimited freedom and opportunity. The Constitution of the United States guarantees life, liberty and the pursuit of happiness to a United States citizen. You are fortunate to have freedom and opportunity to be able to select and to respond to opportunities which utilized your innate ability. You are unable to control your destiny. You must be able to respond in an active manner to chaotic and serendipitous events to achieve a specific goal which requires patience. You must recognize "windows of opportunities" and react to them appropriately. Your ability to make adjustments each day indicates that you will be responsible for making mistakes. Response indicates you will be responsible – no response means that you will not be responsible. Most individuals do not respond to oral or written requests because they do not want to be responsible or accountable for their responses.

Individuals must be able to recognize "windows of opportunities" which occur as chaotic events and to respond to them appropriately. Knowledge for the sake of knowledge provides little reward for the majority of people that I have observed. Dr. Shukri Khuri was a surgeon who was interested in knowledge without a major consideration that financial benefit would be achieved by the knowledge that he obtained. Numerous collaborations between the Naval Blood Research Laboratory and the West Roxbury Veterans Hospital between 1982 to 2008 were done to assess the bleeding diathesis that occurs in patients subjected to extracorporeal bypass surgery. The scientific information that was obtained was utilized solely to improve the care of sick patients.

The behavior of personnel that I was associated with over the past 5 decades was influenced in part by their early upbringing. During the Vietnam War between 1962 to 1974 enlisted corpsmen assigned to the Naval Blood Research Laboratory reflected the influence of the pediatrician, Dr. Spock, who wrote that a baby has the ability to tell right from wrong. Many of the enlisted corpsmen assigned to the Naval Blood Research Laboratory had no conscience, they did not know what was right and what was wrong. Behavior for the Spock children was predicated on what the child wanted. Right was what they wanted to do. Wrong was what they did not want to do. There was not concept of rights and responsibilities. These individuals only had rights with no responsibility. There was no concept that certain behavior was not appropriate and anything that they wanted to do was correct and anything they did not want to do was also correct. Their behavior influenced by Dr. Spock supported the concept that individual could not escape their upbringing which meant they had rights without responsibility.

My stupidity permitted individuals to do what they wanted to do; to praise them for whatever they did and to financially compensate them way above their performance in an attempt to stimulate their involvement in the projects that they were assigned. I overpaid them, over-graded them, and over-evaluated their marginal performance in an attempt to get them committed to their work assignment which was related to collection, processing of blood products and transfusion of healthy volunteers, patients and baboons with blood products. My supervision of these individuals was daily review of their work which was associated with high risk to the healthy volunteers, patients, and baboons. The performance of these individuals was always predictable if they were not supervised daily. If not carefully monitored these employees would cut corners to reduce the work required to carefully execute the assignments which they were responsible to perform correctly. The research on the normal volunteer, patients, and baboons treated with blood products allowed for no errors. The title of the management section of this book was "Only the names changed" because the behavior of the individuals that I carefully and daily supervised was identical. Close supervision was needed for them all, because if I was not present these individuals would reduce any work requirement that was needed to accomplish the assigned project. "Only the names changed" was the description of the behavior of the individuals that I supervised. Their behavior was similar, indifferent and not interested in the work, only how many hours could they avoid to work and how much money did they earn, how much money did the other employees make, where was the supervisor and if he was not present, how could they leave their assignment. This behavior was identical for the majority of the employees – only the names changed but their behavior and performance were identical.

As a government employee for 23 years and 3 months, I became very aware of the 2-year rotation for the military assigned to the Naval Blood Research Laboratory. This rotation period is required because of the predictable behavior of personnel in the same job for longer than 2 years. The best way to characterize the behavior is by tachyphylaxis, that is nonresponsiveness of the individual to the environment and the assigned responsibilities. Personnel need to be replaced every 2 years in order for them not to become nonresponsive and not involved with their assignment. Individuals who are assigned to the same job for two (2) years usually become nonresponsive to the environment and to their assignment.

A major concern with a yearly supported research facility is the need to obtain funds for the research. This requirement for yearly funding by the U.S. Government provides no guarantee that government employees working at a government research facility will have funds to pay their salaries. There is insecurity for government employees working in a research laboratory funded by the U.S. Government.

The maintenance of a work force to perform the research outlined in proposals submitted for funding is the major responsibility of the principal investigator who has submitted the proposal for funding.

In addition, there is a need for proper design of studies which must consider biostatistics, alpha and beta errors, the number of studies to be done; the prospective randomized blinded and crossover studies; the outcome to demonstrate equivalence or nonequivalence; the risk-benefit effects, and the need for multi-center studies to obtain data to support the safety and therapeutic effectiveness of the treatment. At the Naval Blood Research Laboratory the procedures to collect and preserve blood products required that anyone should be able to collect and prepare the blood products using standard operating procedures that were provided. In fact, the major breakthrough in the processing of frozen RBC, frozen platelets, and frozen adult stem cells was provided by visitors who came to Naval Blood Research Laboratory to learn how to freeze RBC, freeze platelets, and freeze pluripotential mononuclear adult stem cells. The visitor asked a) why the supernatant glycerol was frozen with the glycerolized RBC; b) why the supernatant DMSO was frozen with the DMSO-treated platelets; and c) why the supernatant DMSO was frozen with the DMSO-treated pluripotential mononuclear adult stem cells. These questions raised by the visitors were responded to by the removal of the supernatant glycerol from the glycerolized RBC and removal of the supernatant DMSO from the DMSO treated platelets and the removal of the supernatant DMSO from the DMSO treated pluripotential mononuclear adult stem cells prior to freezing in a -80 C mechanical freezer.

The motivation of government employees working in a government supported research laboratory is difficult because the funding to pay their salaries is not guaranteed beyond the year that funds are provided by the U.S. Government. Government employees working in a research laboratory that depends on yearly funds are not happy that their jobs are not guaranteed. The principal investigator is required to prepare research projects that will be approved by the funding agency to obtain the funds which are predicated on the productivity of the work force which consists primarily of disinterested, non-motivated, insecure individuals who are unhappy that they have to work for their salary. The work force would like to be guaranteed their pay with little interest in performing their assignments. The requirement for close supervision of these individuals is great, if they are involved in high risk research performed on healthy volunteers, patients, and baboons.

Northeastern students in the Boston area have a curriculum which provides formal lectures and periods of work in commercial or educational centers to learn how to use the information that they are learning. The Office of Human Resources at Boston University Medical Center (BUMC) from 1965 to 1995 was responsible to recruit work study students from different universities in the Boston area which included Northeastern and other colleges. Formal education does not teach common sense and how information is to be used in daily living – whether in a work environment or non-work environment. The formal education only provides the student with the opportunity to learn how to prioritize and to regurgitate the information provided by the instructor. The student learns how to utilize his or her time related to their assignments. The success of an individual is related to his or her ability to use information to respond to the daily events that occur in their life. Knowledge needs to be used to respond to the daily unpredictable chaotic events that occur. The major issues related to survival in the work place – one must provide service and one must communicate. Communication is done by oral presentation of the issue and by written description of the issues that need to be addressed. The most effective way to communicate is by the written memoranda which discuss the issues and outlines possible solutions like pros and cons presented in the movie Music and Lyrics by Alex Fletcher (Hugh Grant) and his manager. The memoranda should be written to all parties involved with the issue. In 1982 the research performed at the NBRL was site visited by three individuals who were requested to prepare a report on the quality of the research, comment on the results reported; and suggest that the funding agency either continue to support or reject further research. In 1982 the peer review committee was comprised of Dr. Eugene Cronkite, Retired Admiral, MC, USN and senior scientific investigator at the Brookhaven Laboratory located on Long Island, New York; Dr. Tibor Greenwalt, Editor of the Journal Transfusion and Dr. Nathaniel Shulman, Senior Investigator at the National Institute of health (NIH). The NBRL from 1965 to 1985 would have site visits every 2 -3 years for independent evaluation of the research that was being performed. Two (2) important issues were raised at this site visit in 1982. In the 1982 site visit, Dr. Nathaniel Shulman was concerned on how all the projects were documented and monitored. This concern by Dr. Nathaniel Shulman stimulated my writing a daily memorandum which reported all the activities, the interpretation of the results of ongoing studies, the modifications of the studies that were needed, and the assignment of personnel to specific projects. The memoranda were sent to all personnel for information and responses. This approach to write daily memoranda to addresses issues that needed to be resolved existed at the NBRL following the site visit in 1982 to 2008. What I learned from this experience was the meaning of the word "responsibility". I learned over the past 26 years from 1982 to 2008 that the failure to respond to memoranda meant that the persons who received the memoranda which focused on specific issues did not respond because they did not want to be responsible. No responses to the oral request which is usual or the written memoranda which is guaranteed meant that the persons involved with the issue did not want to be responsible for what they said or wrote regarding the issue raised by the oral request or written memoranda. The first lesson learned at law school is that the lawyer never documents any issue. The lawyer at law school is responsible to attend a course on ethics. This course on ethics is to document only "billable hours" with no documentation as to what the billable hours are related to. Lawyers are only interested in billable hours and that the litigation should not be completed quickly which will reduce or eliminate their billable hours and payment for their legal work. Billable hours are usually provided by lawyers without documentation of what was done for the client. Lawyers usually communicate orally but prefer not to respond in writing to a request by their clients because they do not want to be responsible for their actions.

At the 1982 site visit Dr. Eugene Cronkite, the Chairman of the site visit group, reported that the committee had reviewed the research proposal, progress reports, and the publications, but Dr. Cronkite and members of his site committee requested copies of the pre-prints of manuscripts that NBRL would be submitting to peer reviewed journals for publication.

The 1982 site visit focused my attention on documentation of all events that occurred at the NBRL: the administrative issues, the scientific issues, the review of the data, the establishment of standard operating procedures, the need for daily control of measurements, the independent collation and tabulation of the data by individuals who did not perform the measurements, and independent scientific review of the data.

For the period from 1962 to 1965 oral reports were provided to individuals on the severe adverse events (SAEs) observed at the Chelsea Naval Hospital following the transfusion of previously frozen deglycerolized RBC with the occurrence of acute renal failure requiring peritoneal and hemodialysis of the recipients. Whenever this topic was discussed, individuals in attendance would always ask "have you published this information". After repeated questions raised at several meetings I decided that peer reviewed publications would be far superior to publishing abstracts and presenting talks at meetings. The oral presentations were not effective in providing scientific data related to the funded research. This early experience established the requirement that abstracts, oral presentations, chapters and review articles would not be "the goal" of the NBRL. The major activity of the NBRL would be to publish peer reviewed papers with the expectation that the peer reviewed papers would be read!

During the past 45 years, in excess of 500 peer reviewed papers have been published. The peer reviewed papers are extremely difficult to publish, especially when the published data are novel and against the current commercial interest of companies and the financial interest of the status quo.

Publish or perish is the quote that is a given for survival in performing research which is supported by yearly contracts obtained from government and commercial sources. Research funds provided by the government usually require yearly funding and review. The only way to document the quality of your research is to publish the information in peer reviewed journals. The NBRL has been funded from 1965 to 2010, for 45 years, by obtaining government and commercial funds amounting to in excess of $120,000,000 dollars. These funds are justified by the FDA approved blood products that have been provided for use by both military and civilian communities. The requirements for preparing proposals, reports, and peer reviewed publications were the responsibility of Dr. C. Robert Valeri who was assigned to the NBRL which was initially located at Chelsea Naval Hospital and then at Boston University Medical Center as a government owned laboratory operated by a contract with BUMC; a contract between NBRL and the New England Medical Research Institute (NEMRI) at West Roxbury Veterans Administration Hospital; and the NBRL, Inc., a non-profit research laboratory.

The scientific productivity of the NBRL is responsible in part for the significant progress made in Transfusion Medicine that has occurred over the past 45 years. Review of the NBRL accomplishments must be compared to that of other government and non-government supported research laboratories with regards funds provided and the blood products provided and the scientific information published.

The cost for the support of the NBRL and FDA products produced and scientific publications reported in Transfusion Medicine journals document that this government supported laboratory has been one of the most productive government laboratories over the past 45 years.

The recent motion picture Music and Lyrics provide the important message that good scientific observations (Music) need the clearly written papers (Lyrics) to communicate the importance of the science.

The time required to publish a peer-reviewed report is related to "X" – the time to perform the research and collect the data, "X2", the time required to analyze the data, and "X3" the time required to write the paper, revise the paper, and publish the paper in a peer-reviewed journal. The NBRL has been able to reinvent itself over the past 5 decades by the utilization of new technology to address the important issues related to the collection and preservation of blood products, blood substitutes, resuscitation solutions, and hemostatic agents to treat the military and the civilian communities. The issue of "tachyphylaxis" has not infected our research facility which has been able to respond to the important issues related to the safety and therapeutic effectiveness of blood products and treatments for the military and civilian communities. The extensive peer reviewed publications have demonstrated the ability of this government supported research facility to recognize and use new information and new technology to resolve the important issues related to the severe adverse events (SAEs) associated with the use of blood products and to establish the criteria to document the safety and therapeutic effectiveness of blood products, blood substitutes, resuscitation solutions, hemostatic agents, and maintenance of normal temperature of the recipient and the maintenance of normal temperature of the blood products that are infused.

The motivation to perform the research to obtain knowledge is not the current goal of most investigators. Research is currently driven by the need to patent the findings for personal reward and reward for the funding agency whether supported by commercial, academic institutions, or governmental sources. The fundamental problem with patenting research is that the patent may prevent making new observations that may conflict with what you have patented. Patents restrict your freedom to observe and interpret chaotic and serendipitous observations that do not support the patent.

In my experience I never wanted to patent procedures because patents prevent any further development of procedures or concepts that have been patented. The brain may become fixed without any attempt to alter the procedure that was patented. A patent may produce an effect on new thought process equivalent to fixing the brain in formaldehyde.

Fiscal support of a research facility by annual funds provided by the government requires yearly review. Funds to support projects for 3 to 5 years may occur by grants. These funds are provided to support the hypothesis of the investigators. The NIH funding requires that the investigator provides a hypothesis and outlines an approach to support the hypothesis. The investigators are looking for results to support their hypothesis. Hypothesis directed research will not permit the investigation of serendipitous observations that usually result in significant scientific contributions. Serendipitous observations are usually rejected by the investigators who need to provide the information to support their hypothesis. Hypothesis driven research yields results that the investigators are looking for and fail to observe and recognize the chaotic-serendipitous observations unrelated to the hypothesis that the investigators have proposed and the funding agency has supported.

My experience was related to working in the U.S. Navy as a Medical Officer for 23 years and as the scientific director of the NBRL, a government owned and contractor operated laboratory for 20 years to provide safe and therapeutic effective blood products and methods to treat military and civilian casualties. The research was supported by Naval Medical Research and Development Command (NMRDC) of the U.S. Navy's Bureau of Medicine and Surgery, the Office of Naval Research, by the U.S. Congress and by commercial funds. Yearly funds were requested over the 45 year period – which required yearly proposals, progress reports, site visits by independent investigators, and the need to document that funds provided were used to obtain information on the safety and therapeutic effectiveness of blood products, blood substitutes, resuscitation fluids, and hemostatic agents to treat military and civilian casualties.

The success of the research accomplishments of the NBRL from 1965 to 2010 was related to the Vietnam War and the decision by the U.S. Navy to support the preservation of blood products using freezing technology. The accomplishments of the NBRL were possible because from 1965 to 1974, wounded servicemen who returned from South Vietnam were hospitalized at the Chelsea Naval Hospital. These wounded Marines and Naval servicemen were air-evacuated to their U.S. Naval Hospitals close to their homes. These wounded servicemen provided the stimulus for a multiple of studies to assess the optimum method to treat them at Chelsea Naval Hospital and at Danang, South Vietnam using frozen RBC. The research performed was done at a time that there were no institutional review boards (IRB) and no signed informed consents to review and approve the testing and treatment that produced excellent clinical outcome. Two books were written "Blood Banking and the Use of Frozen Blood Products" in 1976 and the other book in 1981 "Hypovolemic Anemia of Trauma – The Missing Blood Syndrome" to report the studies that were performed and the results that were obtained. During the Vietnam War, the first frozen blood bank was deployed at Danang, South Vietnam to provide frozen red blood cells to treat wounded servicemen. The numerous publications provided by the NBRL, an in-house Navy laboratory, supported by the Naval Medical Research and Development Command (NMRDC) of the U.S. Navy's Bureau of Medicine and Surgery and the Office of Naval Research (ONR) were related to the Vietnam War and the need to provide resuscitation fluids, blood products, and surgical and medical care to treat the wounded servicemen.

The U.S. Army supported research to provide fresh whole blood obtained from the troops deployed in South Vietnam and liquid preserved whole blood stored at 4 C; whereas the U.S. Navy supported research to provide frozen RBC to treat wounded servicemen to supplement the liquid method to preserve whole blood. Between 1968 to 1974 the transmission of hepatitis B virus and hepatitis non-A and non-B, subsequently reported as hepatitis C, were the major risks associated with the transfusion of fresh whole blood collected from the troops in combat zones in South Vietnam, i.e. "the walking blood bank".

The freezing of red blood cells with glycerol was thought to reduce the transmission of hepatitis B and hepatitis non-A and non-B by the transmembrane washing of the red blood cells to remove the glycerol prior to transfusion.

The study of methods to preserve red blood cells by freezing procedures permitted the study of autologous RBC in volunteers and allogeneic compatible but identifiable RBC in patients. These studies permitted the assessment of the preservation injury that produced irreversibly damaged RBC that were removed from the circulation and reversibly damaged RBC that circulated and were recovered following transfusion by differential agglutination procedures. The reversibly damaged red blood cells were recovered from the circulation to assess the in vivo restoration of these red blood cells by the following measurements: RBC morphology, RBC deformability, RBC filterability, RBC ATP, DPG, and P50 levels; RBC sodium and potassium levels; RBC coated with B1C globulin; RBC osmotic fragility; and RBC density distribution. The reversible defects in RBC were identified by the recovery of the compatible allogeneic RBC that circulated following transfusion.

In 1965, the method to measure the quantity of nonviable irreversibly damaged red blood cells was debated. Dr. Max Strumia, Chairman of the National Research Council (NRC) introduced the need to quantitate the nonviable preserved RBC by using a double label procedure. One label to detect the preserved RBC and the other label to measure the red blood cell volume of the recipient.

Investigators using the single label procedure have recommended that the extrapolation of the 51Cr radioactivity associated with the preserved red blood cells during the 5, 15, 30 minute post-infusion period to zero time following infusion would provide an accurate estimate of the quantity of nonviable irreversibly damaged RBC that are removed at an accelerated rate during the 24 hour posttransfusion period.

The debate over the single label and double label procedures to assess the quantity of nonviable irreversibly damaged RBC has never been resolved. The NBRL has published numerous papers on the 24-hour posttransfusion survival and lifespan of preserved RBC in the liquid state and frozen state. The second major area of investigation at the NBRL has been the function of preserved RBC in the liquid state and frozen state. This issue was reported in 1954 by Valtis and Kennedy who documented that RBC stored in acid citrate dextrose anticoagulant in the liquid state at 4 C for one week had an increased affinity for oxygen.

The International Committee to Standardize Methods to assess the quality of RBC and platelets has ignored the numerous papers published by the NBRL with regards to the methods used to document that preserved RBC and preserved platelets must circulate and function immediately or shortly after transfusion.

The controversy related to importance of the function of preserved RBC has been debated to a greater degree than the single and double methods to assess the quantity of nonviable RBC. In addition, the adverse effects of the quantity of nonviable compatible red blood cells that are transfused into patients have not been investigated to determine that 25% of the nonviable preserved RBC produce no adverse effects in patients.

The issues related to the survival and function of preserved platelets are identical to those related to the survival and function of preserved RBC. The methods to resuscitate patients subjected to trauma using resuscitation fluids containing Ringer's D,L lactate, Ringer's L lactate, and Ringer's ketone solutions have been investigated. In addition, methods to prepare large volumes of parenteral solutions at the site of need from potable water using membrane technology without heat sterilization have been studied at the NBRL. The parenteral solutions and solutions to deglycerolize red blood cells were made at the NBRL, aboard the LHA USS Saipan ship and at Fort Hood, Texas using membrane technology provided by the Millipore Sterimatics ST-30 instrument.

Over the 45 years of research conducted at the NBRL as an in-house U.S. Navy supported laboratory, as a government-owned and contractor-operated laboratory, and a not-for-profit NBRL research facility, the funding from government sources, commercial sources, and donations have permitted the publication of in excess of 500 peer-reviewed papers on the storage of nonrejuvenated and rejuvenated RBC frozen with 40% W/V glycerol and stored at -80 C for 10 years; and the storage of frozen plasma at -80 C for 7 years. These procedures have been FDA approved.

Extensive data have been published on the freezing of human platelets with 6% DMSO stored at -80 C for at least 2 years, thawed and the washed previously frozen platelets stored in autologous plasma at room temperature (22 C $\pm$ 2 C) without agitation for 5 hours. Recent data have been published to demonstrate that platelets treated with 6% DMSO, can be concentrated to remove the supernatant DMSO prior to freezing and storage in a -80 C mechanical freezer, thawed, diluted and stored in 0.9% sodium chloride or AB plasma at room temperature without agitation for 6 hours. The freezing of platelets treated with 6% DMSO, concentrated to remove the supernatant DMSO prior freezing at -80 C in a mechanical freezer, thawed, diluted with 10 to 20 ml of 0.9% NaCl or a unit of AB plasma and stored at room temperature without agitation for 6 hours prior to transfusion has not been FDA approved.

The technology to freeze human RBC with 40% W/V glycerol with the removal of the supernatant glycerol prior to freezing at -80 C for 10 years, thawed, deglycerolized, and stored in AS-3 at 4 C using the Haemonetics Blood Processor ACP215 instrument; the technology to freeze single donor group O leukoreduced platelets using 6% DMSO, concentrated to remove the supernatant DMSO prior to freezing of the platelets in a -80 C mechanical freezer and group AB plasma frozen at -80 C are now used by the Netherlands military. Frozen group O Rh positive and group O Rh negative RBC, frozen single donor leukoreduced group O platelets concentrated to remove the supernatant DMSO prior to freezing and after thawing resuspended in AB plasma, and frozen AB plasma to treat wounded casualties are now utilized by the Netherlands military. The Netherlands military no longer use non-tested fresh whole blood prior to transfusion to treat military and civilian wounded personnel in the Middle East conflicts. The Netherlands military in combat zones in Afghanistan, Iraq, and Bosnia has reported significant improvement in the survival of wounded casualties who required more than 10 units of red blood cells within 24 hours for resuscitation from 44% to 84% by using frozen blood products at the ratio of 4 units of frozen red blood cells, 3 units of AB plasma, and one unit of single donor group O leukoreduced frozen platelets containing 2.5 to 3.0 X 10^{11} platelets resuspended in a unit of AB plasma with no adverse effects.

In the Middle East conflicts in Iraq and Afghanistan, the fresh whole blood and apheresed platelets used by U.S. Department of Defense (DOD) are not tested for the FDA mandated infectious disease markers prior to transfusion. The technology designed and developed by the NBRL is now available to provide frozen RBC, frozen platelets and frozen plasma all stored in -80 C mechanical freezer to treat military and civilian casualties without the need for fresh whole blood and apheresed platelets which are collected from pre-screened donors and not tested for the FDA mandated infectious disease markers prior to transfusion. The severe adverse events have been reported associated using FDA approved blood products with regards mortality and morbidity associated with transfusion related acute lung injury (TRALI), systemic inflammatory response syndrome (SIRS), posttransfusion infections, myocardial infarction, cerebrovascular insufficiency, and renal insufficiency. The severe adverse events were related to the length of storage of liquid preserved RBC in additive solutions at 4 C for longer than 2 weeks. The FDA is preoccupied with the possible transfusion of infectious diseases associated with the blood products but has failed to recognize the severe adverse events (SAEs) associated with the transfusion of poorly preserved RBC and platelets that do not survive and do not function immediately or shortly after transfusion. The length of storage of liquid preserved platelets at room temperature for longer than 48 hours with agitation adversely affect their survival and hemostatic function immediately and shortly after transfusion.

Severe adverse events related to FDA approved blood products are related to the survival and function of the poorly preserved RBC and poorly preserved platelets and the infusion of biologically active substances present in the preserved RBC and preserved platelets, the presence in the plasma containing antibodies to granulocytes and WBC HLA antigens, and biologically active substances that activate the recipients granulocytes.

The FDA has not licensed a hemoglobin based oxygen carrier (HBOC) because of the severe adverse events observed in patients. The severe adverse events (SAEs) reported in patients who received the hemoglobin based oxygen carriers (HBOC) produced by three commercial companies, Northfield, Biopure, and Hemosol can be explained in part by the Ringer's D,L lactate excipient used to resuspend the hemoglobin based oxygen carrier (HBOC); the resuscitation solution Ringer's D,L lactate that was transfused; and the length of storage of the liquid preserved RBC at 4 C administered with hemoglobin based oxygen carriers (HBOCs).

A major area of active investigation is the evaluation of hemostatic agents to prevent blood loss. Significant progress has been made by Marine Polymer Technologies, Inc. to provide hemostatic agents to control the bleeding site with external pressure. The hemostatic agents activate the platelets and RBC to restore hemostasis. The hemostatic agent also accelerates wound healing and exerts an antimicrobial effect.

The NBRL has been fortunate to assess bleeding associated with surgical causes and bleeding associated with non-surgical causes working in collaboration with Dr. Shukri Khuri at the West Roxbury VA Hospital from 1982 to 2008. A series of serendipitous observations made at the NBRL over the past 26 years in collaboration with Dr. S. Khuri have shown that the non-surgical bleeding diathesis in anemic thrombocytopenic patients is affected by temperature, the hematocrit, the platelet count, platelet size, platelet function, and the plasma clotting proteins. A very important observation has been made which suggests that N-acetylglucosamine is a potent activator of the platelets and RBC to produce thrombin to restore hemostasis in patients with a non-surgical bleeding diathesis. The clinical data suggest that anemic thrombocytopenic patients need to be transfused viable and functional RBC to restore the hematocrit to 35 V%

prior to the transfusion of viable and functional platelets. Anemic and thrombocytopenic patients have dysfunctional platelets which require that the hematocrit of 35 V% is needed to correct the anemia induced reversible platelet dysfunction in the recipient prior to the transfusion of viable and functional platelets.

The NBRL has summarized a 30-year experience in the study of baboon RBC, platelets and plasma clotting proteins. In these studies autologous baboon platelets stored at 22 C for only 48 hours were able to reduce the increased bleeding time produced by aspirin treatment of the baboons. Autologous baboon platelets stored at 22 C with agitation for 3 days and 5 days were not able to reduce the increased bleeding time in the aspirin treated baboons. In the aspirin treated baboon studies which were performed over a 4-year period, the five baboons were infused on eight different occasions in a random manner with autologous fresh, liquid preserved, and cryopreserved baboon platelets. The data in the baboon resolved many of the issues that have been raised over the past 50 years regarding the survival and function of preserved RBC, platelets and plasma clotting proteins.

The recent monograph published in Transfusion supplement (46:1S-42S, 2006) on the NBRL 30-year experience with the study of baboon RBC, platelets and plasma proteins and the 40-year experience reported in another monograph published in Transfusion supplement (47:206S-248S, 2007) on the non-surgical bleeding diathesis in anemic and thrombocytopenic patients: the role of temperature, RBC, platelets, and plasma clotting proteins provided important clinical information on the safety and therapeutic effectiveness of blood products to improve clinical outcome, reduce the severe adverse events (SAEs), and reduce the cost to treat military and civilian personnel.

Major changes in the collection and preservation of blood products are required to provide high quality blood products to reduce adverse events and to reduce cost to the health care system due to the current severe adverse events associated with FDA approved blood products.

The experience at the NBRL for the past 45 years has shown that in vitro testing to predict the therapeutic effectiveness of preserved RBC, preserved platelets, and preserved clotting proteins must identify whether the in vitro measurements represent reversible or irreversible defects related to the survival and function of the red blood cells, platelets, and plasma clotting proteins. The current publications in transfusion medical journals report in vitro measurements which are assumed to correlate to the survival and function of red blood cells, platelets, and plasma clotting proteins.

In vitro measurements to assess the percentage of nonviable irreversibly damaged RBC do not correlate to the 24-hour posttransfusion survival. In our experience at the NBRL, measurements of RBC ATP, filterability, deformability, osmotic fragility, density distribution, morphology, and RBC sodium and potassium levels do not correlate to the RBC 24-hour posttransfusion survival value. These in vitro measurements may detect reversible changes that are corrected following infusion. Our experience shows that RBC function assessed by in vitro measurements of RBC ATP, DPG, and P50 levels do correlate with their in vivo function. In our experience in vivo recovery of preserved platelets one to two hours following transfusion correlated to the in vitro response to hypotonic stress. In vitro testing to assess in vivo platelet hemostatic function has not correlated to platelet aggregation response to specific agonists. It is not known what in vivo agonists stimulate platelets to exert their hemostatic effect. Various agents have been utilized in vitro: calcium, collagen, adenosine diphosphate (ADP), arachidonic acid (AA), thrombin, and dual agonists like AA and ADP, ADP and collagen, ADP and epinephrine and collagen and epinephrine to perform platelet aggregation response with these different agonists. In our experience the aggregation response to the different agonists do not correlate to platelet hemostatic function in vivo assessed by reduction in bleeding time, a reduction in non-surgical blood loss, and an increase in thromboxane at the bleeding time site. In our experience the platelet production of thromboxane A2 following stimulation in vitro with a combination of arachidonic acid (AA) and adenosine diphopshate (ADP) correlated to the function of the preserved platelets to reduce the bleeding time and increase the level of thromboxane at the bleeding time site. The platelets interact with the RBC at the bleeding time site. The agonists produced at the bleeding time site activate the platelets and red blood cells to restore hemostasis by release of the vasoconstrictor substances endothelin from endothelial cells and thromboxane from the platelets; and the activation of the platelets and red blood cells to generate thrombin to convert fibrinogen to fibrin and then activate the lysis of the clot by the stimulation of the fibrinolytic process.

Our data in the study of the hemostatic agents suggest that the subendothelial glycocalyx may release derivatives of polymerized N-acetyl glucosamine which is a potent agonists to activate the platelets and RBC to generate thrombin, and to reduce non-surgical blood loss and to restore hemostasis. The freezing of human and baboon platelets with 6% DMSO by storage in a -80 C mechanical freezer produces a bimodal population of platelets. In the baboon autologous platelets frozen with 6% DMSO, thawed, washed and resuspended in autologous plasma produced a population of Gplb normal and reduced annexin V binding baboon platelets with an in vivo recovery of 48% and a linear lifespan of less than 6 days. The other population of Gplb reduced and increased annexin V binding baboon platelets was rapidly removed from the circulation within 5 minutes and was associated with a reduction in bleeding time.

Significant improvements in the freezing of red blood cells with glycerol and freezing of platelets with DMSO have been observed by the removal of supernatant glycerol from the RBC and the removal of supernatant DMSO from the platelets prior to freezing and storage in a -80 C mechanical freezer.

At the time of biochemical treatment of outdated red blood cells with a solution containing pyruvate, inosine, glucose, phosphate, and adenine (PIGPA) and incubation at 37 C for one hour there was a major concern that bacterial contamination of the RBC would occur during the one hour of incubation at 37 C. To investigate the possibility of bacterial contamination two individuals (DV and JA) who were not trained to collect and process blood products were recruited to biochemically treat outdated red blood cells, freeze the red blood cells with 40% W/V glycerol for storage at -80 C, thaw and deglycerolize the red blood cells. These two (2) individuals who volunteered to process 100 units of outdated RBC by rejuvenation and treatment with 40% W/V glycerol prior to freezing at -80 C, thawed, and washed with 3.2 liters of wash solution in the Haemonetics Blood Processor 15 suggested that the supernatant glycerol should be removed from the RBC prior to freezing and suggested that the RBC should be frozen in the original 600 ml polyvinylchloride plastic bag. The suggestion by the two non-paid volunteers who had no formal training in collection and processing of RBC was incorporated into the method to freeze human RBC in the original polyvinylchloride plastic bag which was increased in volume from 600 to 800 ml and the supernatant glycerol was removed prior to freezing. The removal of the supernatant glycerol of RBC frozen with 40% W/V glycerol permitted the reduction in the volume of wash solution from 3.2 liters to 1.6 liters. The 100 units of red blood cells that were biochemically treated, glycerolized, frozen and washed by the two non-trained volunteers were cultured and all units were negative for aerobic and anaerobic organisms. The cultures of the 100 units of deglycerolized RBC were done by Dr. Charles Ellis at the New England Medical Center, Boston, MA.

The two non-paid volunteers recommended that a shaker should be attached to the Haemonetics Blood Processor 15 to dilute the thawed glycerolized RBC prior to adding the thawed glycerolized RBC to the washing bowl. The Haemonetics Blood Processor 15 was modified by attaching a mixing platform to the Haemonetics Blood Processor 15 to produce the Haemonetics Blood Processor 115 instrument. The Haemonetics 115 utilized external dilution of the thawed glycerolized RBC concentrate prior to washing in the 375 ml disposable polycarbonate bowl. The external dilutions of the thawed glycerolized red blood cells prior to washing in the disposable polycarbonate bowl improved the washing efficiency to remove the glycerol from the red blood cells and the rejuvenation solution used to biochemically treat the red blood cells prior to glycerolization and freezing.

In March 2000, a pathologist, Dr. Hans Ostergaard from Norway visited the NBRL to observe the method to freeze platelets with 6% DMSO, storage in the -80 C mechanical freezer, thawed, and platelets were washed and resuspended in autologous plasma.

Dr. Hans Ostergaard reported that the procedure utilized was too complicated and needed to be simplified. At the time of his visit, the platelets were treated with the 27% DMSO solution in 0.9% NaCl to achieve a final concentration of 6% DMSO and the supernatant DMSO removed by centrifugation of the DMSO treated platelets at 1250 X g for 10 minutes which was the procedure utilized to concentrate the glycerolized RBC to removed the supernatant glycerol prior to freezing. The supernatant DMSO was removed prior to freezing the platelets treated with 6% DMSO in a 300 ml polyvinylchloride plastic bag which was placed in polyester plastic bags and stored in a rigid cardboard container and frozen in a -80 C mechanical freezer. Following thawing the 6% DMSO platelets were diluted with 10 to 20 ml of 0.9% NaCl and stored at room temperature for 6 hours without agitation prior to transfusion. The procedure used to freeze single donor leukoreduced platelets with 6% DMSO and the supernatant DMSO removed before freezing simplified the procedure and eliminated the need for post-thaw washing of the platelets to remove the DMSO prior to transfusion.

The NBRL method to freeze pluripotential mononuclear adult stem cells isolated from peripheral blood treated with 10% DMSO and frozen and stored at -80 C for 1 ½ years, thawed, washed and resuspended in autologous plasma was modified by the removal of the supernatant DMSO prior to freezing the DMSO treated pluripotential mononuclear adult stem cells in a -80 C mechanical freezer. Following thawing, the 10% DMSO pluripotential mononuclear cells were diluted with 10 to 20 ml of 0.9% NaCl and stored at room temperature without agitation prior to transfusion. The procedures to remove the supernatant glycerol from the RBC, the supernatant DMSO from the platelets, and the supernatant DMSO from adult mononuclear stem cells simplified these procedures to freeze RBC, platelets, and adult mononuclear stem cells prior to storage in a -80 C mechanical freezer.

The suggestions of the untrained volunteers recruited to assess the sterility of outdated red blood cells that were biochemically modified prior to glycerolization and freezing and the concern by Norwegian pathologist that the method to freeze platelets with DMSO and need to wash the platelets prior to transfusion were too complex stimulated the studies to remove the supernatant glycerol from the glycerolized red blood cells prior to freezing and the removal of the supernatant DMSO prior to freezing the platelets and the removal of the supernatant DMSO from pluripotential mononuclear adult stem cells prior to freezing in a -80 C mechanical freezer. The freezing of human platelets in the -80 C mechanical freezer was reported by the NBRL in a paper "A simple method for freezing human platelets using 6% DMSO and storage at -80 C" in Blood 43:131-136, 1974.

The progress made in the freezing of human RBC, platelets, pluripotential mononuclear adult stem cells, and plasma at -80 C over the past 45 years was related to serendipitous observations to demonstrate that a -80 C mechanical freezer can freeze group O Rh and group O Rh negative red blood cells with 40% glycerol, concentrated to remove the supernatant glycerol prior to freezing at -80 C for at least 10 years, thawed and deglycerolized in the Haemonetics Blood Processor ACP215 with 1.6 liters of wash solution and stored at 4 C in AS-3 (Nutricel) for 2 weeks, freeze single donor leukoreduced group O platelets with 6% DMSO concentrated to remove the supernatant DMSO and frozen at -80 C for at least 2 years, thawed, diluted with 10 to 20 ml of 0.9% NaCl and stored at room temperature for 6 hours without agitation, freeze ABO, Rh, and

HLA pluripotential mononuclear adult stem cells isolated from peripheral blood with 10% DMSO, concentrated to remove the supernatant DMSO and frozen at -80 C for 1 ½ years, thawed, diluted with 0.9% NaCl and stored at room temperature without agitation for 6 hours prior to transfusion into an ABO, Rh and HLA compatible recipient, and freeze AB plasma from male donors for at least 10 years, thawed and stored at 4 C for 24 hours. The decision to freeze RBC, platelets, adult stem cells, and plasma by storage in a -80 C mechanical freezer was done to avoid the need to freeze these blood products using a controlled rate freezing instrument using liquid nitrogen and storage in the gas phase or liquid phase of liquid nitrogen. Freezing in a -80 C mechanical freezer to maintain a mean temperature of -80 C with a range from -65 C to -90 C produces a freezing rate of 2 to 3 C per minute and produces a bimodal population of platelets treated with 6% DMSO. The thawed previously frozen DMSO platelets contain a population of platelets that circulate and a population of platelets that do not circulate but do exert a hemostatic effect.

The -80 C mechanical freezer contains a dual-cascade air-cooled compressor and is attached to a carbon dioxide tank to be triggered to add the liquid carbon dioxide when the temperature decreases to -65 C because of electrical or mechanical failures can be deployed in combat areas by the military. Frozen blood products consisting of RBC, platelets, and plasma can be transported using dry ice in insulated containers which can maintain the temperature of -65 C to -80 C as documented by the experience of the U.S. Navy during the Vietnam War in 1968 to 1974 period and the Netherlands military in the Middle East combat zones in Iraq, Afghanistan, and Bosnia in 2001 to 2010 period.

The results of 45 years of research revealed that much of the success achieved at the NBRL was due to chaotic serendipitous observations and patience. One cannot emphasize too much the importance of patience to permit the time needed to investigate the chaotic serendipitous observations to achieve the goal of the research to produce safe and therapeutically effective blood products for the civilian and military communities.

Hypothesis driven research requires that investigators find what they are looking for. Serendipitous observations that occur are usually not recognized and if observed are rejected by hypothesis driven research. The need to write progress reports and publish reports may produce the data that cannot be reproduced by the investigators themselves and other investigators. The NBRL was fortunate to be funded for 45 years and to be able to recognize important chaotic and serendipitous observations that were critical to the productivity of our laboratory.

The serendipitous observations that storage of frozen RBC with 40% W/V glycerol, frozen platelets with 6% DMSO, frozen pluripotential mononuclear adult stem cells isolated from peripheral blood frozen with 10% DMSO and plasma frozen in a -80 C mechanical freezer produced freezing rates that permitted acceptable cryopreservation of these blood products without the need of controlled rate freezing instruments using liquid nitrogen and storage in the gas phase of liquid nitrogen or in liquid nitrogen.

Studies in wounded servicemen between 1968 to 1974 that returned from Vietnam provided data to document the maintenance of the central blood volume with the reduction in the peripheral blood volume in chronic hypovolemic anemic patients who were subjected to traumatic wounds. Transfusion of compatible but identifiable allogeneic red blood cells increased the peripheral blood volume to the gastrointestinal tract and muscle, bone and skin of the extremities. The increase in peripheral blood volume occurred following the transfusion of viable red blood cells which increased the shear stress on endothelial cells to produce nitric oxide to vasodilate the hypoperfused areas of the peripheral blood volume. The red blood cells may also have provided nitric oxide bound to the globin portion of hemoglobin (SNOHb) and the red blood cells themselves may produce nitric oxide like the endothelial cells. The studies performed in the chronic hypovolemic anemic patient demonstrated that transfusion of washed viable RBC not only increased the peripheral red blood cell volume but also increased the plasma volume. The red blood cell volume, the plasma volume and total blood volume in these chronic hypovolemic anemic patients were reduced. The normal hematocrit and hemoglobin concentration could not be used to assess the hypovolemic state in these patients with "stress anemia". The red blood cell volume was measured with 51Cr labeled autologous RBC and the plasma volume was measured with radiolabeled cold agglutinin 19S macroglobulin with a molecular weight of 1,000,000. Radiolabeled human albumin could not be used to measure the plasma volume because of the rapid loss of the radiolabeled albumin with molecular weight of 68,000 into the extravascular volume which overestimates the plasma volume. The normal hematocrit in these patients minimized the non-surgical blood loss from open wounds of the extremities which required surgical debridement. The chronic hypovolemic anemic patients subjected to general anesthesia for the debridement of their wounds of the extremities became hypotensive and a few of these wounded servicemen had cardiac arrests associated with the administration of general anesthesia. This observation confirmed that general anesthesia permitted redistribution of the blood volume from the central blood volume in the brain, heart, lungs, and kidneys into the hypoperfused peripheral blood volume which consists of the blood volume in the gastrointestinal tract, muscle, bones and skin of the extremities. The debridement of wounds was associated with minimal non-surgical blood loss in these open wounds of the extremities. This important clinical observation showed that debridement of wounds of the extremities was not associated with significant non-surgical blood loss from the open wounds of the extremities. Another clinical observation was that following the transfusion of RBC which increased the peripheral red blood cell volume to the extremities, the edema in the tissues of the extremities disappeared. The reduction in the tissue edema in the extremities was associated with an increase in the plasma volume.

The transfusion of washed liquid preserved red blood cells and washed previously frozen deglycerolized red blood cells to chronic hypovolemic anemic patients increased the red blood cell volume, the plasma volume, the peripheral red blood cell volume, and the total blood volume. At the same time studies were done to assess whether washed autologous dog red blood cell transfusions would increase the plasma volume in dogs following acute hypovolemic anemia produced by removal of whole blood from the dogs. Dogs subjected to acute hypovolemia with reduced RBC volume and plasma volume were transfused with washed autologous dog red blood cells. Following transfusion, an increase in red blood cell volume and plasma volume occurred. The study demonstrated that plasma volume increased following the transfusion of washed autologous dog red blood cells to acute hypovolemic dogs. These studies indicate that the red blood cell volume regulated the plasma volume and the total blood volume in chronic hypovolemic patients and in acute hypovolemic dogs.

The hematocrit of the chronic hypovolemic anemic patient did not increase following the transfusion of several units of compatible identifiable viable RBC because the plasma volume of the patient increased. Both the red blood cell 2,3 DPG and the RBC creatine levels decreased following the restoration of the red blood cell volume in these patients. The chronic hypovolemic anemic patients had normal hematocrit values and hemoglobin concentration levels but they had significantly elevated RBC 2,3 DPG levels and RBC creatine levels prior to the RBC transfusions. The aggressive transfusions of the chronic hypovolemic anemic patients with extremity injuries to restore the RBC volume, plasma volume and peripheral blood volume prevented the need to amputate extremities in the 300 wounded servicemen hospitalized at the Chelsea Naval Hospital between 1968 to 1974.

The increase in RBC 2,3 DPG level was a compensatory mechanism to permit the reduced volume of red blood cells to improve oxygen delivery to tissue to maintain high p02 tensions in the brain, heart and kidneys without the need to increase cardiac output to increase blood flow to the brain, heart, and kidneys. The hypoperfusion of the muscle permitted leakage of creatine from the muscle which accumulated in the circulating RBC to increase the RBC creatine level. In patients with cardiopulmonary insufficiency increases in RBC 2,3 DPG and creatine levels were observed. Again, the tissue hypoxia produced in patients with cardiopulmonary insufficiency stimulated the RBC to increase the 2,3 DPG level to improve oxygen release from the RBC. The tissue hypoxia that was present in patients with cardiopulmonary insufficiency permitted loss of creatine from the hypoxic muscle which accumulated in the red blood cells. Both the red blood cell creatine and 2,3 DPG levels were elevated in patients with cardiopulmonary insufficiency reflected tissue hypoxia in the non-vital tissues like the muscle of the extremities.

The aggressive transfusion of the patients with traumatic injuries who had multiple fractures of the upper and lower extremities restored the blood volume to these areas. In the study of 300 wounded servicemen from Vietnam who were transferred to Chelsea Naval Hospital from 1968 to 1974, amputation of their extremities rarely occurred. The transfusion of these patients to restore the red blood cell volume, especially the peripheral red blood cell volume to extremities with washed liquid preserved red blood cells and washed previously frozen deglycerolized RBC increased the red blood cell volume, the plasma volume and total blood volume towards their normal values. The multiple fractures of the upper and lower extremities healed without the need to amputate these extremities because of poor perfusion.

Following transfusion in these chronic hypovolemic and anemic patients improvement in appetite occurred together with improvement in their libido and general state of health in these wounded servicemen.

Studies were done to investigate the mechanisms of the red blood cell volume deficits observed in these wounded servicemen with the chronic hypovolemic anemia of trauma. The red blood cell survival of autologous and allogeneic compatible but identifiable red blood cells, the rate of red blood cell production, and the red blood cell destruction were measured. (Valeri CR and Altschule MD. The Hypovolemic Anemia of Trauma: The Missing Blood Syndrome, published by CRC Press, 1981).

Biphasic survival of allogeneic compatible but identifiable red blood cells was observed suggesting the presence and subsequent removal of an extrinsic factor that accelerated the destruction of the donor red blood cells. Treatment of the patient's blood with adrenochrome, a metabolite of epinephrine, showed increase hemolysis which decreased following the RBC transfusions. The extrinsic factor in the patient's blood correlated to the increased hemolysis in their blood following the in vitro treatment with adrenochrome. Red blood cell production was reduced and destruction of preserved compatible nonviable red blood cells was not associated with increases of carbon monoxide, bilirubin, and urinary and fecal urobilinogen excretion. The destruction of preserved allogeneic compatible nonviable red blood cells was associated with porphyrin accumulation in the wounds and increase in urinary excretion of dipyrroles and oxyporphyrin. The data suggested that the degradation of preserved compatible nonviable RBC was through a non carbon monoxide – bilirubin – urobilinogen pathway but instead by degradation into heme and porphyrin derivatives without release of carbon monoxide. In vitro testing to assess the susceptibility of the red blood cells to adrenochrome, a degradation product of epinephrine, suggested a potential mechanism for the degradation of hemoglobin into heme and porphyrin. In vitro testing of blood from the chronic hypovolemic anemic patients showed an increase in hemolysis when treated with adrenochrome.

The subsequent studies of non-surgical blood loss in patients with anemia and thrombocytopenia following cardiopulmonary bypass surgery provided an explanation for the observed reduced non-surgical blood loss from open wounds of the extremities observed in the chronic hypovolemic anemic patients with traumatic injuries. The patients with chronic hypovolemia of trauma had normal or slightly increased hematocrit values. The effect of the normal or slightly increased hematocrit values on the bleeding time and non-surgical blood loss was associated with the decrease in non-surgical blood loss in the open extremity wounds of these patients. The other important observation was that RBC transfusions increased the peripheral red blood cell volume and reduced the edema in the extremities to increase the plasma volume and the total blood volume.

The failure to observe non-surgical blood loss in the edematous extremities of the hypovolemic anemic patients was related to the normal or slightly increased hematocrit values in these patients. The red blood cells have been shown to be extremely important in hemostasis and to the prevention of non-surgical blood loss. There are several mechanisms which have been shown to demonstrate the hemostatic function of RBC. Red blood cells increase shear stress on endothelial cells to stimulate release of ADP from the RBC and RBC may provide arachidonic acid (AA) and adenosine diphosphate (ADP) for the platelet production of thromboxane, a potent vasoconstrictor agent and an agonist to aggregate platelets. Red blood cells release oxygen at high p02 tensions to oxidize the nitric acid produced by the endothelial cells and release endothelin from the endothelial cells to vasoconstrict the microcirculation. The hematocrit value has been correlated to the bleeding time and the volume of blood collected at the bleeding time site. The reduced hematocrit correlated to an increase in the bleeding time, an increased volume of blood collected at the bleeding time site, and a decrease in shed blood thromboxane level at the bleeding time site. In patients following cardiopulmonary bypass surgery the non-surgical blood loss 4 hours following cardiopulmonary bypass surgery has been correlated to the bleeding times and the hematocrit 2 hours following bypass surgery; a reduction in hematocrit and an increased in bleeding time were associated with an increase in non-surgical blood loss. The reduction in hematocrit, like the reduction in temperature, produces a reversible platelet dysfunction characterized by an increase in the bleeding time and a reduction in the thromboxane level in the shed blood collected at the bleeding time site.

The bleeding time correlates to non-surgical bleeding which is affected by temperature, the red blood cells, the platelets and the plasma clotting proteins. The red blood cells interact with the platelets and clotting protein at the bleeding time site to restore hemostasis. At the bleeding time site a derivative of N-acetylglucosamine may be released from the subendothelial glycocalyx of the blood vessel which activates the platelets and red blood cells to produce thrombin to convert fibrinogen to fibrin which then stimulates the fibrinolytic process.

The insights into hemostasis were stimulated by the study of the hemostatic agent provided by Marine Polymer Technologies, Inc. This company isolated polymerized N-acetylglucosamine filamentous material produced by microalgae in a bioreactor. This substance was isolated, purified, and prepared as an endotoxin free hemostatic bandage. The initial fiber length of the filamentous material was long and required that this hemostatic agent (l-NAG) had to be removed from the bleeding site. The filamentous material was shown to vasoconstrict the bleeding site and activates the platelets and the red blood cells to produce thrombin. The filamentous material was reduced in length by gamma radiation and the short filamentous material (s-NAG) can now be applied directly to the bleeding time site with external pressure. The short filamentous material (s-NAG) composed of polymerized N-acetylglucosamine is now degraded at the bleeding time site and this hemostatic agent does not have to be removed. The long filamentous material (l-NAG) has been shown to activate the platelets and interacts with the RBC at the Band 3 site to produce stomatocytes and to expose RBC phosphatidylserine to produce thrombin.

The data suggest that the subendothelial glycocalyx of blood vessel may release derivatives of N-acetylglucosamine which activates the platelets and red blood cells to produce thrombin and to restore hemostasis, accelerate wound healing, and exert an antimicrobial effect. The effects of the polymerized N-acetylglucosamine stimulates release of endothelin from endothelial cells and stimulates the production of thromboxane by platelets from arachidonic acid (AA) and adenosine diphosphate (ADP).

In 1968, whole blood was collected in the acid-citrate-dextrose (ACD) anticoagulant and stored at 4 C for 21 days. During my visit to South Vietnam to observe the frozen blood bank deployed at Danang, South Vietnam and aboard the hospital ship USS Repose located in the China Sea, large numbers of group O Rh positive and group O Rh negative units were outdated and discarded. Subsequently, solutions were prepared at the Naval Blood Research Laboratory to biochemically treat outdated group O Rh positive and group O Rh negative RBC to increase the RBC ATP, DPG level to 1 ½ to 2 times normal prior to freezing. These values for RBC DPG were observed in the RBC of chronic hypovolemic anemic patients who sustained traumatic injuries who returned from South Vietnam to the Chelsea Naval Hospital and the RBC DPG levels of 1 ½ to

2 times normal were observed in patients with cardiopulmonary disorders who were hospitalized at the Chelsea Naval Hospital from 1962 to 1974. Beverly Gabrio and associates have published that adenosine added to blood stored at 4 C increased the RBC level of ATP and improved the storage of RBC in whole blood at 4 C. Similar observations by Strumia and associates reported that adenine and inosine improved the ATP level in liquid preserved RBC. The availability of outdated group O Rh positive and group O Rh negative RBC stored in the acid citrate dextrose (ACD) anticoagulant at 4 C for longer than 21 days stimulated the NBRL to biochemically modify these outdated RBC to increase the RBC ATP, DPG and P50 levels prior to freezing with 40% W/V glycerol and storage at -80 C and freezing with 20% W/V glycerol and storage at -150 C in the gas phase of liquid nitrogen.

The serendipitous observations that the chronic hypovolemic anemic patients who returned from South Vietnam to Chelsea Naval Hospital had significantly elevated RBC 2,3 DPG levels and patients hospitalized at the Chelsea Naval Hospital from 1962 to 1974 with cardiopulmonary insufficiency had increased RBC DPG levels of 1 ½ to 2 times normal stimulated the study to biochemically modify outdated RBC to increase the RBC ATP, DPG, and P50 levels similar to those observed in wounded servicemen with chronic hypovolemic anemia and patients with cardiopulmonary disorders hospitalized at the Chelsea Naval Hospital.

The design and development of solutions containing substances like pyruvate, inosine, phosphate, and adenine with and without glucose were studied. The presence of glucose was associated with carmelization of these solutions following heat sterilization and the glucose was eliminated to prevent the carmelization. These solutions were utilized to increase the RBC ATP, DPG and P50 levels in RBC stored in ACD and CPD at 4 C for 3 to 6 days as indated rejuvenated RBC and in RBC stored in ACD and CPD at 4 C for 22 to 25 days as outdated rejuvenated RBC to increase the ATP, DPG and P50 values to 1 ½ to 2 to 3 times normal by incubation at 37 C for one hour prior to glycerolization and freezing. The rejuvenation solution containing pyruvate, inosine, phosphate, and adenine (PIPA) was studied and was shown to be stable at room temperature for 18 months.

There was a major investigation to control the temperature during incubation from 4 C to 20 C during the first 30 minutes and from 20 C to 32 C during the second 30 minute period of incubation in a water bath maintained at 37 C. The 50 ml of rejuvenation solutions were added to the

RBC in the 600 ml or 800 ml polyvinylchloride plastic bag and placed in double plastic bags and then placed in a water bath maintained at 37 C.

Liquid preserved RBC stored in ACD, CPD, CPDA-1 and in the additive solutions CPD AS-1, CP2D AS-3, and CPD AS-5 for 3 to 6 days were glycerolized, the supernatant glycerol removed and were frozen as nonrejuvenated RBC and then stored at -80 C. Liquid preserved RBC stored in ACD, CPD, CPDA-1 and in the additive solutions CPD/AS-1, CP2D/AS-3, CPD/AS-5 for 3 to 6 days were biochemically treated with PIPA and then glycerolized, the supernatant glycerol removed, and frozen at -80 C as indated rejuvenated RBC. Liquid preserved RBC stored in the additive solutions CPDA-1 for 36 to 38 days and in CPD AS-1, CP2D AS-3, and CPD AS-5 for 42 days were biochemically treated with PIPA then glycerolized, the supernatant and glycerol removed, and frozen as outdated rejuvenated RBC. The nonrejuvenated glycerolized RBC frozen at -80 C, thawed and deglycerolized in the Haemonetics Blood Processor ACP215 and stored in the additive solution AS-3 for 2 weeks have 24 hour posttransfusion survival of 75% and red cell oxygen affinity which is moderately increased and less than 1% hemolysis. The indated rejuvenated glycerolized frozen RBC following thawing, deglycerolization in the Haemonetics Blood Processor ACP215 and stored at 4 C in AS-3 for 2 weeks have 24 hour posttransfusion survival of at 75% and red cell oxygen affinity which is normal or slightly decreased and less than 1% hemolysis. The outdated rejuvenated glycerolized frozen RBC following thawing, deglycerolization in the Haemonetics Blood Processor ACP215 and stored at 4 C in AS-3 for at least 24 hours have 24 hour posttransfusion survival of 75% and normal or slightly increased red blood cell oxygen affinity and less than 1% hemolysis. The biochemical treatment of indated RBC and outdated RBC produce RBC with ATP, DPG and P50 levels similar to those of patients with chronic hypovolemic anemia of trauma and patients with cardiopulmonary disorders. The serendipitous observations made in patients with chronic hypovolemic anemia of trauma and in patients with cardiopulmonary insufficiency provided the levels of RBC, ATP, DPG, and P50 that were achieved to biochemically modify indated and outdated RBC prior to glycerolization and freezing. The biochemical treatment produced RBC with function that patients with chronic hypovolemic anemia of trauma and with cardiopulmonary insufficiency produce to provide optimum oxygen delivery to the brain, heart, and kidneys without the need to increase blood flow. Studies demonstrated that the freezing of non-rejuvenated glycerolized RBC containing 40% W/V glycerol and stored at -80 C for 37 years maintained the RBC ATP, DPG and P50 levels in the thawed, deglycerolized RBC washed using electrolyte solutions that were present at the time of freezing.

Washing glycerolized RBC in the bowl designed by Allan Latham showed that continuous flow washing in the Latham disposable polycarbonate bowl designed to isolate plasma, platelets and red blood cells did not efficiently wash the glycerolized RBC to remove the glycerol in the Haemonetics Blood Processor 15. The procedure was modified by dilution of the glycerolized RBC prior to washing in the Haemonetics Blood Processor 15. The modified instrument to deglycerolize the RBC with the integrally attached mixer was named the Haemonetics Blood Processor 115 instrument. The Haemonetics Blood Processor 115 instrument removed the glycerol from the glycerolized RBC using a 375 ml washing bowl with an internal seal. Both the procedures for adding the glycerol and removing the glycerol from the RBC using the Haemonetics Blood Processor 115 were open systems using the modified 375 ml disposable polycarbonate bowl with an internal seal designed by Mr. Allan Latham.

In 1998, the Haemonetics Blood Processor 115 instrument was modified to produce the Haemonetics Blood Processor 215 automated cell processor (ACP 215) to permit both the glycerolization and deglycerolization of RBC using a functionally closed system by the use of the sterile docking device, in-line 0.22 micron filters to add the glycerol to the RBC with a pump; the deglycerolization of the RBC using the 275 ml or the 325 ml disposable polycarbonate washing bowl with an external seal. The bowl was spun at 5,800 rpm and the bowl stopped during the washing procedure several times to permit mixing within the bowl; the thawed glycerolized RBC are diluted on two separate occasions using an integrally attached mixer and in-line 0.22 micron filter to add the solutions to deglycerolize the thawed glycerolized RBC using 50 ml of 12% NaCl and 2.0 liters of 0.9% NaCl and 0.2 gm% glucose. The Haemonetics Blood Processor ACP215 is a functionally closed instrument to add and to remove glycerol from RBC, so that the washed RBC can be stored at 4 C for 2 weeks in the additive solution AS-3 (Nutricel) with less than 1% hemolysis. The high separation 275 ml disposable bowl which is FDA approved for performing plateletpheresis procedure was attached to the disposable glycerolization sets to simplify both the glycerolization of non-rejuvenated RBC and rejuvenated RBC. The high separation 275 ml bowl attached to the disposable glycerolizing set permitted concentration of the RBC prior to glycerolization and the concentration of the glycerolized RBC to remove the supernatant glycerol prior to freezing in the 1,000 ml polyvinylchloride plastic bag. The attachment of the high separation bowl to the disposable glycerolizing set has completely automated the concentration of the non-rejuvenated and rejuvenated red blood cells prior to glycerolization and the removal of the supernatant glycerol prior to freezing in the 1,000 ml polyvinylchloride plastic bag. The attachment of the high separation bowl to the disposable glycerolizing set reduces the labor to concentrate the RBC prior to and following glycerolization and removal of the supernatant glycerol prior to freezing of nonrejuvenated, indated rejuvenated and outdated rejuvenated RBC using the Haemonetics Blood Processor ACP215 instrument.

The NBRL has published procedures to freeze RBC as nonrejuvenated and indated and outdated rejuvenated RBC, freeze platelets, freeze plasma, and freeze peripheral blood mononuclear pluripotential adult stem cells to produce safe and therapeutically effective frozen RBC, frozen platelets, frozen plasma and frozen adult mononuclear stem cells for use by the military and civilian communities.

To preserve red blood cell viability and function, platelet viability and function, and the viability and function of plasma proteins methods were reported by the NBRL to measure in vivo survival and function of preserved red blood cells, preserved platelets and preserved plasma proteins. Currently the FDA requires data on the in vitro recovery of preserved red blood cells, the 24 hour posttransfusion RBC survival and RBC lifespan, the hemolysis, the sterility and the testing for infectious disease markers in the RBC products. There are no requirements that the preserved red blood cells function in vivo. In a similar manner the FDA requires data on the in vitro recovery of preserved platelets, the in vivo recovery of the platelets one to two hours after transfusion, the platelet lifespan, pH of the platelets, sterility and the testing for infectious disease makers in the preserved platelets. Like preserved RBC, there is no testing required by FDA to document that the preserved platelets function in vivo. Over the past 45 years the NBRL has evaluated procedures to assess the in vitro recovery of preserved red blood cells, the hemolysis, the 24 hour posttransfusion survival of preserved red blood cells and the RBC lifespan using radioisotope and non-radioisotope procedures, the RBC function to carry and to release oxygen at high tissue oxygen tension to the brain, heart and kidneys, to restore hemostasis by the reduction in the bleeding time, and the reduction in nonsurgical blood loss, sterility and the testing for infectious disease markers. The NBRL has evaluated methods to assess the in vitro recovery of the preserved platelets, radioisotope and non-radioisotope procedures to measure the in vivo recovery of the preserved platelets 1-2 hours following infusion and the platelet lifespan, platelet hemostatic function to reduce the increase bleeding time in stable thrombocytopenic patients and in normal volunteers and baboons treated with aspirin, the pH of the platelets, sterility, and the testing for the infectious disease markers.

The NBRL has reported on the measurement of RBC ATP, DPG, p50 levels to assess both in vitro and in vivo function of the preserved RBC. The NBRL has reported on preserved platelet response to hypotonic stress to assess the in vivo recovery of preserved platelets one to two hours after transfusion and platelet production of thromboxane A2 following stimulation with a combination of arachidonic acid (AA) and adenosine diphosphate (ADP) to assess the in vitro hemostatic function of the preserved platelets.

The survival and function of preserved RBC, preserved platelets, and preserved plasma proteins were assessed during the study of wounded servicemen who returned from South Vietnam to the Chelsea Naval Hospital and patients at the Chelsea Naval Hospital who required blood transfusion; patients at the University Hospital at Boston University Medical Center (BUMC) who were subjected to cardiopulmonary bypass surgery and thrombocytopenic patients; patients at the West Roxbury Veterans Hospital subjected to cardiopulmonary bypass surgery, and a population of baboons and dogs housed at the New England Medical Center (NEMC) and at Boston University Medical Center (BUMC) from 1972 to 2002.

The NBRL peer reviewed papers published the methods that were designed and developed to assess in vitro recovery and in vivo survival and function of the preserved RBC, preserved platelets and preserved plasma proteins in normal volunteers, patients, and baboons.

The data collected demonstrated that in vitro testing of the preserved red blood cells, platelets, and plasma proteins did not determine whether these measurements represented reversible or irreversible changes in these blood products.

Studies performed at the NBRL in normal volunteers, patients and baboons documented that preserved red blood cells and preserved platelets that circulated had reversible defects that were repaired following transfusion. The preserved red blood cells and preserved platelets that were removed from the circulation had irreversible defects which could not be identified because these irreversibly damaged cells could not be isolated following transfusion and studied. Whereas recovery of the circulating reversibly damaged RBC and platelets permitted the study of the in vivo restoration of the RBC and platelets following transfusion.

Studies were performed at the NBRL in healthy volunteers, patients, baboons, and dogs. In addition, studies were performed to assess isolated bovine eyes, isolated rat kidneys, isolated dog hearts, and isolated rabbit hearts perfused with human RBC with decreased and increased affinity for oxygen at normothermic and hypothermic temperatures. These studies were designed and the data tabulated and analyzed by independent personnel not involved in the testing procedures.

The research performed at the NBRL was influenced by the following individuals:

1. Admiral H. Richover, the father of nuclear power to operate nuclear submarines for the past 50 years stimulated our study of the preservation of adult pluripotential mononuclear stem cells isolated from peripheral blood of healthy volunteers between 1980 to 1985. These studies demonstrated that autologous peripheral blood pluripotential mononuclear adult stem cells collected from normal volunteers frozen with 10% DMSO and stored at -80 C in mechanical freezers for 1 ½ years can be used to repopulate the bone marrow of military personnel exposed to radiation injury.
2. Mr. Howard Hughes, the father of aviation and responsible for the safety of air travel, stimulated the NBRL commitment to provide safe and therapeutically effective blood products
3. Mr. William Gates, the inventor of the computer technology has globalized communication. Our laboratory has utilized his contribution to tabulate and analyze the data that were collected and published at the NBRL over the past 45 years.
4. Dr. Mark David Altschule participated in the study of the wounded servicemen who returned from Vietnam who were hospitalized at the Chelsea Naval Hospital. Dr. Altschule is the "gold standard" of what a physician should be. Dr. Mark David Altschule was the best clinician at the Harvard Medical School between 1954 to 1958. Dr. Altschule reported that the study of the wounded servicemen

and their response to treatment taught him what convalescence was "the period of time required for the patient to restore to normal the peripheral blood volume". Dr. Mark Altschule was convinced that transfusion of red blood cells to restore to normal the peripheral blood volume in patients with chronic hypovolemia was responsible for the repair of the multiple fractures of the extremities without the need for amputations in the patients hospitalized at the Chelsea Naval Hospital from 1968 to 1974.

5. Dr. Thomas Durant, a close friend who recognized the existence of global health problems and worked to solve them and he was a very important supporter of the research conducted at the NBRL and convinced Congressman J. Joseph Moakley to provide congressional funds to support the NBRL
6. The late Congressman J. Joseph Moakley from Massachusetts, a politician who was a "true servant of the people" who worked to improve the welfare of people and not his own self interest. Congressman J. Joseph Moakley provided congressional funds from 1998 to 2003 at the time of his death to support the research that was performed at the NBRL.
7. The NBRL was fortunate to work with Dr. Edward Merrill, Dr. Max Strumia, and Dr. Shukri Khuri. Professor Edward Merrill was the Carbon and Petroleum (CP) Dubbs Professor of Chemistry at MIT who designed and provided the Merrill viscometer to the NBRL to measure blood viscosity. Dr. Max Strumia, the Chairman of the National Research Council between 1965 to 1970 provided the criteria to evaluate the safety and the therapeutic effectiveness of blood products. The NBRL collaborated closely with Dr. Shukri Khuri at the West Roxbury VA Hospital from 1982 to 2008. Dr. Khuri did not accept information without critical and independent review of the study design and the data that were collected and analyzed by numerous biostatisticians. Dr. Khuri wanted to place his finger in the site where the nails were driven in the hands of the crucified individual. Shukri did not believe anything unless he carefully designed the studies and independently reviewed and analyzed the data with biostatisticians. Dr. Khuri was responsible for the studies performed at the West Roxbury VA Hospital to define the optimum treatment of nonsurgical blood loss in patients subjected to cardiopulmonary

bypass surgery in addition to his seminal research on myocardial pH and the assessment of surgical care and clinical outcome at the Veterans Administration Hospital.

As Niccolo Machiavelli reported in the book "The Prince", we learned at the NBRL the following:

1. Better to be lucky than smart.
2. Man is good and honest until for self interest, becomes bad and dishonest.
3. To control society, the subjects must fear the leader.
4. If the people do not fear the leader, then the leader must punish the disobedient subjects.
5. Religion and laws of the society are used to control the behavior of the people.
6. Lawyers now provide the fear of litigation that Machiavelli stated is needed to control the behavior of people.
7. Patents, biostatisticians, affirmative action, and lawyers now restrict freedom and opportunity to perform research to improve the health of the military and civilian communities.

   We agree with the Noble Laureate Charles Brenton Huggins, M.D. who reported the following:

   Do not go to meetings, they are a waste of time.

   Do not write books.

   Do not go to the library.

   Work on new and important problems that are not in books. Discover first. Do library research to find out whether you can connect your new discovery to established ideas. Then publish a concise paper.

   We support the statement by Maurice Maeterlinck:

   At every crossway on the road that leads to the future, every progressive spirit is opposed by a thousand men appointed to guard the past.

In 1958, I obtained a medical degree and selected a training program in internal medicine and a clinical fellowship in hematology. My interest was to practice internal medicine and hematology and I was not interested in performing research. I was told by several of my supervisors that I would never practice medicine and I was offered research fellowships which I rejected. I was asked in undergraduate school at Tufts College, Medford, MA by Dr. Russell Carpenter, Chairman and Professor of Biology to apply to Harvard Medical School and he arranged my interview with Dr. Roy Greep, Dean of Harvard Dental School. The interview with Dr. Greep was pleasant and I remember telling Dr. Greep that I really did not know what I wanted to do, whether to go to law school or medical school. Today, after 45 years in the study of blood products to treat patients, I have not yet decided what I want to do. The study of blood products was never considered to be work. I considered the study of blood products to be an extension of my training before I would get a real job. My mother was ill with a terminal illness and I decided to select medical school rather than law school. Within 24 hours of my interview with Dean Greep I was selected to become a student at Harvard Medical School with a scholarship. My father Angelo and his brother Dominic were both born in Italy and following the death of their mother, my father and his brother, together with my grandfather, immigrated to the United States. My father married my mother Domenica Piccolomini who was born in Pittsburgh, PA and they met and married in Leominster, MA. I was one of five children, three boys and two girls and my father, without any opportunity to go to school, opened a restaurant in Shirley, MA, a small town adjacent to Leominster and near Fort Devens, where a U.S. Army base was located in Central Massachusetts.

I attended public schools in Leominster and upon graduation Mr. Robert Laserte, the Guidance Counselor was responsible to suggest where the student should apply to college. I remember Mr. Robert Laserte telling me I could be a big fish in a small pond or a small fish in a big pond. I selected Tufts College in Medford, MA where I received a Bachelor of Science (BS) degree in 1954. I never knew what I wanted to do. I enjoyed Political Science courses and I was always concerned that the course was not Science-Politica. The emphasis on how the politics influenced the science was of concern to me in undergraduate school and then in medical school.

In my medical training, I spent two years at Boston City Hospital and one year at the New England Medical Center where Dr. Sam Progrer directed the Pratt Clinic which cared for relatively healthy patients compared to the very sick patients that were admitted to the Boston City Hospital. At the BCH, the intern did all the testing on the patients that he was responsible for, i.e. blood testing, urinalysis, lumbar punctures, cultures of blood and urine and took the electrocardiograms. There were no intensive care units and very sick patients were on the ward and the responsibility of the interns and the residents. At the New England Medical Center the senior resident cared for relatively healthy patients and I learned from Dr. Sam Progrer, the Director of the Hospital, how he recruited patients for annual physical examinations. During my stay at the New England Medical Center I spent a lot of time with Dr. Ted Astwood and his staff in the Department of Endocrinology. I was not interested in the ongoing endocrine research projects that I learned about in Dr. Astwood's laboratory. As a second year medical student I visited Dr. Astwood to obtain his opinion on a thyroid nodule that I detected at the end of the second year at medical school. Dr. Astwood did not recommend surgery to remove the thyroid nodule which was recommended by Dr. Francis Moore, Chief of Surgery at Peter Bent Brigham Hospital. Dr. Astwood suggested that I treat the thyroid nodule by taking thyroid medication whereas Dr. Francis Moore recommended surgical removal of the thyroid nodule. Dr. Moore prevailed and the nodule was removed followed by a postoperative hemorrhage into my neck because surgical hemostasis was not achieved and the wound site was not drained. The postoperative bleeding that occurred was an important event that stimulated my interests in hemostasis related to surgical and nonsurgical bleeding. Dr. Andrew Jessiman ligated the severed artery after examining my neck which the surgical residents did not do. Prior to the visit of Dr. Andrew Jessiman I was sent to a psychiatrist because the surgical residents thought I was emotionally unstable. Dr. Francis Moore had left the hospital to attend the marriage of his daughter and my postoperative care was the responsibility of the surgical residents. To this day, I am very grateful to Dr. Andrew Jessimen and Dr. Leroy Van Dam, Chief of Anesthesia for my care as a patient at the Peter Bent Brigham Hospital in 1956 for the surgical removal of the thyroid nodule.

I returned to the BCH for a one year fellowship in clinical hematology working with Dr. W.C. Moloney and Dr. Jane Desforges. In 1961, I was on a rotation in pediatrics working for Dr. Jane Desforges. A four year old boy with lead intoxication was a patient that I was assigned to consult on. The peripheral blood smear showed stippling of his red blood cells which I was interested in and prepared a protocol to study the mechanism of RBC stippling in patients with lead intoxication. The proposal was submitted and $10,000 was provided by the Massachusetts Public Health Department to study the stippling of red blood cells by lead.

Rabbits were purchased and housed at the Holy Ghost Hospital in

Cambridge, MA. The rabbits were injected with lead and stipple cells were produced and studies were performed to document that the lead was shown to aggregate the messenger ribonucleic acid (RNA) in the reticulocytes that were produced in the rabbit. Of interest, in 2006, Dr. Craig Mello at the University of Mass received the Noble Prize in Medicine related to inhibitors of messenger RNA. The pediatric patient with lead intoxication and stipple cells in his peripheral smear was studied in 1961 and in 2006 Dr. Craig Mello received a Nobel prize to study inhibitors of messenger RNA. Our studies performed in 1961 which were not published demonstrated that the RBC stippling was due to the aggregation of messenger RNA in the reticulocytes of rabbits treated with lead.

My experience has been that research performed at the NBRL was stimulated by the clinical events that were recognized as chaotic and serendipitous observations. Data were collected to help to interpret the observation without a preconceived hypothesis. At the Chelsea Naval Hospital, all the patients who received blood transfusions were studied. There was no institutional review board (IRB) and no written informed consent (IC) between 1962 to 1974.

My experience in the research funded by the Naval Medical Research and Development Command of the U.S. Navy's Bureau of Medicine and Surgery, the Office of Naval Research (ONR) and the Congress of the United States was related to chaotic serendipitous observations which stimulated extremely productive research on the preservation of blood products, blood substitutes, resuscitation solutions, and hemostatic agents to treat military and civilian personnel.

In 1962, I volunteered to join the U.S. Navy and I was assigned to the Blood Research Laboratory at the Chelsea Naval Hospital. In July 1962 four (4) patients were in acute renal shutdown following transfusion of deglycerolized red blood cells at the Chelsea Naval Hospital and I was responsible as the hematologist to investigate the etiology of these severe adverse events associated with the transfusion of the previously frozen deglycerolized RBC.

Patients at the Chelsea Naval Hospital had been transfused with previously frozen deglycerolized RBC from 1958 to 1962. I reviewed all the records of patients who were transfused frozen deglycerolized RBC at the Chelsea Naval Hospital. I did not learn anything from the review of these records except that posttransfusion infections were not reported. Deglycerolized RBC were stored at 4 C in autologous plasma or in an albumin-glucose medium for as long as 21 days prior to transfusion. The open method to glycerolize and deglycerolize RBC using the Cohn Blood Fractionator was not associated with contamination of the deglycerolized red blood cells stored at 4 C in autologous plasma or the albumin-glucose medium for 21 days. The Office of Naval Research (ONR) had funded Dr. Edward J. Cohn to design and develop the Cohn Blood Fractionator to isolate albumin from fresh whole blood to produce a 25% concentrate of albumin as the first blood substitute to treat patients in hemorrhagic shock. Cohn and his associates decided to salvage the fresh RBC in the Cohn Blood Fractionator which were routinely discarded using the procedure reported by Dr. Audrey Smith to freeze human red blood cells with glycerol and storage of the glycerolized RBC in dry ice and alcohol at -79 C.

The intracellular glycerol used to freeze the RBC needed to be removed prior to transfusion to avoid intravascular hemolysis of the glycerolized RBC. Tullis and his associates speculated that transmembrane washing of the red blood cells to remove the glycerol would reduce or eliminate the hepatitis virus. The deglycerolized RBC were resuspended in an albumin-glucose medium to avoid the use of plasma which transmits hepatitis. Unfortunately aggregation of the outdated albumin in the glucose medium occurred and was responsible for the severe adverse events (SAEs) that were observed in four patients in July 1962 when I arrived at the Chelsea Naval Hospital. The acute renal insufficiency in the four patients who received deglycerolized RBC in the outdated albumin-glucose medium was due to the aggregated albumin in the resuspension medium. The use of high concentrations of glycerol and freezing by storage in a -80 C mechanical freezer was one approach to long-term frozen preservation of RBC which required the post-thaw washing to reduce the intracellular cryoprotective agent glycerol prior to resuspension in plasma or the albumin-glucose medium and storage at 4 C.

The alternative approach to long-term frozen storage of blood utilized extracellular additives like hydroxyethyl starch and pentastarch and freezing using liquid nitrogen with storage in the liquid phase of liquid nitrogen at -197 C or in the gas phase of liquid nitrogen at -150 C. The liquid nitrogen approach was promoted by Linde Corporation and investigators in the Netherlands while the -80 C mechanical freezer approach was promoted by the U.S. Navy. The containers used to freeze the red blood cells and the transportation of the frozen RBC were issues that were investigated together with the length of the storage of the RBC in the frozen state at -80 C, -120 C, -150 C, -197 C, and storage in the liquid state at 22 C and 4 C following thawing together with the breakage of the frozen units.

Dr. Richard L. Veech, with whom I have collaborated from 1985 to the present, has been concerned that I was never trained to perform research. Dr. Veech had received his MD from Harvard Medical School and PhD working with Dr. Hans Krebs, a Nobel Prize laureate. I had attended Harvard Medical School where I did not learn anything except the medical words like hematocrit, bleeding time and nonsurgical blood loss but it took me 45 years to learn the meaning of these words. Dr. Veech does not understand how I accomplished what I did working at the NBRL without a mentor like Hans Krebs. I was fortunate to have worked with Dr. Mark David Altschule, the best clinician at the Harvard Medical School from 1954 to 1958 who worked at the NBRL for 15 years prior to his death. Mark Altschule told me that I had monumental ignorance and was a C minus performer, whereas Dr. Richard Veech and Dr. Richard Wurtman were two students at Harvard Medical School that were A performers. Dr. Altschule investigated my background and learned that what I did was all in my genes. My mother was a Piccolomini, her name was Domenica Piccolomini Valeri. Dr. Altschule reported that Pope Gullio the 2nd was a Piccolomini who sponsored Michelangelo. Dr. Altschule reported my genes were responsible for the creative research that our laboratory performed in the field of transfusion medicine for which I had no formal training. Dr. Altschule would tell me that it was all in the Piccolomini genes. Dr. Richard Veech now disagrees with the Piccolomini genes and reported that the Valeri genes were responsible. Dr. Veech reported that the Valeri family was one of the seven (7) most important families in the Republic of Rome 300 BC and recently the excavation beneath the Vatican in Rome discovered the Valeri tomb adjacent to the tomb of St. Peter.

Both Dr. Altschule and Dr. Veech failed to recognize that my accomplishments in science were due to my mother and my father. My mother Domenica Piccolomini Valeri provided me the following rules of life: a) it is better to be lucky than smart; b) be happy with what you have today because tomorrow you may have nothing; c) take the bad with the good; d) you can always do better; e) associate yourself with people from whom you can learn; f) at the time of your death you will be lucky to have friends equal to the number of fingers on one hand, excluding your thumb.

My father, Angelo Valeri, told me repeatedly: a) don't give a sucker an even break; b) I would always be a student and referred to me as "the student". Dr. Mark Altschule taught me clinical medicine and that medicine was an "art" and not a "science" and Dr. Veech taught me the chemistry of life, i.e. biochemistry. I owe much of my success to Dr. Mark Altschule and Dr. Richard Veech.

But my success is directly due to my mother who died at a very young age and my father who lived a long and good life. In addition, my wife Cynthia Ann Campbell provided the lyrics for all the music reported in the scientific reports that NBRL has published over the past 45 years related to Transfusion Medicine. I have been very lucky to have been born and grown up when affirmative action and political correctness were not the controlling preoccupation in our society. I had unlimited opportunity and freedom to investigate chaotic observation and serendipitous events without the restriction that are now present because of affirmative action, political correctness, governmental regulations, lawyers and biostatisticians that control our society and limit freedom and opportunity.

I was fortunate to live long enough to learn that patience is needed before serendipitous observations can be used in clinical medicine to improve the health care of our society. The utilization of the -80 C mechanical freezers to freeze RBC, platelets, and plasma by the Netherlands military has demonstrated the safety and therapeutic effectiveness of these frozen blood products without the need for fresh whole blood to treat military and civilian casualties requiring more than 10 units of red blood cells within a 24-hour period in combat zones in Afghanistan and Iraq. It is of interest that the Netherlands investigators pioneered the use of liquid nitrogen to freeze blood products but now utilize -80 C mechanical freezers to freeze red blood cells, platelets and plasma to treat combat casualties. The use of the -80 C mechanical freezer which is the temperature of dry ice and alcohol was used with high concentration of glycerol (40% W/V) to salvage fresh RBC that were discarded by Dr. Edwin J. Cohn who isolated plasma from fresh whole blood to prepare albumin using the Cohn Blood Fractionator. The fresh RBC that were routinely discarded were frozen with high concentration of glycerol (40% W/V) and frozen in dry ice and alcohol at -79 C using the protocol reported by Audrey Smith, a veterinarian who initially froze bull spermatozoa and then human red blood cells. The Cohn Blood Fractionator procedure was funded by the Office of Naval Research (ONR) to produce human albumin concentrates as the first blood substitute which has been shown to be a poor plasma volume expander and should not be used to resuscitate patients in hemorrhagic shock. The salvaging of the routinely discarded fresh red blood cells using 40% W/V glycerol and storage at -80 C has been shown to provide safe and therapeutically effective RBC following storage at -80 C for at least 21 years. The deployment of -80 C mechanical freezers to freeze RBC for at least 10 years, platelets for at least 2 years, and plasma for at least 10 years to treat wounded patients in a combat zone by the Netherlands military demonstrated that chaotic observations, serendipity and patience have provided blood products for clinical use that are safe and therapeutically effective to treat wounded military and civilian patients in a combat zone without the need for fresh whole blood.

The abstract which was reported by Dr. John Badloe and Dr. Femke Noorman from the Ministry of Defense, Military Blood Bank, Leiden, Netherlands at the Annual meeting of the AABB, San Diego, CA October 22-25, 2011 confirm the procedures provided by the NBRL, Boston, MA to freeze human RBC, platelets, and plasma in -80 C mechanical freezers. This report by the Netherlands military demonstrates that fresh whole blood advocated by Colonel William Crosby and the U.S. Army is no longer used by the Netherlands military and the fresh whole blood can be replaced by frozen RBC, frozen plasma and frozen platelets at a ratio of 1:1:1 to significantly improved survival of the massively transfused patients.

The following abstract was published in the Transfusion Supplement Volume 51, No. 35, and reported at the annual meeting of the American Association of Blood Banks (AABB) at San Diego, CA from October 22-25, 2011.

51-030D

The Netherlands Experience with Frozen −80°C Red Cells, Plasma and Platelets in Combat Casualty Care

J Badloe1 (jf.badloe@mindef.nl), F Noorman1.

1Ministry of Defense, Military Blood Bank, Leiden, Netherlands

Background/Case Studies: Since 1987 the Netherlands Military Blood Bank has worked closely with Dr CR Valeri for the production of −80°C frozen blood products. With the procedures of his Naval Blood Research Laboratory the Netherlands Military is able to provide frozen red cells since 1993 and frozen plasma and platelets since 2001 for peacekeeping and peace enforcing missions abroad the Netherlands. With the availability of these −80°C frozen blood products the "walking blood bank" and its potentially unsafe blood products are obsolete and this concept is thus safely abolished in 2001 by the Netherlands military. Since the introduction of 4°C storage of thawed red cells in 2004, the Netherlands military mainly use-80°C frozen blood products to cover operational needs. Here we describe the experiences with these products of NLD blood bank facilities in Afghanistan, from Aug 2006-April 2010.

Study Design/Methods: All −80°C frozen products are leukodepleted and of universal donor type, produced in the Netherlands, shipped at −80°C (dry ice) and stored in theatre at −80°C.Products are thawed on demand (red cells, plasma and platelets) or for 4°C storage after thaw (red cells 14 days and plasma 7 days). Occasionally, non frozen liquid red cells are sent as a supplement to cover (expected) higher usage. All products are in compliance with international regulations and guidelines.

Results/Findings: During the past 4.7 years, 1002 patients (83% Afghan) were transfused with 6164 −80°C frozen blood products (2168 red cell units, 2953 plasma units and 1043 platelet units) and 876 units liquid red cells. On one location, where all blood products were provided by the Netherlands Military Blood Bank, blood usage and survival were further analyzed. It showed that >95% of the transfused patients were trauma patients, of which 14% (48 out of 341) required more than 10 red cell units within 24 hours. In these massively transfused patients survival improved from 44% (N16) to 84% (N32) after the introduction of the new transfusion policy in Nov 2007. No walking blood bank was required and no shortages or transfusion reactions were reported.

Conclusion: Fully tested, frozen blood products, readily available after thaw proved to be a safe, available, effective and efficient blood support for combat casualty care and together with the use of a 1:1:1 ratio increased survival in MT patients significantly.

The recent paper by Henkelman S and Rakhorst G in Transfusion (52:2272-2273, 2012) "Does modern combat still need fresh whole blood transfusions?" reported that the Dutch military blood bank eliminated the use of fresh whole blood on site and implemented the routine use of frozen group O Rh positive and group O Rh negative deglycerolized RBC, frozen single donor leukoreduced group O platelets, and fresh frozen AB plasma to treat wounded casualties in war zones. The RBC, platelets, and plasma are frozen and stored at -80 C in mechanical freezers. Following thawing, the deglycerolized RBC are stored at 4 C in the additive solution AS-3 for 14 days, the thawed AB plasma stored at 4 C for 7 days, and the thawed single donor leukoreduced platelets containing 2.5 to 3.0 X 10^{11} platelets in AB plasma stored at room temperature without agitation for 6 hours.

The authors reported that fresh whole blood (FWB) needs to be tested for infectious transmissible disease and bacterial contamination. In addition, the transfusion of FWB has been associated with a high incidence of febrile non-hemolytic transfusion reactions, human leukocyte antigen alloimmunization and transfusion associated graft versus host disease (TA-GVHD). They reported that Gilstad C and associates published in Transfusion (52:930-935, 2012) clinical symptoms of TA-GVHD in a trauma patient who was resuscitated with FWB. The Dutch army has demonstrated that frozen RBC, frozen platelets, and frozen plasma are safe, effective, and provide an adequate blood supply. The authors report that infusion of FWB in military settings is not recommended because of the possible risk of disease transmission, burden to the military personnel, and is not justified when frozen blood products are utilized. The authors were concerned with the potential that FWB can produce graft versus host disease in recipients. The frozen RBC containing 40% W/V glycerol and stored at -80 C can be treated with 2500 cGy of gamma radiation to inactivate the immunocompetent lymphocytes in the deglycerolized RBC as reported by Valeri CR and associates in Transfusion 41:545-549, 2001. The data show similar acceptable results for RBC frozen with 40% W/V glycerol at -80 C and treated in the frozen state with 2500 cGy of gamma radiation and for RBC that were not irradiated, all of which were washed and then stored in a sodium chloride glucose solution for 3 days before autologous transfusion.

Universal donor group O Rh positive and group O Rh negative RBC, group O single donor leukoreduced platelets resuspended in AB plasma, and AB plasma eliminate the need to provide fresh whole blood to treat patients subjected to traumatic injuries in war zones. The current method to freeze single donor leukoreduced platelets containing 2.5 – 3 X 10^{11} platelets equivalent to the number of platelets isolated from 4 to 6 units of fresh whole blood treated with 5% DMSO in 0.9% NaCl, the supernatant DMSO removed prior to freezing and storage in a -80 C mechanical freezer, and following thawing resuspended in a unit of AB plasma eliminate the need to provide the platelets in fresh whole blood and single donor leukoreduced apheresed platelets stored at room temperature for 5 days with agitation.

Employment of personnel to work at the U.S. NBRL was limited by the number of civil servants available to work at the U.S. NBRL located at Chelsea Naval Hospital and then at the 615 Albany Street site. Personnel were hired at University Hospital by Dr. C. P. Emerson principal investigator on a contract between University Hospital and U.S. NBRL from 1965 to 1979 to work at the U.S. NBRL on this contract. Dr. C.P. Emerson and Mr. Howard Buzzee, the Director of Personnel at University hospital approved the policy that University Hospital personnel working at the U.S. NBRL were able to accrue both annual and sick leave without limitation. Ongoing studies required that University Hospital personnel who worked at the U.S. NBRL located at Chelsea Naval Hospital and then at the 615 Albany Street site were permitted to accrue annual leave and sick leave without limitation because of the requirement that UH personnel were needed each day, including Saturday and Sunday, to treat and evaluate normal volunteers, patients, and baboons that were being studied.

Mr. Howard Buzzee, who was the Director of Personnel at University Hospital (UH) recognized the rights and responsibility of both the employees and the principal investigator. Dr. C. P. Emerson at UH who was contracted by the U.S. NBRL between 1965 to 1979 to perform specific studies and employed personnel at UH to work at the U.S. NBRL. The Office of Human Resources was concerned with the formation of a union at BUMC and made every effort to support the employees. To avoid the employment of personnel at NBRL from 1986 to 2004, independent contractors were hired to perform specific assignments and when the assigned projects were completed the independent contractors were terminated.

Dr. Valeri had the responsibilities to perform the research that was approved for funding by government contracts and by commercial companies.

BU needed the 615 Albany Street site for their real estate development and evicted the NBRL on December 31, 2004. The frozen blood products, the equipment, the publications, the personnel records, the research records, FDA data, and all the material at the NBRL and in its storage area were removed and sent to two storage facilities which the NBRL, Inc., a not for profit research laboratory, had to pay for. BUMC refused to provide funds to store the frozen blood products produced by the NBRL over the past 20 years when NBRL was a contract laboratory with Boston University Medical Center. The 615 Albany Street building was secured on September 1, 2004 and entrance into this building was permitted by the BUMC Security Department.

In September 1999 and in November 1999, Dr. Aram Chobanian did not want to negotiate the Congressional contracts between the NBRL and BUMC because the 615 Albany Street site was needed for BUMC as part of the Biomedical Research Complex development on Albany Street. The NBRL was supported by Congressman J. Joseph Moakley who was responsible for the Congressional funds provided to the NBRL in 1998, 1999, 2000, 2001, 2002 and 2003. The U.S. Navy negotiated with UH to accept for one dollar the 615 Albany street building donated to University Hospital by New England Nuclear Company in 1974. The U.S. Navy decontaminated the building and renovated the building for 2 million dollars. The NBRL utilized this government renovated building from April 1974 until BU evicted the NBRL on December 31, 2004. Dr. Valeri was evicted from the 615 Albany Street site because the building was reverted to BUMC on December 31, 2003. Dr. Valeri had negotiated the congressional funds provided by the U.S. Congress for October 1, 2003 to September 30, 2004 with New England Medical Research Institute (NEMRI) at West Roxbury VA Hospital and the NBRL, Inc., a not for profit research laboratory and performed government and commercial supported research on frozen platelets, resuscitation solutions and hemostatic agents at the Brigham and Women's Hospital in collaboration with Dr. Dennis Orgill and Dr. Herbert Hechtman from 2003 to 2010.

The NBRL was relevant to Boston University Medical Center (BUMC) from July 1965 to December 31, 2003 and provided at least 10 million dollars in overhead to BUMC and a renovated building at 615 Albany Street, Boston, MA. The NBRL become irrelevant on December 31, 2003 when BU needed the building for the Bioterrorism Center which the U.S. Congress provided $215,000,000 to build and a 2.4 billion dollar contract to operate.

CHAPTER 1 REFERENCES

1. Valeri CR, Henderson ME. Recent difficulties with frozen glycerolized blood. JAMA 188:1125-1131, 1964.
2. Valeri CR, Fortier NL. Red-cell 2,3-diphosphoglycerate and creatine levels in patients with red-cell mass deficits or with cardiopulmonary insufficiency. NEJM 81:1452-1455, 1969.
3. Valeri CR, Altschule MD, Pivacek LE. The hemolytic action of adrenochrome, an epinephrine metabolite. J Med 3:20-40, 1972.
4. Valeri CR, Cooper AG, Pivacek LE. Limitations of measuring blood volume with iodinated I 125 serum albumin. Arch Int Med 132:534-538, 1973.
5. Valeri CR, Altschule MD. Hemolysis in vitro of blood obtained from patients with traumatic injuries. J Trauma 13:678-686, 1973.
6. Valeri CR, Feingold, H, Marchionni LD. A simple method for freezing human platelets using 6% dimethylsulfoxide and storage at -80 C. Blood 43:131-136, 1974.
7. Valeri CR, Feingold H, Marchionni LD. The relation between response to hypotonic stress and the ^{51}Cr recovery in vivo of preserved platelets. Transfusion 14:331-337, 1974.
8. Valeri CR. Hemostatic effectiveness of liquid-preserved and previously frozen human platelets. NEJM 290:353-358, 1974.
9. Valeri CR. Oxygen transport and viability of preserved red blood cells. J Med 5:278-291, 1974.
10. Valeri CR, Valeri DA, Anastasi J, Vecchione JJ, Dennis RC, Emerson CP. Freezing in the primary polyvinylchloride plastic collection bag: A new system for preparing and freezing nonrejuvenated and rejuvenated red blood cells. Transfusion 21:138-149, 1981
11. Melaragno AJ, Carciero R, Feingold H, Talarico L, Weintraub L, Valeri CR. Cryopreservation of human platelets using 6% dimethylsulfoxide and storage at -80C. Effects of 2 years of frozen storage at -80C and transportation in dry ice. Vox Sang 49:245-258, 1985.
12. Valeri CR. Cryopreservation of human platelets and bone marrow and peripheral blood totipotential mononuclear stem cells. Ann N Y Acad Sci 459:353-66, 1985.

13. Carciero R, Valeri CR. Isolation of mononuclear leukocytes in a plastic bag system using Ficoll-Hypaque. Vox Sang 49:373-380, 1985.
14. Valeri CR, Donahue K, Feingold HM, Cassidy GP, Altschule MD. Increase in plasma volume after the transfusion of washed erythrocytes. Surg Gynec Obstet 162:30-36, 1986.
15. Vogel WM, Dennis RC, Cassidy GP, Apstein CS, Valeri CR. Coronary constrictor effect of stroma-free hemoglobin solutions. Am J Physiol (Heart Circ Physiol 20) 251:H413-H420, 1986.
16. Valeri CR, Feingold H, Melaragno AJ, Vecchione JJ. Cryopreservation of dog platelets with dimethylsulfoxide: Therapeutic effectiveness of cryopreserved platelets in the treatment of thrombocytopenic dogs, and the effect of platelet storage at -80 C. Cryobiology 23:387-394, 1986.
17. Vogel WM, Lieberthal W, Apstein CS, Levinsky N, Valeri CR. Effects of stroma-free hemoglobin solutions on isolated perfused rabbit hearts and isolated perfused rat kidneys. Biomater Artif Cells Artif Organs 16(1-3):227-35, 1988.
18. Valeri CR, Pivacek LE, Gray AD, Cassidy GP, Leavy ME, Dennis RC, Melaragno AJ, Niehoff J, Yeston N, Emerson CP, Altschule MD. The safety and therapeutic effectiveness of human red cells stored at -80C for as long as 21 years. Transfusion 29:429-437, 1989.
19. Crowley JP, Valeri CR, Metzger J, Gray A, Schooneman F, Man NK, Merrill E: The estimation of whole blood viscosity by a porous bed method. Am J Clin Pathol 96:729-737, 1991.
20. Khuri SF, Wolfe JA, Josa M, Axford TC, Szymanski I, Assousa S, Ragno G, Patel M, Silverman A, Park M, Valeri CR. Hematologic changes during and after cardiopulmonary bypass and their relationship to the bleeding time and nonsurgical blood loss. J Thorac Cardiovasc Surg 104:94-107, 1992.
21. Valeri CR, Khabbaz K, Khuri SF, Marquardt C, Ragno G, Feingold H, Gray AD, Axford T. Effect of skin temperature on platelet function in patients undergoing extracorporeal bypass. J Thorac Cardiovasc Surgery 104:108-116, 1992.
22. Cordts PR, LaMorte WW, Fisher JB, DelGuercio C, Niehoff J, Pivacek LE, Dennis RC, Siebens H, Giorgio A, Valeri CR, Menzoian JO. Poor predictive value of hematocrit and

hemodynamic parameters for erythrocyte deficits after extensive elective vascular operations. Surg Gynec Obstet 175:243-248, 1992.

23. Rosenblatt MS, Hirsch EF, Valeri CR. Frozen red blood cells in combat casualty care: Clinical and logistical considerations. Milit Med 159:392-397, 1994
24. Valeri CR, MacGregor H, Cassidy G, Tinney R, Pompei F. Effects of temperature on bleeding time and clotting time in normal male and female volunteers. Crit Care Med 23:698-704, 1995.
25. Khuri SF, Valeri CR, Loscalzo J, Weinstein MJ, Birjiniuk V, Healey NA, MacGregor H, Doursounian M, Zolkewitz MA. Heparin causes platelet dysfunction and induces fibrinolysis before cardiopulmonary bypass. Ann Thorac Surg 60:1008-14, 1995.
26. Thompson A, Valeri CR, Lieberthal W. Endothelin receptor A blockade alters hemodynamic response to nitric oxide inhibition in rats. Am J Physiol 269 (Heart and Circ Physiol 38):H-743-H748, 1995.
27. Valeri CR, Pivacek LE. Effects of the temperature, the duration of frozen storage, and the freezing container on in vitro measurements in human peripheral blood mononuclear cells. Transfusion 36:303-308, 1996.
28. Valeri CR, Crowley JP, Loscalzo. The red cell transfusion trigger: has a sin of commission now become a sin of omission? Transfusion 38:602-610, 1998.
29. Michelson AD, Barnard MR, Khuri SF, Rohrer MJ, MacGregor H, Valeri CR. The effects of aspirin and hypothermia on platelet function in vivo. Br J Haematol 104:64-68, 1999.
30. Khuri SF, Healey N, MacGregor H, Barnard MR, Szymanski IO, Birjiniuk V, Michelson AD, Gagnon DR, Valeri CR. Comparison of the effects of transfusions of cryopreserved and liquid-preserved platelets on hemostasis and blood loss after cardiopulmonary bypass. J Thorac Cardiovasc Surg. 117:172-184, 1999.
31. Barnard MR, MacGregor H, Mercier R, Ragno G, Pivacek LE, Hechtman HB, Michelson AD, Valeri CR. Platelet surface p-selectin, platelet-granulocyte heterotypic aggregates, and plasma soluble p-selectin during plateletpheresis. Transfusion 39:735-741, 1999.
32. Valeri CR, Pivacek LE, Cassidy GP, Ragno G. Posttransfusion

survival (24-hour) and hemolysis of previously frozen, deglycerolized RBCs after storage at 4 C for up to 14 days in sodium chloride alone or sodium chloride supplemented with additive solutions. Transfusion 40:1337-1340, 2000.

33. Valeri CR, Pivacek LE, Cassidy GP, Ragno G. The survival, function, and hemolysis of human RBCs stored at 4 C in additive solution (AS-1, AS-3, or AS-5) for 42 days and then biochemically modified, frozen, thawed, washed, and stored at 4 C in sodium chloride and glucose solution for 24 hours. Transfusion 40:1341-1345, 2000.

34. Valeri CR, Ragno G, Pivacek LE, Cassidy GP, Srey R, Hansson-Wicher M, Leavy ME. An experiment with glycerol-frozen red blood cells stored at –80 C for up to 37 years. Vox Sang 79:168-174, 2000.

35. Valeri CR, Ragno G, Pivacek LE, Srey R, Hess JR, Lippert LE, Mettille F, Fahie R, O'Neill EM, Szymanski IO. A multicenter study of in vitro and in vivo values in human RBCs frozen with 40-percent (wt/vol) glycerol and stored after deglycerolization for 15 days at 4 C in AS-3: assessment of RBC processing in the ACP 215. Transfusion 41:933-939, 2001.

36. Valeri CR, Ragno G, Pivacek LE, O'Neill EM. In vivo survival of apheresis RBCs, frozen with 40-percent (wt/vol) glycerol, deglycerolized in the ACP 215, and stored at 4 C in AS-3 for up to 21 days. Transfusion 41:928-932, 2001.

37. Valeri CR, Cassidy G, Pivacek LE, Ragno G, Lieberthal W, Crowley JP, Khuri SF, Loscalzo J. Anemia-induced increase in the bleeding time: implications for treatment of nonsurgical blood loss. Transfusion 41:977-983, 2001.

38. Valeri CR, Pivacek LE, Cassidy GP, Ragno G. Volume of RBCs, 24- and 48-hour posttransfusion survivals, and the lifespan of (51)Cr and biotin–X-N-hydroxysuccinimide (NHS)-labeled autologous baboons RBCs: effect of the anticoagulant and blood pH on (51)Cr and biotin-X-NHS elution in vivo. Transfusion 42:343-8, 2002.

39. Valeri CR, MacGregor H, Giorgio A, Ragno G: Circulation and hemostatic function of autologous fresh, liquid preserved, and cryopreserved baboon platelets transfused to correct an aspirin-induced thrombocytopathy. Transfusion 42:1206-16, 2002.

40. Valeri C: Status report on the quality of liquid and frozen red blood cells. Vox Sang 83(1):193-196, 2002.
41. Valeri CR, Giorgio A, MacGregor H, Ragno G. Circulation and distribution of autotransfused fresh, liquid-preserved and cryopreserved baboon platelets. Vox Sang 83:347-351, 2002.
42. Koustova E, Rhee P, Hancock T, Chen H, Inocencio R, Valeri CR, Alam HB. Ketone and pyruvate Ringer's solutions decrease pulmonary apoptosis in a rat model of severe hemorrhagic shock and resuscitation. Surgery 134(2):267-274, 2003.
43. Valeri CR, Srey R, Lane JP, Ragno G. Effect of WBC reduction and storage temperature on PLTs frozen with 6 percent DMSO for as long as 3 years. Transfusion 43(8):1162-7, 2003.
44. Valeri CR, MacGregor H, Giorgio A, Srey R, Ragno G. Comparison of radioisotope methods and a non-radioisotope method to measure the RBC volume and RBC survival in the baboon. Transfusion 43(10): 1366-1373, 2003
45. Valeri CR, Srey R, Tilahun D, Ragno G. In vitro effects of polymerized N-acetyl glucosamine (NAG) on the activation of platelets in platelet rich plasma with and without red blood cells. J Trauma (suppl) 57:S22-S25, 2004.
46. Valeri CR, Srey R, Tilahun D, Ragno G. In vitro quality of red blood cells frozen with 40% w/v glycerol at -80C for 14 years, deglycerolized with the Haemonetics ACP 215, and stored at 4C in additive solution-1 or additive solution-3 for up to 3-weeks. Transfusion 44:990-995, 2004.
47. Valeri CR, MacGregor H, Giorgio A, Ragno G: Comparison of radioisotope methods and a non-radioisotope method to measure platelet survival in the baboon. Transfusion and Apheresis Science 32(3):275-81, 2005.
48. Valeri CR, MacGregor H, Ragno G. Correlation between in vitro aggregation and thromboxane A2 production in fresh, liquid-preserved, and cryopreserved human platelets: Effects of agonists, pH, and plasma and saline resuspension. Transfusion 45:596-603, 2005.
49. Valeri CR, Ragno G, Khuri S. Freezing human platelets using 6% DMSO with removal of the supernatant solution prior to freezing and storage at -80C without post-thaw processing. Transfusion 45(12):1890-1898, 2005.

50. Valeri CR, Ragno G, van Houten P, Rose L, Rose M, Egozy E, Popovsky M. Automation of the glycerolization of the RBC using the high separation bowl in the Haemonetics ACP215 instrument. Transfusion 45(10):1621-1627, 2005.
51. Valeri CR, Ragno G. The effect of storage of fresh frozen plasma at -80 C for as long as 14 years on plasma clotting proteins. Transfusion 45(11):1829-1830, 2005.
52. Valeri CR, Dennis RC, Ragno G, MacGregor H, Menzoian JO, Khuri SF. Limitations of the hematocrit to assess the need for RBC transfusion in hypovolemic anemic patients. Transfusion 46:365-371, 2006.
53. Valeri CR, Ragno G, Veech RL. Effects of the resuscitation fluid and the HBOC excipient on the toxicity of the HBOC: Ringer's D,L-lactate, Ringer's L-lactate, Ringer's ketone solutions. Art Cells, Blood Substitutes and Biotechnology 34(6):601-606, 2006.
54. Jaskille A, Koustova E, Rhee P, Britten-Web J, Chen H, Valeri CR, Kirkpatrick JR, Alam HB. Hepatic apoptosis following hemorrhagic shock in rats can be reduced through modifications of conventional Ringer's solution. J Am Coll Surg 202(1):25-35, 2006.
55. Valeri CR, Ragno G. Cryopreservation of human blood products. Trans Apher Sci 34:271-287, with an editorial on pages 267-269, 2006.
56. Valeri CR, Ragno G. The effects of preserved red blood cells on the severe adverse events observed in patients infused with hemoglobin based oxygen carriers. Artificial Cells, Blood Substitutes and Biotechnology 36 (1):3-18, 2008.
57. Valeri CR, Ragno G, Veech RL. Severe adverse events associated with hemoglobin based oxygen carriers: role of resuscitative fluids and liquid preserved RBC. Trans Apher Sci 39:205-211, 2008.
58. Valeri CR, Ragno G. Role of nitric oxide in the prevention of severe adverse events associated with blood products. Trans Apher Sci 39:241-245, 2008.
59. Pietramaggiori G, Yang HJ, Scherer SS, Kaipainen A, Chan RK, Alperovich M, Newalder J, Demcheva M, Vournakis JN, Valeri CR, Hechtman HB, Orgill DP. Effects of poly-N-acetyl

glucosamine (pGLcNAc) patch on wound healing in db/db mouse. J Trauma 64(3):803-808, 2008.

60. Fischer TH, Valeri CR, Smith CJ, Scull CM, Merricks EP, Nichols TC, Demcheva M, Vournakis JN. Non-classical processes in surface hemostasis: mechanisms for the poly-N-acetyl glucosamine-induced alteration of red blood cell morphology and surface prothombogenicity. Biomed Mater 3(1):1-9, 2008.
61. Pietramaggiori G, Scherer SS, Mathews JC, Lancerotto L, Gennaoui A, Ragno G, Valeri CR, Orgill DP. Quiescent platelets stimulate angiogenesis and diabetic wound repair. J Surg Res 160:169-177, 2008.
62. Scherer SS, Pietramaggiori G, Matthews J, Perry S, Assmann A, Carothers A, Demcheva M, Muise-Helmericks RC, Seth A, Vournakis JN, Valeri CR, Fischer TH, Hechtman HB, Orgill DP. Poly-N-acetyl glucosamine (pGlcNAc) nano-fibers: a new bioactive material to enhance diabetic wound healing by cell migration and angiogenesis. Ann Surg 250:322-330, 2009.
63. Valeri CR, Ragno G. An approach to prevent the severe adverse events associated with transfusion of FDA-approved blood products. Trans Aph Sci 42:223-33, 2010.
64. Valeri CR, Vournakis JN. The mRDH bandage for surgery and trauma. J. Trauma 71:S162-S166, 2011.
65. Fischer TH, Hays WE, Valeri CR. Poly-N-acetyl glucosamine materials accelerate hemostasis in patients treated with anti-platelet drugs. J. Trauma 71:S176-S182, 2011.
66. Scherer SS, Pietramaggiori G, Matthews J, Gennaoui T, Demcheva M, Fischer TH, Valeri CR, Orgill DP. Poly-N-acetyl glucosamine membranes induce angiogenesis and wound healing in ADP inhibitor treated diabetic mice. J. Trauma 71:S183-S186, 2011.
67. Erba P, Adini A, Demcheva M, Valeri CR, Orgill DP. Poly-N-acetyl glucosamine fibers are synergistic with vacuum assisted closure in augmenting the healing response of diabetic mice. J. Trauma 71:S187-S193, 2011.
68. Valeri CR, Ragno G. Letter regarding Stinner DJ and associates "Prevalence of late amputations during the current conflicts in Afghanistan and Iraq". Military Medicine 176:11, 2011.
69. Veech RL, Valeri CR, VanItallie TB. The mitochondrial permeability transition pore provides a key to the diagnosis and

treatment of traumatic brain injury. IUBMB Life 64(2):203-207, 2012.

70. Valeri CR, Veech RL. The unrecognized effects of the volume and composition of the resuscitation fluid used during the administration of blood products. Trans Aph Sci 46:121-123, 2012.
71. Valeri CR. Blood Banking and the Use of Frozen Blood Products. Chemical Rubber Company, Boca Raton, FL, 1976.
72. Valeri CR, Altschule MD. Hypovolemic Anemia of Trauma: The Missing Blood Syndrome, Chemical Rubber Company, Boca Raton, FL, 1981.
73. Valeri CR. Use of rejuvenation solutions in red blood cell preservation. CRC Crit Rev Clin Lab Sci 17:299-374, 1982.
74. Valeri Cr, Ragno G: The survival and function of baboon red blood cells, platelets and plasma proteins: a review of the experience from 1972 to 2002 at the Naval Blood Research Laboratory, Boston, Massachusetts. Transfusion 46(8):1-42, 2006.
75. Valeri CR, Khuri S, Ragno G. Non-surgical bleeding diathesis in anemic thrombocytopenic patients: role of temperature, RBC, platelets, and plasma clotting proteins. Transfusion 47:206S-248S, 2007.
76. Henkelman S, Rakhorst G: Does modern combat still need fresh whole blood transfusions? Transfusion 52:2272-2273, 2012.
77. Gilstad C, Roschewski M, Wells J, Delmas A, Lackey J, Uribe P, Popa C, Jardeleza T, Roop S. Fatal transfusion-associated graft-versus-host disease with concomitant immune hemolysis in a group A combat trauma patient resuscitated with group O fresh whole blood. Transfusion 52:930-935, 2012.
78. Valeri CR, Pivacek LE, Cassidy GP, Ragno G. In vitro and in vivo measurements of gamma-radiated, frozen, glycerolized RBCs. Transfusion 41:545-549, 2001.
79. Lelkens CCM, Koning JG, de Kort B, Floot IBG, Noorman F: Experiences with frozen blood products in the Netherlands military. Trans Aph Sci 34:289-296, 2006.
80. Noorman F, et al: Frozen -80 C red cells, plasma and platelets in combat casualty care. Transfusion Suppl 49(35):28A, 2009.
81. Badloe J, Noorman F: The Netherlands experience with frozen

-80 C red cells, plasma and platelets in combat casualty care S1-301. Transfusion Suppl 51(35), 2011.

82. Noorman F, Badloe JF. -80 C frozen platelets - efficient logistics, available, compatible, safe and effective in the treatment of trauma patients with or without massive blood loss in military theatre. AABB Oct 2012, Transfusion practice/clinical case studies, S53-030H, oral presentation ; Transfusion 52(Suppl) S53-030H, 53A, 2012.
83. Noorman F, Strelitski R, Badloe JF. -80 C frozen platelets are activated compared to 24 hour liquid stored platelets and quality of frozen platelets is unaffected by a quick preparation method (15 min) which can be used to prepare platelets for the early treatment of trauma patients in military theatre. AABB Oct 2012, Components and component processing: Platelets ; SP23, Poster presentation, Transfusion 52(Suppl) SP23, 62A, 2012.
84. Noorman F, Strelitski R, Badloe JF. Lyophilized plasma, an alternative to 4 C stored thawed plasma for the early treatment of trauma patients with (massive) blood loss in military theatre. AABB Oct 2012, Components and component processing: Plasma, SP7, Poster presentation, Transfusion 52(Suppl) SP7, 55A, 2012.
85. Noorman F, Badloe JF. -80 C frozen red blood cells, plasma and platelets : efficient logistics, available, compatible, safe and effective in the treatment of trauma patients with or without massive blood loss in military theatre. AABB Oct 2012, SP383 ; Poster presentation. Transfusion 52(Suppl) SP383, 198A, 2012.

CHAPTER 2

DECADES AND EVENTS: WHAT OCCURRED AND WHAT WAS LEARNED. PART 1.

This book was written to review the chaotic-serendipitous observations that were made at the NBRL during the past 45 years which stimulated the research that was supported by Naval Medical Research and Development Command of the Navy's Bureau of Medicine and Surgery, Office of Naval Research, the Congress of the United States and commercial funds. The serendipitous observations were made working with the numerous academic and commercial collaborators who provided technology and expertise to conduct the research. These investigations were made using the new technologies that were made available over the past 5 decades. This book describes just plain old luck - serendipity that Horace Walpole in 1789 defined as making discoveries by accident and from chaotic-serendipitous observations like the examples of Alfred Noble and dynamite, Marie Curie and radium, and Alexander Fleming and penicillin.

The Cohn Blood Fractionator was designed and developed by Dr. Edward J. Cohn to isolate albumin from fresh whole blood as the first blood substitutes and the fresh red blood cells were discarded. Dr. Edward J Cohn, Dr. James Tullis and their associates utilized the protocols reported by Dr. Audrey Smith to salvage red blood cells isolated in the Cohn Blood Fractionator by treatment with a high concentration of glycerol and freezing at -79C, the temperature of dry ice and alcohol. Harris Refrigeration Company in the Boston area designed water-cooled, mechanical freezers to maintain temperatures at -80C and -120 C to freeze the human RBC containing 40-45 W/V glycerol. Two serendipitous events occurred: 1) the utilization of the Cohn Blood Fractionator to isolate albumin from plasma obtained from fresh blood and 2) the salvaging of the fresh red blood cells by treatment with high concentrations of glycerol and storage at -80C using the protocol reported by Dr. Audrey Smith initially to freeze bull spermatozoa and then to freeze human red blood cells. (Smith AU, Polge C, Vet Res 62:115-116, 1950 and Smith AU, Lancet 2:910-911, 1950.)

Glycerol is an intracellular cryoprotectant which requires that the thawed glycerolized RBC need to be washed to reduce the residual glycerol concentration to less than 1% prior to transfusion to prevent intravascular hemolysis. The Cohn Blood Fractionator was utilized to deglycerolize the RBC and to resuspend the deglycerolized RBC in autologous plasma or in an albumin-glucose medium. The speculation by Tullis and his associates was that transmembrane washing to remove the glycerol would reduce or eliminate the hepatitis virus and the resuspension of the deglycerolized RBC in the albumin-glucose resuspension medium instead of the autologous plasma would prevent the transmission of hepatitis. Hepatitis was the major disease transmitted by blood transfusion in the 1950 and 1960 period. These two serendipitous observations were involved in the development of freezing human RBC with 40% W/V glycerol and storage at -80C: a) the Cohn Blood Fractionator to isolate plasma from fresh whole blood to prepare albumin which was the first blood substitute and b) the salvaging of fresh RBC by Dr. Audrey Smith report that human RBC glycerolized to 40-45% W/V could be frozen at -80C in mechanical freezers.

In 1965, Dr. David Danon from the Weizman Institute of Science, Rehovoth, Israel was on his way to visit Dr. Paul Marks at Columbia University to provide Dr. Paul Marks with the fragiligraph that Dr. Danon designed and developed to measure continuous osmotic fragility of RBC. In addition to the fragiligraph, Dr. Danon provided to the U.S. Naval Blood Research Laboratory (NBRL) the phthalate esters to measure the density distribution of the RBC. Instead of providing the fragiligraph and the phthalate esters to Dr. Paul Marks, Dr. David Danon left the fragiligraph and the phthalate esters at the NBRL. These two procedures were utilized to measure the osmotic fragility and the density distribution of fresh and preserved RBC. The osmotic fragility and density distribution of RBC were utilized at the NBRL to correlate these measurements to the in vivo survival of the RBC preserved in the liquid state and the frozen state (Danon D, J Lab Clin Path 16:337-342, 1963; Danon D and Markikovsky Y, J Lab Clin Med 64:668-674, 1964.

In 1954, Valtis DJ and Kennedy AC reported the defective gas transport of RBC stored at 4C in acid citrate dextrose anticoagulated whole blood for one week. Dr. Alistair Bellingham from England visited the NBRL in 1968. Dr. Alistair Bellingham provided the Bellingham and Huehns tonometer which was attached to a cuvette used in the Unicam spectrophotometer to measure the RBC affinity for oxygen (RBC P50 value) - p02 at which 50% of the hemoglobin was oxygenated for fresh, liquid preserved, and frozen RBC. Over the past 4 decades the RBC p50 has been measured at the NBRL using the Unicam spectrophotometer and the Bellingham and Huehns tonometer, the IL282 Co-oximeter, the Hemoscan and the Hemoxanalyzer instruments to assess the function of fresh and preserved RBC (Valtis DJ, Kennedy AC, Lancet 1:119-124, 1954; Bellingham AJ and Huehns ER, Forvarsmedicin 5:207-211, 1969; Festa RS, Asakura T, Transfusion 19:107-113, 1979; Asakura T, Reilly MP, In Nicolau C, editor Oxygen Transport Red Blood Cells, New York, Pergamon Press, 1986, pgs 57-75; Dennis RC, Valeri CR, Clin Chem 26:1304-1308, 1980).

Dr. Fabian Lionetti, Dr. Normand Fortier and Dr. Alan Keitt established the methods to measure serum haptoglobin level and RBC ATP, DPG, and creatine levels in the 1970 period. In addition, glutathione level, glutathione stability, glucose-6-phosphate dehydrogenase, hexokinase, gluthatione reductase, and lipid levels in fresh and preserved RBC were measured. Dr. Eugene Cronkite and Dr. James Bond from Brookhaven National Laboratory, Long Island, NY assisted the NBRL to count platelets by phase microscopy and by the Coulter counter technology using the Coulter Model B, Model JT, Model STKS, and ZBI counter with the H4 channelyzer attachment (Cronkite EP, Transfusion 6:18-22, 1966; Handin RI et al, Am J Clin Path 56:661-664, 1971; Lionetti FJ et al, Transfusion 6:116-123, 1966; Valeri CR, McCallum LE, Nature 205:561-563, 1965; Valeri CR et al, Clin Chem 11:581-588, 1965).

Dr. Jonathan Costa at the National Institute of Health (NIH) and Dr. David Shepro at the Biology Department of Boston University quantitated the platelet dense bodies by electron microscopy. Dr. Lee Rodkey at the Naval Medical Research Institute (NMRI), Bethesda, MD assayed blood samples for carboxyhemoglobin concentration to assess the degradation of the nonviable compatible RBC to increase the carboxyhemoglobin level in the blood (Costa J et al, FEBS Letters 99:141-146, 1979; Sweetman HE et al, Thrombosis Res 17:55-61, 1980).

In 1965 Dr. Richard Rosenfield and Dr. I.O. Szymanski introduced the Technicon Autoanalyzer to perform automated differential agglutination to measure the survival of compatible but identifiable RBC. Dr. CP Emerson and Ms. Rose Aloia collaborated with the NBRL to measure the survival of compatible but identifiable RBC by the manual procedure using light microscopy. The NBRL has utilized the automated differential agglutination procedure using the Technicon Autoanalyzer and the Coulter Counter to measure the survival of compatible but identifiable RBC using anti-A, anti-B, and anti-M antibodies. (Szymanski IO et al, Transfusion 8:65-73, 1968; Szymanski IO, Valeri CR, Transfusion 8:74-83, 1968; Valeri CR et al, Vox Sang 49:195-205, 1985).

The NBRL collaborated with Dr. Mark D. Altschule and Dr. Edwin Gordy, the last post-doctoral fellow of Dr. David Drabkin at the University of Pennsylvania to measure the percent oxygen saturation of hemoglobin, percent carboxyhemoglobin, percent methemoglobin and total hemoglobin by the cyanmethemoglobin method using the IL282 co-oximeter. In addition, the Lex02Con Instrument was used to measure oxygen content in blood using the galvanic fuel cell and by the Van Slyke procedure. Nitric oxide was measured using chemi-luminescence and photolysis in collaboration with Dr. J. Stamler and Dr. J. Loscalzo. Dr. E. Merrill C.P. Dubbs Professor of Chemistry at M.I.T. provided the porous bed viscometers to measure blood viscosity at shear rate of 19 inverse seconds. Working in collaboration with Dr. A Michelson and M. Barnard the flow cytometer was utilized to assess platelets, platelet microparticles, and platelet and WBC heterotypic aggregates in fresh and preserved blood products and in normal volunteers, baboons, and patients prior to and following transfusion. F. Pompei provided the NBRL with the infrared laser instrument to measure the temperature of the skin adjacent to the template bleeding time site. (Merrill EW, Physiol Rev 49:863-888, 1969; Valeri CR et al, J. Lab Clin Med 79:1035-1040, 1972; Dennis RC, Valeri CR, Clin Chem 26:1304-1308, 1980; Merrill EW et al, Lab Medica Int 10:19-24, 1993; Crowley JP et al, Am J Clin Pathol 96:729-737, 1991; Stamler JS et al: Proc Natl Acad Sci US: 89:7674-7677, 1992; Kestin AS et al, Blood 82:107-117, 1993; Michelson AD et al, J Thromb Haemos 5:633-640, 1994; Valeri CR et al, Crit Care Med 23:698-704, 1995).

In 1965, Dr. M. Strumia, Chairman of the National Research Council, reported the need to measure the RBC volume of the recipient at the time of the transfusion of autologous RBC labeled with radioisotopes and allogeneic compatible identifiable RBC measured by differential agglutination using anti-A, anti-B, anti-M antibodies in the Technicon Autoanalyzer and in the Coulter Counter. Dr. M Strumia reported that radiolabeled human albumin can be used in normal volunteers to estimate the plasma volume. Dr. Strumia reported that the total blood volume and the RBC volume in normal volunteers can be estimated from the plasma volume measured using radiolabeled human albumin and total body hematocrit. However, Dr. Strumia reported that Evans blue dye and radiolabeled albumin used in the Volemetron Instrument could not used to accurately measure the plasma volume in patients. (Strumia MM et al, Blood 13:128-145, 1958; Strumia MM et al, Blood 19:115-122, 1962; Strumia MM et al, Transfusion 8:197-209, 1968; Valeri CR et al, Arch Int Med 132:534-538, 1973).

Dr. Max Strumia reported that in patients radiolabeled albumin with a molecular weight of 68,000-7S molecules overestimated the plasma volume because of its rapid distribution into the extravascular space. At the NBRL, Dr. A. Cooper, who worked with Sir John Dacie in England isolated, purified and radiolabeled cold agglutinin with a molecular weight of 1,000,000 - 19S molecules with 125I and 131I radioisotopes to measure the plasma volume in normal volunteers and patients. Dr. M. Strumia advocated the need to utilize a dual method to assess the 24-hour posttransfusion survival and lifespan of preserved RBC. Radioisotopes 51Cr, 99mTc and 111In oxine and non-radioisotope biotin X-NHS hydroxysuccinimide were utilized to measure the RBC volume and the survival of fresh and preserved RBC in normal volunteers, patients and baboons at the NBRL (Valeri CR, Ragno G, Transfusion 46(Suppl):1S-42S, 2006; Valeri CR et al, Transfusion 47(Suppl): 206S-248S, 2007). Dr. M. Strumia reported that the single radioisotope label procedure using 51Cr to measure the 24-hour posttransfusion survival and lifespan of preserved RBC would detect "eternal RBC". Dr. M. Strumia reported that nonviable compatible preserved RBC are rapidly removed from the circulation during the mixing of the preserved RBC in the blood volume of the recipient. The rapid loss of nonviable compatible RBC during the 5 to 30 minute post-infusion period allow the remaining viable RBC to have prolonged lifespan which Dr. Strumia referred to as "eternal RBC". The observations of Strumia and his associates have been confirmed by numerous studies performed at the NBRL in normal volunteers, baboons and patients. These findings demonstrate that albumin is too small a molecule to remain in the plasma volume in patients and albumin is a very poor oncotic protein to increase the plasma volume and should not be used as a blood substitute or as a resuscitation solution. Human and bovine albumin were isolated from fresh blood by Dr. E.J. Cohn and his associates to resuscitate wounded patients in hemorrhagic shock during World War II. Dr. E. Churchill reported in the New York Times that bovine albumin infused into wounded servicemen in North Africa during World War II produced anaphylactic shock and should not be used instead of human albumin. Dr. Churchill also advocated the use of blood and not human albumin and plasma to resuscitate wounded servicemen. The therapeutic benefit of human albumin to treat patients in hemorrhagic shock has never been demonstrated and should not be used. (Farrugia A, Transf Med Rev 24:53-63, 2010; The Safe Study Investigators, N Engl J Med 350:2247-2256, 2004).

In July 1962, four (4) patients who received previously frozen deglycerolized RBC at the Chelsea Naval Hospital developed acute renal insufficiency and were treated with peritoneal dialysis and hemodialysis. The records of the patients who received previously frozen deglycerolized RBC using the Cohn Blood Fractionator revealed that patients who developed acute renal failure received deglycerolized RBC resuspended in an outdated albumin-glucose medium. The aggregates of outdated albumin in the medium measured by acrylamide gel electrophoresis, nephelometry, and ultra-centrifugation were identified to be the cause of the acute renal insufficiency. The albumin was obtained from an outdated source and then reprocessed prior to preparation of the outdated albumin-glucose medium. The review of the records of the 1327 patients revealed that deglycerolized RBC stored in autologous plasma and the albumin-glucose medium at 4 C for 21 days were not associated with bacterial contamination of the RBC product which was processed in an open system to add and to remove the glycerol using the Cohn Blood Fractionator. A total of 2,324 units of previously frozen deglycerolized RBC were transfused without any report of an adverse event. Captain L. Haynes, MC, USN and Dr. J. Tullis and their associates reported that the use of previously frozen deglycerolized RBC resuspended in the albumin glucose medium would eliminate the potential transmission of hepatitis. However, Haugen RK in 1979 reported in the NEJM (301:393-395, 1979) hepatitis following the transfusion of previously frozen deglycerolized RBC and washed red cells in patients. Alter HJ and associates in NEJM (298:637-

642, 1978) reported the transmission of hepatitis in chimpanzees transfused with frozen deglycerolized RBC. In 1954, Valtis DJ and Kennedy AC reported the respiratory defect in RBC stored in acid-citrate-dextrose (ACD) anticoagulated whole blood at 4 C for one week. The increase in RBC affinity for oxygen was related to the decrease in the RBC organic phosphate compound 2,3 DPG. Gardos G in 1966 had reported that RBC DPG was a negatively charged molecule in RBC which neutralized the high concentration of the positively charged RBC potassium ion. Benesch R and Benesch RE and Chanutin A and Curnish RF in 1967 reported that the RBC 2,3 DPG and ATP levels affected the RBC affinity for oxygen. Studies were performed at the NBRL to measure not only 24-hour posttransfusion survival and lifespan of the preserved RBC but also the RBC function assessed by in vitro and in vivo measurement of RBC ATP, DPG and P50 levels (the partial pressure of oxygen at which 50% of the hemoglobin is oxygenated). (Valtis DJ, Kennedy AC, Lancet 1:119-124, 1954; Haynes LL et al, JAMA 173:1657-1663, 1960; Valeri CR and Henderson ME, JAMA 188:1125-1131, 1964; Tullis JL et al, Vox Sang 8:100-101, 1963; Valeri CR, Transfusion: 5:25-35, 1965; Valeri CR, Transfusion 5:36-53, 1965; Haugen RK, N Engl J Med 301:393-395, 1979; Alter HJ et al, New Engl J Med 298:637-642, 1978; Benesch R, Benesch RE, Biochem Biophys Res Commun 26:162-167, 1967; Chanutin A, Curnish RF, Arch Biochem Biophys 121:96-102, 1967; Gardos G, Acta Biochem Biophys Acad Sc Hung 1:139-148, 1966).

In 1965, Dr. C. Huggins, the son of the Nobel Prize laureate, Dr. C. Brenton Huggins, designed and developed a method to deglycerolize RBC frozen with 40% W/V glycerol by agglomeration of the glycerolized RBC in a low ionic medium without the need for a centrifuge. The US Navy was interested in utilizing this approach to deglycerolize the RBC to replace the Cohn Blood Fractionator. Dr. C. Huggins was hired as a consultant to the US NBRL located at Chelsea Naval Hospital to supervise the use of his technology to freeze glycerolized RBC in a low ionic medium and deglycerolize the RBC by agglomeration using 6.8 liters of low ionic wash solution and the deglycerolized RBC were resuspended in 0.9% NaCl solution. The volume of 0.9% NaCl used to disaggregate the RBC agglomerated by the Huggins method was 75 to 100 ml of 0.9% NaCl. The first two (2) in vivo 24-hour posttransfusion survival measurements performed in healthy male volunteers at the NBRL yielded values of 60% and 85% using the 125I/51Cr double radioisotope procedure.

In discussions of these data with Dr. C. Huggins, he reported that the 85% 24-hour posttransfusion survival value indicated that the procedure was successful. He was unable to interpret the poor 24-hour posttransfusion survival of 60%. Studies at the NBRL demonstrated that the volume of 250 ml of 0.9% NaCl was needed to restore the physical-structural parameters of the deglycerolized RBC to normal, whereas the volume of 75 to 100 ml of 0.9% NaCl solution produced crenated RBC with decreased mean corpuscular volume (MCV), increased mean corpuscular hemoglobin concentration (MCHC), reduced potassium level and RBC with increased resistance to osmotic fragility. Some of the dense RBC were rapidly removed from the circulation and were sequestered in the spleen of the normal volunteer and released into the circulation during the 3 to 5 day postinfusion period. (Huggins CE, Monogr Surg Sci 3:133-173, 1966; Valeri CR, Transfusion 6:247-253, 1966; Valeri CR, Bond JC, Transfusion 6:254-262, 1966; Valeri CR et al, Transfusion 6:543-553, 1966; Valeri CR et al, Transfusion 6:554-564, 1966).

Dr. Morton Grove Rasmussen, the Director of the Massachusetts General Hospital Blood Bank, performed the compatibility testing of the deglycerolized RBC by the Huggins method by screening of the recipient's serum for atypical antibodies and typing the donor and recipient RBC for group ABO antigens and the Rh antigens without performing the antiglobulin-Coombs cross-match. The Coombs crossmatching of the donor deglycerolized RBC processed by the Huggins procedure was not done at the Mass General Hospital. NBRL observed that RBC treated with a glycerol solution in a low ionic medium prior to freezing and deglycerolized in a low ionic medium produced Coombs positive RBC. RBC in low ionic medium were coated with B_1C globulin and the addition of Na2EDTA to the glycerol solution was needed to prevent the Coombs positive RBC that were transfused to patients at the Mass General Hospital prior to the evaluation of the Huggins procedure to freeze human RBC at the NBRL. (Valeri CR: Transfusion 6:247-253, 1966).

The thawed glycerolized RBC were washed by dilution with low ionic solutions and by agglomeration of the glycerolized RBC. Dr. C. Huggins reported that the agglomerated RBC should be disaggregated with 75 to 100 ml of 0.9% NaCl. Studies performed at the NBRL demonstrated that a volume of 250 ml of 0.9% NaCl was needed to restore the physical structural parameters of the deglycerolized RBC. The disaggregation of the agglomerated RBC with 250 ml of 0.9% NaCl and storage at 4C for only

24 hours was required because both the addition of the glycerol solution prior to freezing and the thawing and deglycerolization of the RBC were done in open systems which required that the deglycerolized RBC could be stored at 4 C for only 24 hours. In addition, the deglycerolized RBC stored at 4 C accumulated hemolysis which required that the deglycerolized RBC had to be concentrated by centrifugation to remove the supernatant solution containing the hemolysis and to adjust the hematocrit of the deglycerolized concentrated RBC to a hematocrit of 80 V% prior to transfusion. The need to concentrate the deglycerolized RBC to remove the high levels of hemolysis associated with dilution agglomeration procedure and resuspension in 250 ml of 0.9% NaCl required that a refrigerated centrifuge was needed in addition to the Huggins Cytoglomerator to prepare RBC frozen by the Huggins method.

The original concept that washing of the deglycerolized RBC by dilution agglomeration and resuspension in 0.9% NaCl would eliminate the need for a centrifuge was not correct. Deployment of -80C water-cooled mechanical freezers, Huggins Cytoglomerator, wash solutions, and refrigerated centrifuges were needed to provide frozen RBC processed by the Huggins method. Studies at NBRL required modification of the procedure introduced by Dr. Huggins by addition of Na2EDTA to the glycerol solution, disaggregation of the diluted-agglomerated RBC using 250 ml of 0.9% NaCl, and the centrifugation of the deglycerolized RBC to prepare RBC concentrates with hematocrits of 80 V% prior to transfusion to eliminate the hemolysis present in the deglycerolized agglomerated resuspended RBC prepared by the Huggins procedure. In addition, studies performed at NBRL documented that RBC frozen by the Huggins procedure with 40% W/V glycerol and washed by the dilution-agglomeration procedure could be stored at -80C for less than 2 years. Studies performed at NBRL demonstrated that washing of the Huggins frozen RBC using electrolyte solutions in the Haemonetics Blood Processor 115 and the IBM Blood Processor 2991-1 and 2991-2 instruments permitted storage of the Huggins frozen RBC with Na2EDTA for at least 21 years at -80 C with acceptable in vitro and in vivo survival values. The NBRL modifications of the Huggins procedure to freeze human group O Rh positive and group O Rh negative RBC permitted the deployment for the first time of a frozen blood bank utilizing -80C mechanical freezers, Huggins cytoglomerator, refrigerated centrifuges, and wash solutions at Danang, South Vietnam between 1968-1974 and aboard the hospital ships

USS Sanctuary and USS Repose to supplement fresh whole blood and the liquid preserved whole blood stored in acid citrate dextrose anticoagulant at 4C for 21 days to successfully treat wounded casualties. (Brodine CE et al, J Surg Res 7:545-548. 1967; Valeri CR et al, Bibl Haematol 29:735-738, 1968; Valeri CR et al, Transfusion 10:102-112, 1970; Valeri CR et al, Transfusion 29:429-437, 1989).

The deployment of the first frozen blood bank at Danang, South Vietnam in 1968 demonstrated that the liquid preserved RBC in acid-citrate-dextrose anticoagulated whole blood at 4C for 21 days were received in South Vietnam at or near their outdating. The RBC were group O Rh positive and group O Rh negative universal donor RBC with low titer Anti-A and Anti-B whole blood. This observation stimulated the procedure to biochemically modify these outdated group O Rh positive and group O Rh negative RBC to increase the RBC ATP, DPG and p50 values prior to glycerolization and freezing. Studies were conducted at the NBRL to prepare and test solutions to salvage outdated universal donor group O Rh positive and group O Rh negative RBC using solutions containing pyruvate, inosine, phosphate and adenine with and without glucose. Dr. Beverly Gabrio and associates and Dr. Max Strumia and associates had reported that addition of nucleosides to RBC stored at 4C increase the ATP level. Over the past 4 decades, studies were done at NBRL to biochemically treat outdated RBC stored in acid-citrate-dextrose (ACD), citrate-phosphate-dextrose (CPD), CPD supplemented with adenine (CPDA-1) and CPD/AS-1, CP2D/AS-3, and CPD/AS-5 RBC stored at 4C prior to glycerolization and freezing. (Gabrio BW et al, J Clin Invest 34:1509-1512, 1955; Strumia MM et al, J Lab Clin Med 76:907-914, 1970; Valeri CR and Zaroulis CG, NEJM 287:1307-1313, 1972; Valeri CR et al, Transfusion 22:102-106, 1982; Valeri CR et al, Transfusion 40:1341-1345, 2000).

In 1980 the American Red Cross was utilizing the additive solutions AS-1 (ADSOL) to store RBC in CPD/AS-1 at 4C for 49 days which was approved by the Food and Drug Administration (FDA). The American Red Cross Northeast was interested in biochemical treatment of outdated group O Rh positive and group O Rh negative RBC stored in CPD/AS-1 at 4C for 49 days. NBRL was requested by ARC Northeast to study whether or not RBC stored in CPD/AS-1 at 4C for 49 days could be biochemically modified and frozen. Studies done in 1980 showed that autologous RBC stored in CPD/AS-1 for 49 days, biochemically modified to increase the RBC ATP, DPG, and p50 values, frozen with 40% W/V glycerol and stored at -80C, thawed and deglycerolized and stored at 4C in sodium chloride glucose phosphate solution for 24 hours had 24-hour posttransfusion survival of 60%. These observations required that control studies were needed to assess the 24-hour posttransfusion survival of autologous and allogeneic RBC stored in CPD/AS-1 for 49 days. The 24-hour posttransfusion survival for autologous RBC was measured using the double radioisotope procedure: 125I albumin to measure the plasma volume and the total body hematocrit (peripheral venous hematocrit multiplied by 0.89) to measure the RBC volume of the healthy male volunteer and 51Cr labeling of the autologous liquid preserved RBC. At the same time, the 24-hour posttransfusion survival values of allogeneic compatible identifiable RBC stored at 4C in CPD/AS-1 for 49 days transfused into stable anemic patients were measured using the automated differential agglutination procedure using the Technicon Autoanalyzer and the RBC volume of the patient was measured using 51Cr-labeled autologous RBC. The method utilized to assess the allogeneic compatible identifiable RBC stored at 4C in CPD/AS-1 for 49 days produced 24-hour posttransfusion survival values of 55% which was similar to the 24-hour posttransfusion survival values observed with autologous RBC stored in AS-1 (ADSOL) for 49 days measured by the 125I/51Cr double radioisotope procedure. (Valeri CR et al, Surg Gynec Obstet 166:33-46, 1988).

The questions raised by the American Red Cross Northeast stimulated studies by the NBRL to assess the survival and function of autologous and allogeneic RBC stored in CPD/AS-1 for as long as 49 days. The results of the NBRL studies demonstrated that 24-hour posttransfusion survival of autologous and allogeneic RBC preserved in CPD/AS-1 could be stored at 4C for only 35 days with a 24-hour posttransfusion survival of 75% and the RBC function was preserved at 4C for only 2 weeks. These findings were in contrast to several other investigators who reported that the autologous RBC stored in CPD/AS-1 at 4C for 56 days had 24-hour posttransfusion survival of at least 70% and the FDA had approved storage of the RBC stored in CPD/AS-1 at 4C for 49 days.

Dr. Joseph Bove, who was the Chairman of the Standards Committee of the AABB, suggested that our observations should be reported as a letter to the editor of the New England Journal of Medicine. In 1985, the FDA conducted a meeting to review the length of storage of human RBC in CPD/AS-1 at 4C and the information reported resulted in reducing the length of storage of CPD/AS-1 RBC from 49 days to 42 days but not to 35 days. (Valeri CR, NEJM 312:377-378, 1985; Valeri CR et al, Surg Gynec Obstet 166:33-46, 1988).

The question raised by the American Red Cross Northeast in 1980 whether CPD/AS-1 RBC stored at 4C for 49 days could be biochemically modified to increase the RBC ATP, DPG and p50 values prior to freezing stimulated the controversy regarding the measurement of nonviable RBC in preserved RBC products. Dr. M. Strumia in 1965 had reported that a double label procedure was needed to detect the quantity of nonviable RBC in the preserved RBC product. Two radioisotopes were utilized, one radioisotope to measure the RBC volume of the normal recipient and the other radioisotope to measured the preserved 51Cr labeled autologous RBC. At the NBRL, the 125I albumin and 51Cr method, the 99mTC and 51Cr method, and the 111In-oxine and 51Cr method and the double 51Cr method were evaluated in normal volunteers and in baboons to assess 24-hour posttransfusion survival and lifespan of preserved autologous RBC. Alternatively, methods were used in patients to measure the survival of compatible but identifiable allogeneic RBC using anti-A, Anti-B, and Anti-M antibodies using the Technicon Autoanalyzer and the Coulter Counter and 51Cr to measure the RBC volume of the patient. If RBC are well preserved, then the single method utilized using 51Cr to label the preserved RBC without independent measurement of the RBC volume of the recipient will provide similar data to the double isotope methods which measures independently the red cell volume of the recipient. If the preserved RBC contain a quantity of nonviable RBC the single 51Cr method will not detect these irreversibly damaged RBC that are rapidly removed from the circulation during the 5, 10, 15 and 30 minute postinfusion period. The single 51Cr method has been reported in the majority of the published reports because it is a simple procedure compared to the double radioisotope procedure to measure the survival of autologous preserved RBC. Unfortunately, the single 51Cr method to measure the 24-hour posttransfusion survival values overestimates the 24-hour posttransfusion survival when nonviable RBC are rapidly removed during the mixing time of the small aliquot in the recipient's blood volume.

The request by the American Red Cross Northeast that the NBRL investigate the salvaging of universal donor O positive and O negative RBC stored in CPD/AS-1 for 49 days resurrected the controversy discussed by Dr. M. Strumia in 1965 as to the method to determine the quantity of nonviable RBC in the preserved RBC products which contain a quantity of irreversibly damaged preserved RBC.

In October 1997, NBRL did not receive the funds for the 3rd year of a contract with the Naval Medical Research and Development Command and the Office of Naval Research. The reduction in funds required that the NBRL obtain funds from a commercial company to perform three (3) research projects that were requested by the commercial company. Without these commercial funds provided by the Haemonetics Corporation and its Chief Operating Officer, John White and Chief Fiscal Officer, Jim Rice, the NBRL would have been forced to reduce its staff and curtail its research without the commercial funds provided by Haemonetics Corporation. The commercial funds were allocated to perform the following three (3) studies:

1) comparing the storage of autologous human RBC obtained by the manual collection of a unit of blood to isolate the RBC and a unit of RBC collected by apheresis procedure using the Haemonetics MCS Instrument from the same donor with storage of the RBC in the additive solution AS-3 (Nutricel) at 4C for 56 days; 2) the safety of the collection of two (2) units of RBC (180 ml of RBC on two occasions for a total volume of 360 ml of RBC) from normal male and female blood donors by the Haemonetics RBC apheresis procedure; 3) the evaluation of the Haemonetics Blood Processor ACP215 instrument to glycerolize and deglycerolize RBC with storage at 4C in AS-3 (Nutricel) for 3 weeks in this functionally closed system.

These three (3) studies funded by Haemonetics Corporation in 1998 provided fundamental information on the superiority of the additive solution AS-3 (Nutricel) compared to AS-1 (ADSOL) as an additive solution to preserve deglycerolized RBC stored at 4C for 2 weeks processed in the functionally closed Haemonetics Blood Processor ACP215 instrument. The studies demonstrated that autologous RBC in CPD/AS-1 can be stored at 4C for 42 days with a posttransfusion survival of 70%, whereas autologous RBC stored in CP2D/AS-3 can be stored at 4C for 56 days with a 24-hour posttransfusion of 70%. Autologous RBC stored in CP2D/AS-3 at 4C can be stored for 2 weeks longer than autologous RBC stored in CPD/AS-1 obtained from the same donor when measured by a 24-hour posttransfusion survival measurement of 70% assessed by the double isotope labeling procedure using the 125I albumin and 51Cr method. These studies demonstrated that the removal of 2 units of RBC from female and male donors by the RBC apheresis procedure produced a significant increase in bleeding time and that viable and functional RBC had a greater effect on the bleeding time than viable and functional platelets. The three (3) studies performed at NBRL were done to justify the commercial funds that were needed to support the NBRL when the government funds were not received to support the third year of a contract between NBRL and the Naval Medical Research and Development Command (NMRDC) and the Office of Naval Research (ONR) in 1997 (Valeri CR et al, Transfusion 41:933-939, 2001; Valeri CR et al, Transfusion 41:928-932, 2001; Valeri CR et al, Transfusion 41:977-983, 2001; Valeri CR et al, Vox Sang 80:48-50, 2001, Valeri CR et al, Transfusion 44:990-995, 2004).

Questions regarding whether membrane oxygenators produced less damage to RBC than bubble oxygenators and whether centrifugal pumps produced less damage to RBC than roller pumps during cardiopulmonary bypass surgery were investigated at the NBRL in collaboration with Dr. Shukri Khuri at the West Roxbury VA Hospital. Mr. Fred Levan, a salesman at Terumo Company, stimulated and supported these studies. Studies were done to assess the safety and therapeutic effectiveness of nonwashed autologous shed blood collected during orthopedic and cardiopulmonary bypass surgery were stimulated and supported by Mr. Robert Cotter and the Solcotrans Company. The safety and therapeutic effectiveness of autologous shed nonwashed blood obtained during cardiopulmonary bypass surgery and autologous shed nonwashed blood collected from patients during orthopedic surgical procedures were evaluated. Studies were conducted at the New England Baptist Hospital by Dr. J. McCarthy and associates; at the Lahey Clinic by Dr. W. Healey and associates; at Children's Hospital, Boston, MA by Dr. FT Blevins and associates to assess nonwashed autologous orthopedic shed blood. Dr. S. Khuri and associates at West Roxbury VA Hospital studied the safety and therapeutic effectiveness of nonwashed shed blood collected from patients following cardiopulmonary bypass surgical procedures. A volume of 10 to 15% of the patient's blood volume was autotransfused as nonwashed shed blood following surgery produced clinical benefits without any adverse events in these patients subjected to orthopedic surgical procedures and cardiopulmonary bypass surgical procedures. The survival, function and hemolysis of shed nonwashed autologous RBC and shed washed autologous RBC were similar to fresh autologous RBC in dogs subjected to 100% exchange transfusions. In addition, studies were performed in baboons to assess the survival and function of autologous RBC subjected to clotting, lysis by treatment with urokinase, and washing did not adversely affect their survival and function. These studies in dogs and baboons simulated the treatment of autologous nonwashed and washed shed blood collected from patients subjected to cardiopulmonary bypass surgery and to orthopedic surgical procedures. (Faris PM et al, J Bone Joint Surg 73A:1169-1178, 1991; Blevins FT et al, J Bone Joint Surg 75A:363-371, 1993; Healy WL et al, Clin Ortho Rel Res 286:15-17, 1993; Healy WL et al, Clin Ortho Rel Res 99:53-59,1994; Axford TC et al, Ann Thorac Surg 57:615-622, 1994; Valeri CR et al, Transfusion 41:1384-1389, 2001; Valeri CR et al, Ann Thorac Surg 72:1598-1602, 2001; Khuri SF et al, J Thorac Cardiovasc Surg 104:94-107, 1992; Valeri CR et al, Perfusion 21:291-296, 2006).

A study was done to determine whether or not autotransfusion of nontreated plasma and plasma treated with urokinase with and without aprotonin affected hemostasis in healthy baboons. The infusion of a volume of urokinase or urokinase-aprotonin treated autologous plasma equivalent to 15% of the blood volume was not associated with a bleeding diathesis in healthy baboons. (Valeri CR et al, J Card Surg 21:565-571, 2006).

In the 1980 to 1985 period, the Naval Medical Research and Development Command of the U.S. Navy Bureau of Medicine and Surgery, and Office of Naval Research were interested in the methods to treat hypothermic patients subjected to traumatic injury and blood loss in the cold environment of North Atlantic and in the Scandinavian countries. The effects of hypothermia on platelet function and hemostasis were evaluated in normal volunteers at the NBRL, in baboons housed at New England Medical Center (NEMC) and Boston University Medical Center (BUMC), and in patients at the West Roxbury VA Hospital subjected to cardiopulmonary bypass surgery. Baboons were cooled externally to reduce their core temperature and skin temperature and normal volunteers were subjected to local hypothermia. The effect of hypothermia on platelet function was documented by the extension of the standardized bleeding time and the reduction in thromboxane, a potent vasoconstrictor substance, in the shed blood collected at the bleeding time site. Both in vitro and in vivo studies were done in baboons and normal volunteers to document that hypothermia produced a reversible platelet dysfunction. Rewarming of the baboons and patients restored platelet function. Studies were done in collaboration with Dr. S. Khuri at the West Roxbury VA Hospital to document that during cardiopulmonary bypass surgery in the same patient warming of one extremity reduced the bleeding time and increased the shed blood level of thromboxane collected at the bleeding time site compared to the other cold extremity in which the bleeding time was increased and the level of shed blood thromboxane was reduced at the bleeding time site. The effects of temperature on platelet function and the clotting proteins in normal volunteers were studied at the NBRL. In collaboration with Dr. S. Khuri and his associates at the West Roxbury VA Hospital the temperature of the operating room was increased and the patients were warmed with heating blankets to reduce the bleeding time and to reduce nonsurgical blood loss. (Valeri CR et al, Ann Surg 205:175-181, 1987; Valeri CR et al, J Thorac Cardiovasc Surg 104:108-116, 1992; Michelson AD et al, J Thrombosis and Haemostasis 5:633-640, 1994; Valeri CR et al, Crit Care Med 23:698-704, 1995; Valeri CR and Ragno G, Transfusion 46:1S-42S, 2006; Valeri CR et al, Transfusion 47:206S-248S, 2007).

Dr. Shukri Khuri, Chief of Thoracic and Cardiovascular Surgery and Chief of Surgery at the West Roxbury VA Hospital was interested in methods to reduce nonsurgical blood loss during cardiopulmonary bypass surgery. From 1982 to 1999 numerous collaborative studies were performed between NBRL and WRVA Hospital to assess the effects of temperature, heparin, hemodilution, platelets, plasma clotting proteins and RBC on nonsurgical blood loss in patients subjected to cardiopulmonary bypass surgery. The results of these studies were reported in the monograph "Nonsurgical Bleeding Diathesis in Anemic, Thrombocytopenic Patients: Role of Temperature, Red Blood Cells, Platelets and Plasma Proteins". (Valeri CR et al, Transfusion 47:206S-248S, 2007).

In 1998, funds were obtained from Marine Polymer Technologies, Inc. to support the NBRL to study a highly purified polymerized, endotoxin free, sterile N-acetylglucosamine substance isolated from a microalgae as a hemostatic agent. Numerous studies were done to demonstrate that this polymerized beta N-acetylglucosamine substance activated platelets and RBC to reduce nonsurgical blood loss. (Vournakis JN et al, J Trauma [Suppl] 57:S2-S6, 2004; Thatte HS et al, J Trauma [Suppl] 57:S7-S12, 2004; Thatte HS et al, J Trauma [Suppl] 57:S13-S21, 2004; Valeri CR et al, J Trauma [Suppl] 57:S22-S25, 2004; Valeri CR and Vournakis JN, J Trauma [Suppl] 71:S162-S166, 2011).

A prospective randomized double-blind study was performed at WRVA Hospital by Dr. S. Khuri and his associates to evaluate the hemostatic agent containing polymerized N-acetylglucosamine bandage to reduce the bleeding time at the arterial site where the catheter was removed in patients undergoing catheterization. The pressure applied to the bleeding time site with the hemostatic agents was controlled in this prospective and randomized study. The N-acetylglucosamine bandage produced significant reduction in bleeding time compared to the control bandage when the external pressure was controlled. (Najjar SF et al, J Trauma [Suppl] 57:S38-S41, 2004).

In studies performed by Ikeda Y and associates and Favuzza J and Hechtman HB, the role of endothelin released from endothelial cells was documented to be important in the mechanism to restore hemostasis at the catheter site when the hemostatic agent containing N-acetylglucosamine and pressure were applied. Working in collaboration with Dr. D. Orgill and his staff at the Brigham and Women's Hospital, Boston, MA studies in diabetic mice showed that the polymerized N-acetylglucosamine bandage was not only a potent hemostatic agent but also accelerated wound healing. In addition to the hemostatic effect and the acceleration of wound healing, Linder HB and associates have reported that polymerized N-acetylglucosamine bandage exerts an antimicrobial effect. (Favuzza J, Hechtman HB, J Trauma [Suppl] 57:S542-S544, 2004; Ikeda Y et al, J Surg Res 102:215-220, 2002; Pietromaggiori G et al, J Trauma 64:803-808, 2008; Scherer SS et al, Ann Surg 250:323-330, 2009; Linder HB et al, PloSI 6:e18996, 2011).

In September 1998, Dr. RL Veech, a senior NIH investigator reported at the Institute of Medicine meeting on resuscitation fluids that isolated rat hearts perfused with Ringer's ketone solution had improved myocardial function and reduced oxygen consumption compared to isolated rat hearts perfused with Ringer's D,L lactate. This critical observation by Dr. R. L. Veech stimulated the NBRL to investigate the therapeutic effectiveness of Ringer's D,L lactate, Ringer's L lactate and Ringer's ketone solution as parenteral solutions to resuscitate animals subjected to hemorrhagic shock. The toxicity of Ringer's D,L lactate reported by Dr. R.L. Veech and associates was observed in patients treated during the Vietnam War who developed acute respiratory distress syndrome following the infusion of the Ringer's D,L lactate solution. Studies were conducted in animals to demonstrate that Ringer's ketone solution was superior to Ringer's D,L lactate to resuscitate animals subjected to blood loss by Dr. Hasan B. Alam, Dr. Peter Rhee and associates (Valeri CR et al, Art Cells, Blood

Subst Biotech 34:601-606, 2006; Valeri CR et al, Trans Aph Sci 39:205-211, 2008).

During the Vietnam War, NBRL had to provide the research team located at Danang, South Vietnam solutions to deglycerolize frozen RBC. The procedure that was utilized at Danang, South Vietnam, aboard the hospital ships USS Repose and USS Sanctuary and at Clark Air Force Base required 6.8 liters of wash solutions to deglycerolize the RBC using agglomeration of RBC in low ionic media. NBRL, working in collaboration with Millipore Corporation, Bedford, MA, studied the Sterimatics instrument (ST-30) to produce from potable water, sterile water for injection using membrane technology without the need for final heat sterilization. Studies were supported by the NBRL to produce a Ringer's ketone solution to provide a substrate resuscitation solution using the ketone NaD betahydroxybutyrate instead of Ringer's D,L lactate solution which has been shown to produce toxicity in patients and in animals. These studies were stimulated by the observation reported by Dr. Richard L. Veech and Dr. Kieran Clarke in isolated perfused rat hearts. Solution containing the ketone NaD betahydroxybutyrate improved myocardial function and reduced oxygen consumption compared to perfusion of the isolated rat hearts with Ringer's D,L lactate. In studies conducted at University Hospital at Boston University Medical Center between 1980 and 1985 with Dr. H.B. Hechtman, Dr. R. Berger, Dr. R. Dennis, Dr. N. Yeston and Dr. R. Weisel, myocardial function and oxygen consumption in patients were studied following cardiopulmonary bypass surgery. The effect of RBC oxygen delivery was studied to compare liquid preserved nonwashed RBC with 70% DPG and normal affinity for oxygen to biochemically modified previously frozen and washed RBC with 150% DPG and reduced affinity for oxygen. These studies showed that RBC with a 150% DPG level and reduced affinity for oxygen, improved the volumes loading response of the hearts in patients following cardiopulmonary bypass surgery. These studies demonstrated that human hearts consumed more oxygen if more oxygen was provided to the heart. The observations of Dr. R. Veech and Dr. K. Clarke that perfusion of isolated rat hearts with Ringer's ketone solution improved myocardial function but reduced oxygen consumption was significantly different from our clinical observation that improved myocardial function occurred with increased oxygen consumption by the hearts in patients following cardiopulmonary bypass surgery transfused with human RBC with 150% DPG level and reduced affinity for oxygen.

In these patients the more oxygen you provided to the heart, the more oxygen was consumed to improve cardiac response to volume loading by the increase in cardiac output and cardiac work. Ketones in the perfusion fluid improved myocardial function of isolated perfused rat hearts but did not increase the oxygen consumption of the heart. This is an extremely important observation demonstrating that ketones improve myocardial function at a reduced oxygen consumption. (Dennis RC et al, Surgery 77:741-747, 1975; Valeri CR et al, Prog Clin Biol Res 21:597-616, 1978; Dennis RC et al, Ann Thoracic Surg 26:17-26, 1978; Valeri CR et al, Trans Aph Sci 39:205-211, 2008).

Dr. R. Veech's observation with regards to the degradation of hemoglobin in vitro to release iron and protoporphyrin was a major stimulus to investigate the destruction of RBC in patients subjected to traumatic injuries sustained during the Vietnam War and who developed chronic hypovolemic anemia of trauma and the missing blood syndrome. (Veech RL et al: Biochem J 105:1209-1217, 1967).

The missing blood syndrome observed in chronic hypovolemic patients subjected to traumatic injury was investigated at NBRL between 1968 to 1974. The studies on The Hypovolemic Anemia of Trauma: The Missing Blood Syndrome which was published by CRC Press, Boca Raton, FL, 1981, provide data on the degradation of RBC through a non carbon monoxide and non bilirubin-urobilinogen pathway. The data suggest that epinephrine degradation through a plasma pathway may produce products like adrenochrome and adrenolutin that degrade hemoglobin into porphyrin and heme and not into bilirubin and urobilinogen with the release of carbon monoxide. Drs. R. Bannerman and M. Kreimer-Birnbaum measured the urinary and fecal dipyrroles and Dr. D. Borg measured urinary oxyporphyrins in these patients with chronic hypovolemia of trauma. Studies of RBC response to adrenochrome was performed with the help of Z. L. Hegedus.

In 1980, the Naval Medical Research and Development Command of the U.S. Navy's Bureau of Medicine and Surgery, and the Office of Naval Research were concerned with the potential exposure of U.S. Naval personnel to nuclear radiation which was utilized by Admiral Hyman Rickover to power nuclear submarines. The NBRL demonstrated that autologous dog pluripotential mononuclear adult stem cells isolated from peripheral blood frozen with 10% DMSO in -80C mechanical freezers, thawed, washed and reinfused repopulated the bone marrow in beagle dogs subjected to lethal radiation. Studies were done to demonstrate that human blood mononuclear cells (PBMC) adult stem cells can be frozen with 10% DMSO and stored at -80C for 1 1/2 years and following thawing and washing had in vitro recovery that was 90% that of fresh PBMC's, viability that was 90% and growth in CFU-GEMM (granulocyte-erythroid-monocyte-megakaryocytic) tissue culture assay that was similar to that of fresh PBMC's. The measurements of CD34 antigen, CFU-GEMM (granulocyte-erythroid-monocyte-megakary-ocytic) tissue culture assay and 7 amino actinomycin D (7AAD) test were used to assess the viability and function of fresh, liquid preserved, and frozen pluripotential mononuclear human adult stem cells isolated from peripheral blood. (Hunt SM et al, Blood 57:592-598, 1981; Hunt SM et al, Transfusion 23:387-390, 1983; Carciero R, Valeri CR, Vox Sang 49:373-380, 1985; Valeri CR, Ann NY Acad Sci 459:353-366, 1985; Valeri CR In Treatment of Radiation Injuries, Ed. D. Browne et al, Plenum Press, New York, pps 19-27, 1990; Valeri CR et al, Transfusion 36:303-308, 1996; O'Murchadha ET et al, Exp Hematol 16:235-239, 1988; Valeri CR et al, Technical Report 85-01).

The composition of the resuspension medium for the RBC preserved in the liquid state at 4C and deglycerolized RBC stored at 4 C is critical to maintain the survival and function, reduce hemolysis and maintain the physical-structural and metabolic integrity of the preserved RBC. Studies conducted at NBRL have demonstrated that composition of the anticoagulant-preservative solution to collect and store the RBC at 4C affect their survival, function and hemolysis. Studies were done to assess the quality of RBC preserved in the anticoagulant-preservative solution acid-citrate-dextrose (ACD), citrate-phosphate-dextrose (CPD) and RBC stored in additive solutions (CPD/AS-1 (ADSOL), CP2D/AS-3 (Nutricel) and CPD/AS-5 (Optisol). Studies were done to assess the quality of deglycerolized RBC stored in autologous plasma, deglycerolized RBC stored in 5% indated albumin and glucose solution; deglycerolized RBC stored in 5% outdated-albumin and glucose solution, deglycerolized RBC stored in AS-1 (ADSOL) solution and deglycerolized RBC stored in AS-3 (Nutricel) solution. The composition of the resuspension media had significant effects on their survival, function and hemolysis of the deglycerolized RBC. The best resuspension medium for liquid preserved RBC stored at 4C and deglycerolized RBC stored at 4C is the additive solution AS-3 (Nutricel) which contains glucose, citrate, citric acid, phosphate, adenine, and 0.9% sodium chloride. (Valeri CR, Transfusion 5:25-35, 1965; Valeri CR, Transfusion 5:36-53, 1965; Valeri CR, Crit Rev Clin Lab Sci 17:299-374, 1982; Valeri CR et al, Transfusion 40:1337-1340, 2000; Valeri CR et al, Transfusion 40:1341-1345, 2000; Valeri CR et al, Transfusion 41:928-932, 2001; Valeri CR et al, Transfusion 41:933-939, 2001; Valeri CR et al Transfusion, 44:990-995, 2004.

To fund the NBRL from 1962 to 2010, required annual proposals, site visits, publications, and progress reports to support the research. In 1962, the blood research laboratory at the Chelsea Naval Hospital was supported by a contract with the Protein Foundation located at Jamaica Plain, MA; Capt L. Haynes, MC, USN and Commander Mary Sprout, MSC, USN worked with Dr. J. Tullis at the Protein Foundation, Jamaica Plain, MA to utilize the Cohn Blood Fractionator an instrument to glycerolize RBC prior to freezing at -80C and to deglycerolize the RBC. Two thousand three hundred and twenty four (2324) units of previously frozen deglycerolized RBC were transfused to 1327 patients prior to July 1962 when 4 patients with acute renal shutdown occurred. From July 1962 to June 1965 studies were performed to determine the cause of the renal shutdown and papers were published to document that the albumin-glucose medium used to resuspend the deglycerolized RBC was responsible for the renal shutdown. The albumin was outdated and was reprocessed and resuspended in glucose which produced aggregated albumin and the aggregated albumin was responsible for the four (4) cases of renal shutdown and the cause of severe adverse events observed with frozen glycerolized blood. (Valeri CR and Henderson MR, JAMA 188:1125-1131, 1964). In June 1965, the U.S. Naval Blood Research Laboratory was established at Chelsea Naval Hospital by the Naval Medical Research and Development Command (NMRDC) of the U.S. Bureau of Medicine and Surgery. In January 1974, the Chelsea Naval Hospital renamed the Boston Naval Hospital was disestablished and the U.S. NBRL was relocated to the Talbot building at BUMC and then to 615 Albany Street site, a government own building purchased from University Hospital for one dollar and used by the U.S. NBRL from April 1974 to January 1979. The New England Nuclear Company had donated this building to University Hospital at BUMC. In 1979, The U.S. NBRL was disestablished and the NBRL was operated as a government-owned, contractor operated facility with a contract negotiated with BUMC from January 1979 to September 30, 2003. In fiscal year October 1, 2003 to September 30, 2004, the contract was negotiated with New England Medical Research Institute (NEMRI) at West Roxbury Veterans Administration Hospital (WRVA) and NBRL Incorporated, a non-profit laboratory. On December 31, 2003 the 615 Albany Street building was reverted to BUMC and NBRL was evicted from the site on December 31, 2004 by BUMC.

The NBRL was funded from 1965 to 2010 by contracts and grants with Naval Medical Research and Development Command of the U.S. Navy's Bureau of Medicine and Surgery, the Office of Naval Research, the Congress of the U.S., and by commercial funds. From 1998 to 2004, the NBRL received funds from the U.S. Congress supported by the late Congressman J. Joseph Moakley. As noted above, the 3rd year of funding of the contract for year 1997 was not received and NBRL had to obtain commercial funds from Haemonetics Corporation, Solcotrans, Genzyme, Zymequest, VI Technologies, and Marine Polymer Technology Inc. to support the research.

NBRL studies were performed at the Chelsea Naval Hospital from 1962 to 1974, at the University Hospital at BUMC from 1974 to 1985, and at the West Roxbury VA Hospital from 1982 to 2008. Baboons and dogs were studied housed at New England Medical Center (NEMC) and at Boston University Medical Center (BUMC) from 1972 to 2002, and diabetic mice housed at the Brigham and Women's Hospital were studied from 2003 to 2010.

The movie "Music and Lyrics" underscores eight important issues that are applicable to research activity conducted at the NBRL from 1965 to 2010.

a. Pandering was needed to obtain funds from government and commercial sources to perform the research for the 45-year period;
b. Tone deaf responses were received to the oral presentation and published papers provided by the NBRL to the military and civilian communities;
c. Several peer-reviewed papers were rejected by journals to protect the self-interest of the civilian and military organizations responsible to provide the safe and therapeutically effective blood;
d. Dinner and not dessert was provided to the funding agencies. Naval Medical Research and Development Command of the U.S. Navy's Bureau of Medicine and Surgery; Office of Naval Research, the Congress of the U.S. and the commercial companies which provided funds to support the research projects at NBRL.

e. Failure of the NBRL to provide the lyrics to the music provided by the 500 plus peer-reviewed articles which demonstrated the superiority of frozen RBC, frozen platelets, fresh frozen plasma without the need for fresh whole blood to treat the military and civilian casualties with safe and therapeutically effective blood products and to prevent severe adverse events (SAEs) associated with FDA approved blood products.
f. Need of a catalyst "like Cee" to utilize frozen blood products to supplement the liquid blood products to prevent the severe adverse events observed with the current FDA approved blood products.
g. To perform the research, the NBRL had to deal with "Jerks" as discussed by Alex Fletcher (Hugh Grant) and Sophie Fischer (Drew Barrymore).
h. The chaotic and unpredictable daily issues were resolved, predicated on the pros and cons of each issue as discussed by Alex Fletcher (Hugh Grant) and his agent.

The following chaotic and serendipitous observations stimulated research projects.

a. Cohn Blood Fractionator designed to isolate albumin from plasma permitted the freezing of fresh RBC with 40% W/V glycerol and storage at -80C in mechanical freezers using the protocol reported by A. Smith (Smith AU, Lancet:2:910-211, 1950).
b. Albumin is an unacceptable resuscitation solution because it distributes rapidly into the extravascular space. Iodinated albumin cannot accurately measure the plasma volume in patients using the Volemetron and albumin is a poor plasma volume expander for the treatment of hemorrhagic shock. (Valeri CR et al, Arch Int Med 132:534-538, 1973; Valeri CR, Altschule MD, Chemical Rubber Company Press, Boca Raton, FL, 1981; Farrugia A, Trans Med Rev 24:53-63, 2010; The SAFE Study Investigators, N Engl J Med 350:2247-2256, 2004).
c. Dr. D. Drabkin's observations published in 1935 that nitric oxide and sulfur bind to hemoglobin. Nitric oxide binds to hemoglobin to produce NOHB and SNOHb and is a potent vasodilator that increases blood flow and inhibits platelet function. Hydrogen sulfide is also a vasodilator and binds to hemoglobin to produce sulfhemoglobin (Drabkin DL, Austin JR, J Biol Chem 112:51-65, 1935; D'Emmanuale di Villa Bianca R et al, PNAS Early Edition 1-6, 2008).
d. Dr. D. Danon from the Weizman Institute, Israel provided the

fragiligraph and the phthalate esters to measure RBC osmotic fragility and RBC density distribution. (Danon D, J Clin Path 16:337-342, 1963; Danon D, Markikovsky Y, J Lab Clin Med 64:668-674, 1964).

e. Dr. Charles Huggins introduced agglomeration of RBC which produced Coombs positive RBC by accumulation of BlC globulin on the RBC, the disaggregation of the agglomerated RBC with 75 ml to 100 ml of 0.9% NaC1 produced osmotically resistant, dense RBC and RBC with decreased potassium levels. The dense osmotically resistant RBC were sequestered in the spleen and released into the circulation over the 3 to 5 days posttransfusion in normal volunteers (Huggins CE, Monogr Surg Sci 3:133-173, 1966; Valeri CR, Transfusion 6:247-253, 1966; Valeri CR, Bond JO, Transfusion 6:254-262, 1966; Valeri CR et al, Transfusion 6:543-553, 1966; Valeri CR et al, Transfusion 6:554-564, 1966).

CHRONOLOGY OF ACCOMPLISHMENTS AT NAVAL BLOOD RESEARCH LABORATORY, BOSTON, MASSACHUSETTS

1956: Establishment of the Blood Research Laboratory at Chelsea Naval Hospital, Chelsea, MA

1957: Successful freeze preservation of red blood cells using the Cohn Blood Fractionator.

1960: Establishment of the first frozen blood bank for rare blood types at the Blood Research Laboratory at Chelsea Naval Hospital, Chelsea, MA.

1962: Studies initiated to evaluate the quality of preserved red blood cells and the cause of the severe adverse events following the transfusion of previously frozen deglycerolized RBC processed in the Cohn Blood Fractionator

1965: Evaluation of the Huggins cytoagglomerator procedure to deglycerolize RBC frozen with 40% W/V glycerol and stored at -80 C.

1965: Establishment of the U.S. Naval Blood Research Laboratory by the Naval Medical Research and Development Command at Chelsea Naval Hospital.

Successful freeze-preservation of red blood cells using the Huggins procedure.

Clinical evaluation of frozen RBC processed by the Huggins method at the Chelsea Naval Hospital.

Establishment of frozen blood banks at Danang, South Vietnam; Clark

Air Force Base in the Philippine Islands, and aboard the hospital ship USS Repose.

1967: Establishment of a frozen blood bank aboard the hospital ship USS Sanctuary

Establishment of red blood cell collection and freezing centers at Oakland Naval Hospital, Oakland, CA and at the National Naval Medical Center, Bethesda, MD

Evaluation of the Arthur D. Little reusable stainless steel bowl to wash frozen glycerolized red blood cells.

1965-1967: Development of an automated differential agglutination (ADA) technique to measure allogeneic compatible identifiable red blood cells survival using the Technicon Auto analyzer

Freezing and evaluation of the low concentration (20% W/V) glycerolized red blood cells stored at -150 C in the gas phase of liquid nitrogen.

1968-1974: Extensive clinical evaluation, of the chronic hypovolemic anemia of trauma, and investigation of the missing blood syndrome.

1968-1974: Hemoglobin degradation into heme and porphyrin. Convalescence is the time required to restore to normal the peripheral blood volume to the gastrointestinal tract and to the muscle, bones and skin of the extremities.

1969-1971: Establishment and evaluation of the procedures to salvage outdated human red blood cells by biochemical modification prior to freeze-preservation with 40% W/V glycerol at -80 C in mechanical freezers and with 20% W/V glycerol at -150 C in the gas phase of liquid nitrogen.

1971-1973: Use of LexO2 Con fuel cell to measure the oxygen content in blood.

1971-1973: The assessment of the hemostatic effect of human autologous liquid-preserved and cryopreserved platelets to correct the increased bleeding time in aspirin treated normal volunteers.

1971-1973: Baboon studies to evaluate the safety and therapeutic effectiveness of cryopreserved red blood cells with normal or 1 ½ to 2 times normal 2,3 DPG levels.

Establishment of procedures to treat indated liquid-stored human red blood cells to increase 2,3 DPG levels to 1 1/2 to 2 to 3 times normal before cryopreservation with the high and low glycerol methods.

1971-1973: Studies to establish methods to isolate and preserve autologous red blood cells, platelets, and granulocytes in the baboon.

1972-1973: The in vivo survival and function of platelets isolated from a single unit of human blood by serial differential centrifugation, platelet cryopreservation with 6% DMSO at 2- 3 C per minute by storage in a mechanical freezer at -80 C for at least 8 months.

1973: Freeze preservation of indated and outdated rejuvenation human red blood cells with improved capacity to deliver oxygen.

1973-1975: Physiologic studies in baboons demonstrating: (a) an increase in cardiac output during the 2 to 6 hours after transfusion of preserved baboon red blood cells with low 2,3 DPG and an increased affinity for oxygen; and (b) a decrease in cardiac output during the 2 to 6 hours after transfusion of preserved baboon red blood cells with 1 1/2 times normal 2,3 DPG and a decreased affinity for oxygen.

1973-1978: Liquid and. freeze preservation of dog and human platelets.

1974: Studies to show how red blood cell washing alone, glycerolization and washing, and glycerolization, freezing and washing remove white blood cells, platelets and plasma from the red blood cells.

1974: Study of the plasticizer, di-2-ethylhexyl phthalate (DEHP) in whole blood, platelet concentrate, and platelet-poor plasma.

Collaborative study with Naval Submarine Research Laboratory, Groton, CT to study platelet survival in volunteers subjected to hyperbaric exposure. The decreased platelet count was due to decreased platelet production in normal volunteers subjected to hyperbaric exposure.

1974-1978: Study of formation and removal of microaggregates in previously frozen washed red cells.

1975-1977: Establishment of methods to salvage human outdated red blood cells by biochemical treatment prior to freeze-preservation with 40% W/V glycerol at -80 C for 3½ to 4 years. The RBC had excellent in vitro recovery, acceptable 24-hour posttransfusion survival, and normal oxygen transport function.

1975-1978: 51Cr labeling of human lymphocytes, granulocytes, monocytes, and platelets using velocity sedimentation through a phosphate-buffered bovine albumin gradient. Establishment of a method to assess the in vivo survival and function of autologous human red blood cells perfused through a blood oxygenator.

1976: Studies to evaluate granulocytes isolated from human and baboon blood by elutriation.

1976-1978: Studies to demonstrate that human red blood cells with increased 2,3 DPG levels (1½ to 2 to 3 times normal) improve myocardial function immediately following extracorporeal bypass.

Attenuation of the increased affinity of human red blood cells during hypothermia by transfusion of biochemically modified red cells with 2,3 DPG levels increased to 1 1/2 to 3 times normal.

Freeze-preservation of human red blood cells in the primary polyvinyl chloride (PVC) plastic bag.

1977: In vitro assay of platelet function by measuring thrombin-induced release of 5- hydroxytryptamine (serotonin).

1978: A new system for freeze-preservation of human red cells with higher quality at lower cost in the primary polyvinyl chloride (PVC) plastic collection bag.

Successful freeze-preservation of human platelets isolated by plateletpheresis from a single donor using 6% DMSO and storage in a -80 C mechanical freezer.

1979: Feasibility studies at the PACOM Blood Program Office, Okinawa, Japan to evaluate nonrejuvenated and outdated-rejuvenated human red blood cells frozen in the polyvinylchloride plastic primary collection bag and stored at -80 C in mechanical freezers.

Evaluation of the Haemonetics Blood Processor 30 and the IBM Blood Processor for isolation of platelets from normal volunteers.

1980: Feasibility studies at the Frozen Blood Bank Module deployed at the Mobile Fleet Hospital, Twenty- Nine Palms, CA: Evaluation of nonrejuvenated and outdated- rejuvenated human red blood cells frozen with 40% W/V glycerol in the polyvinylchloride plastic primary collection bag and stored at -80 C in mechanical freezers.

Collection of 6 to 8 units of platelets by apheresis procedures, and freeze-preservation with 6% DMSO in polyvinylchloride plastic bags at -80 C.

Evaluation of the Fenwal CS-3000 to isolate platelets from normal volunteers.

1981: Feasibility studies at the Frozen Blood Bank Module deployed at the Mobile Fleet Hospital, Bridgeport, CA: Evaluation of nonrejuvenated and outdated-rejuvenated human red blood cells frozen in the polyvinylchloride plastic primary collection bag and frozen at -80 C. Deployment of the Haemonetics Blood Processor 115 to deglycerolize the frozen RBC by Armed Services Blood Program of the Department of Defense during the Persian Gulf War.

Evaluation of a protocol for collecting 6 to 8 units of platelets from a single donor by apheresis procedures, freeze-preservation with 6% DMSO and storage at -80 C, and transportation in the frozen state in dry ice.

Evaluation of long-term preservation of plasma and cryoprecipitate stored at -20 C and at -80 C.

Frozen red blood cells containing 40% W/V glycerol and frozen platelets containing 6% DMSO stored at -80 C in mechanical freezers.

1981: Feasibility studies at the PACOM Blood Program Office, Okinawa, Japan: Evaluation of protocols for pooling 6 to 8 units of platelets from single units of blood and by plateletpheresis of a single donor, and freeze-preservation with 6% DMSO and storage at-80 C.

Evaluation of long-term preservation of plasma and cryoprecipitate at -20 C and at -80 C on the fibrinogen, factors V and VIII, and fibronectin levels.

1985: Hypothermia produces a reversible platelet dysfunction.

1998: The ketone NaD betahydroxybutyrate produced improved myocardial function and reduce oxygen consumption in isolated perfused rat hearts compared to Ringer's D,L lactate reported by Dr. K. Clarke and Dr. R.L. Veech.

1998: Dr. Ernest Ray Pariser, Dr. John Vournakis and Sergio Finkielsztein at Marine Polymer Technologies, Inc. isolated polymerized N-acetylglucosamine from microalgae. This hemostatic substance activates platelets and RBC to restore hemostasis, accelerates wound healing, and exerts an antimicrobial effect. The glycocalyx in the subendothelial area of blood vessels may produces substances similar to polymerized N-acetylglucosamine which is a potent activator of the hemostatic mechanisms to reduce the bleeding time and reduce nonsurgical blood loss. Endothelin released from endothelial cells and thromboxane produced by platelets are two potent vasoconstrictor substances involved in the restoration of hemostasis and. reduction in nonsurgical bleeding.

1985-1998: Studies by Dr. Shukri Khuri and Dr. Francis Moore, Jr., demonstrated the effect of complement activation during cardiopulmonary bypass surgery reported by Dr. John Kirklin was inhibited by hypothermia, heparin, and hemodilution. Bleeding diathesis in patients subjected to cardiopulmonary bypass surgery was studied in collaboration with Dr. S. Khuri and associates demonstrated that heparin increased the bleeding time by producing a platelet dysfunction related to the decreased thromboxane production at the bleeding site and by the increase in fibrinolytic activity.

Heparin, hypothermia, and anemia increase the bleeding time, decrease thromboxane production at the bleeding site, and increase nonsurgical blood loss. Increased nonsurgical blood loss 4 hours following cardiopulmonary bypass was correlated to the increase in bleeding time and the reduction in the hematocrit 2 hours after the bypass procedure. The hematocrit of 35V% was associated with decreased bleeding time, reduced nonsurgical blood loss, and reduced need for allogeneic RBC and allogeneic fresh frozen plasma in patients subjected to cardiopulmonary bypass surgery. Platelets frozen with 6% DMSO and stored at -80 C, washed and resuspended in. autologous plasma reduced nonsurgical blood loss and the need for allogeneic RBC and allogeneic fresh frozen plasma compared to liquid preserved platelets stored with agitation at room temperature for a mean of 3.4 days in patients following cardiopulmonary bypass surgery.

1988: The blood viscosity measured using the Merrill viscometer at a shear rate of 19 inverse seconds correlated to the cube of the hematocrit and square of the fibrinogen level in the blood.

2001: Anemia produces a reversible platelet dysfunction.

2003: The nonsurgical bleeding diathesis in anemic thrombocytopenic patients needs to be treated by the transfusion of viable and functional RBC to achieve a hematocrit of 35 V% prior to transfusion of viable and functional platelets.

2003: Severe adverse events related to hemoglobin based oxygen carriers (HBOCs) were associated with the length of storage of liquid RBC at 4 C that were administered, the excipient

Ringer's D,L lactate used to resuspend the HBOC, and. the infusion of Ringer's D,L lactate solution at the time the HBOCs were infused.

2003: Dr. John Gibson's lesion of collection was due to the elution of 51Cr from the red blood cells and not to the removal of nonviable red blood cells from the circulation.

2003: Freezing rate at 1 C per minute using instruments to control the rate of freezing using liquid nitrogen is not necessary; freezing rates at 2-3 C per minute by storage in -80 C mechanical freezers provide acceptable freezing rates for human RBC, platelets, plasma, and pluripotential mononuclear adult stem cells isolated from peripheral blood.

2003: Freezing at 2-3 C per minute by storage of human platelets treated with 6% DMSO in a -80 C mechanical freezer produces a bimodal population of human and baboon platelets. One population of platelets that circulates and the other population of platelets that function to reduce the bleeding time.

2003: The removal of supernatant glycerol from the RBC containing 40% W/V glycerol; the removal of supernatant DMSO from platelets containing 6% DMSO; and the removal of supernatant DMSO from pluripotential mononuclear adult stem cells isolated from peripheral blood containing 10% DMSO simplified the freezing of these cells stored at -80 C in mechanical freezers.

2003: In vitro testing has been performed to predict the in vivo 24 hour posttransfusion survival and lifespan of preserved RBC. No in vitro tests have been shown to quantitate the irreversibly damaged nonviable autologous and allogeneic red blood cells that are removed during the 24-hour posttransfusion survival period. Preservation of RBC produces reversible injury which can only be assessed by the in vivo recovery of the donor RBC during the posttransfusion period. The 24-hour posttransfusion survival of autologous RBC has to be measured using the double radioisotope methods using 1251 albumin and 51Cr, 99mTc and 51Cr, 111In-oxine and 51Cr, and the double 51Cr methods. The 24-hour posttransfusion survival and lifespan of compatible allogeneic RBC can be measured using differential agglutination of compatible but identifiable RBC using the Technicon Autoanalyzer and 51Cr method and the Coulter counter and the 51Cr method. Reversibly damaged RBC transfused into compatible donors and recovered by manual differential agglutination using anti-A and anti-B antibodies and by the biotin-avidin procedure demonstrate in vivo restoration of RBC morphology, RBC osmotic fragility, RBC density distribution, and increase in RBC ATP, DPG, p50 and K+ levels and decrease in RBC Na+ level during the posttransfusion period. The reduction in vitro of the nitric oxide binding to the sulfhydryl group of globin (SNOHb) in fresh and preserved RBC has been reported. However, no in vivo data were reported that the reduction in SNOHb in red blood cells following transfusion is irreversible.

2005: In vitro testing of preserved RBC ATP, DPG, p50, K+ levels, filterability, deformability, osmotic fragility, and density distribution, does not predict the quantity of the irreversibly damaged RBC and 24-hour posttransfusion survival value.

2005: The lifespan of the autologous and allogeneic preserved RBC is dependent on the in vivo environment of the recipient. Allogeneic preserved RBC must be compatible with the recipient. The preservation procedure cannot modify RBC antigenicity which may be recognized by the recipient as incompatible red blood cells. The environment into which the compatible preserved RBC are infused determines their lifespan and affects the physical-structure and metabolic state of the donor red blood cells.

2005: In vitro testing of RBC ATP, DPG, and p50 correlate with the oxygen transport function of the preserved RBC.

2005: In vitro testing of fresh and preserved platelets is utilized to detect the quantity of irreversibly damaged platelets that are removed during the 1-2 hour posttransfusion period. The response of platelets to hypotonic stress correlated to the in vivo recovery of the platelets following transfusion but not to the lifespan of the platelets. In vitro testing of platelet aggregation to single and dual agonists like ADP and collagen and collagen and epinephrine does not correlate with the in vivo recovery and function of the preserved platelets and to the lifespan of the preserved platelets. The lifespan of the autologous platelets is dependent upon whether or not platelet antigenicity has been modified by the preservation procedure and/or the disinfection procedure.

2005: Platelet production of thromboxane in vitro to the dual agonists adenosine diphosphate (ADP) and arachidonic acid (AA) correlates to the in vivo function of the platelets better than the platelet aggregation response to the dual agonists ADP and AA. In vivo restoration of platelet aggregation response to agonists occurs during the posttransfusion period of liquid preserved platelets stored at 22$\pm$2 C for 24 and 48 hours.

2005: In vivo testing of platelets should provide data on the survival and function of the platelets. Platelet in vivo recovery and lifespan can be assessed by 111-In labeled platelets and the recipient's blood volume. The increase in platelet count in patients can be used to assess the in vivo survival of platelets from the number of platelets transfused and the blood volume of the recipient. The function of fresh and preserved platelets can be assessed by the reduction of the increased bleeding time in normal volunteers and baboons produced by aspirin treatment 18 to 24 hours prior to the transfusion of autologous and compatible allogeneic non-aspirin treated preserved platelets. In normal volunteers and baboons in vivo survival and function of autologous non-aspirin treated platelets can be assessed by 111-In-oxine labeled platelets, the recipient's blood volume, and the bleeding time reduction in aspirin-treated normal volunteers and baboons.

2005: The in vivo function of allogeneic compatible platelets can be assessed in patients by the reduction in bleeding time, reduction in nonsurgical blood loss, and the reduction in the need for allogeneic RBC and fresh frozen plasma.

2005: The hematocrit has a greater effect on the bleeding time and nonsurgical blood loss than do platelets and plasma clotting proteins.

2005: The plasma clotting protein factor V, factor VIII, the von Willebrand's factor, and fibrinogen level can be used to assess the function of plasma clotting protein in fresh frozen plasma and in cryoprecipitate stored at -20 C and at -80 C.

2005: Using the Haemonetics Blood Processor ACP215 instrument, outdated RBC stored at 4 C in CPDA1 for 36 to 38 days, CPD/AS-I, CP2D/AS-3, and CPD/AS-5 RBC stored at 4 C for 42 days can be biochemically modified with a solution containing pyruvate, inosine, phosphate and adenine (PIPA) by storage at 37 C for 60 minutes, treated with 40% W/V glycerol and the supernatant glycerol solution removed and then frozen at -80 C, thawed, washed and stored in AS-3 additive solutions for at least 24-hours with 24-hour posttransfusion survival of 75%, RBC with normal or slightly to moderately increased affinity for oxygen and less than 1% hemolysis.

2005: Using Haemonetics Blood Processor ACP2I5 RBC in CPDA1, CPD/AS-l, CP2D/AS-3 and CPD/AS-5 stored at 4 C for 3 to 6 days can be biochemically modified with a solution containing pyruvate, inosine, phosphate and adenine (PIPA) by storage at 37 C for 60 minutes, treated with 40% W/V glycerol and the supernatant glycerol removed and then frozen at -80 C, thawed, washed and stored at 4 C in AS-3 for 2 weeks with 24-hour posttransfusion survival of 75%, normal to improved oxygen transport function, and less than 1% hemolysis.

2005: Limitation of the peripheral venous hematocrit to assess whether the individual is hypervolemic, normovolemic, or hypovolemic. The hematocrit alone should not be used as the transfusion trigger to assess whether or not the patient should be transfused. Clinical signs and symptoms should be used together with the hematocrit to assess whether the total blood volume is normal or reduced and the quality and quantity of RBC that are needed. The quality of the RBC survival and function needs to be considered in the treatment of the anemic normovolemic patients and the

anemic hypovolemic patients. Most patients are hypovolemic and anemic with falsely elevated hematocrit values.
2005: The clinical assessment of the central blood volume and the peripheral blood volume of the recipient is needed to determine the quality and quantity of the RBC that are needed.
2005: The central blood volume is assessed by the arterial blood pressure, heart rate, cardiac output, the arterial blood p02, pCO2, and pH and urine output. The peripheral blood volume is assessed by the postural changes in vital signs. The temperature of the big toe and sexual function are important assessments of the peripheral blood volume. The appetite is an important clinical assessment of blood flow to the gastrointestinal tract. Transfusion of the recipient to restore the peripheral blood volume will improve the physical, sexual and psychological well being of the patient. The transfusion trigger of the patient should not be solely dependent on the peripheral venous hematocrit but on the physical and general state of health of the recipient.
2005: Prophylactic and therapeutic platelet transfusions depend upon platelet count, platelet size and platelet hemostatic function and the hematocrit. The hematocrit has a greater effect on the bleeding time and the non-surgical blood loss than do the platelet count, platelet size and platelet hemostatic function and the plasma clotting proteins: factor V, factor VIII, von Willebrand's factor and fibrinogen levels.
2005: Temperature has an important effect on function of red blood cells, platelets and plasma clotting protein. Hypothermia alone and anemia alone produce a reversible platelet dysfunction. Nonsurgical bleeding diathesis needs to be treated by warming the patient, warming the resuscitation solutions and the blood products during infusion using warming devices to 40 C, transfusion of viable and functional RBC to achieve a hematocrit of 35 V% and by the transfusion of viable and functional platelets and clotting proteins. Surgical bleeding needs to be treated by surgical intervention and by hemostatic agents applied with external pressure.
2005: Using the Haemonetics Blood Processor ACP215 instrument, RBC stored in the liquid state in CPDA-l and in the additive solutions (CPD/AS-1, CP2D/AS-3, and CPD/AS-5) at 4 C for 3 to 6 days treated with 40% W/V glycerol and the supernatant glycerol solution removed and then frozen at -80 C for at least 10 years, thawed, washed and stored at 4 C in AS-3 for 2 weeks provide RBC with 24 hour posttransfusion survival of 75% that function shortly after transfusion to restore hemostasis.

Single donor leukoreduced platelets stored in the liquid state at 22 C with agitation for 24-hours treated with 6% DMSO, concentrated to remove the supernatant DMSO, frozen and stored at -80 C, thawed, and diluted with 10 to 20 ml of 0.9% sodium chloride stored at room temperature without agitation for 4 hours have in vivo recovery of 25 to 30% 1 to 2 hours after transfusion and lifespan of 7 days.

Single donor previously frozen, thawed, washed platelets reduced the nonsurgical blood loss and reduced the need for allogeneic RBC and allogeneic fresh frozen plasma in patients following cardiopulmonary bypass surgery better than single donor or pooled preserved platelets stored with agitation at room temperature for a mean of 3.4 days.

2005: Fresh plasma frozen and stored at -80 C for at least 14 years maintains the function of the plasma clotting proteins factor V, VIII and fibrinogen.

2005: As suggested by current practice guidelines, clinical judgment, not the hematocrit, should be the ultimate factor in determining the need for RBC transfusions. Tachycardia, shortness of breath, pallor, decreased tissue turgor, postural hypotension, lightheadedness or dizziness, decreased appetite, weakness, and fatigue are important signs and symptoms. Moreover, a patient properly apprised of the potential risks and. benefits of a transfusion should be allowed a voice in this important decision.

2005: In 1968, about 40 years ago, the U.S. Navy deployed -80 C water-cooled mechanical freezers in South Vietnam to freeze red blood cells to supplement fresh whole blood and liquid preserved whole blood to treat casualties. Recently, 40 years later, the -80 C air-cooled mechanical freezers have been deployed in combat zones in the Middle East by the Netherlands military under the direction of Dr. Charles Lelkens, Dr. Femke Noorman, and Dr. John Badloe and their associates to freeze red blood cells, platelets and plasma all frozen and stored at -80 C in mechanical freezers without the need for fresh whole blood to successfully treat both military and civilian casualties.

2005 to 2008: Forty-five years of research at the NBRL supported by the U.S. Navy's Bureau of Medicine and Surgery, Naval Medical Research and Development Command, the Office of Naval Research (ONR), the Congress of the United States and by commercial companies provided the following information on the safety and therapeutic effectiveness of frozen RBC, frozen platelets, frozen plasma and frozen pluripotential mononuclear stem cells obtained from peripheral blood to treat military and civilian patients.

a. Using the Haemonetics Blood Processor ACP2I5, non-leukoreduced, nonrejuvenated group 0 Rh positive and group 0 Rh negative RBC stored at 4 C in CPDA-l, CPD/AS-l, CP2D/AS-3, and CPD/AS-5 for 3 to 6 days, treated to achieve 40% W/V glycerol, the supernatant glycerol removed and the glycerolized RBCs frozen in polyvinyl chloride plastic bags at -80 C for at least 10 years, thawed, deglycerolized can be stored in the additive solution AS-3 (Nutricel) at 4 C for 2 weeks;

b. Using the Haemonetics Blood Processor ACP215 non-leukoreduced indated rejuvenated group 0 Rh positive and group 0 Rh negative RBC stored at 4 C in CPDA-1, CPD/AS-l, CP2D/AS-3, and CPD/AS-5 for 3 to 6 days, biochemically treated with pyruvate, inosine, phosphate, and adenine (PIPA), treated to achieve 40% W/V glycerol and the supernatant glycerol removed, and then frozen and stored at -80 C for at least 10 years, thawed, deglycerolized can be stored in the additive solution AS-3 at 4 C for 2 weeks;

c. Using the Haemonetics Blood Processor ACP215 non-leukoreduced outdated rejuvenated group 0 Rh positive and group 0 Rh negative RBC stored at 4 C in CPDA-1 for 36 to 38 days and CPD/AS-l, CP2D/AS-3, and CPD/AS-5 for 42 days, biochemically treated with pyruvate, inosine, phosphate, and adenine (PIPA), treated to achieve 40% W/V glycerol and the supernatant glycerol removed, and then frozen and stored at -80 C for at least 10 years, thawed, deglycerolized can be stored in the additive solution AS-3 at 4 C for at least 24 hours;

d. Leukoreduced single donor group O platelets treated to achieve 6% DMSO, the supernatant DMSO removed and the DMSO platelets frozen and stored at -80 C for at least 2 years, thawed, diluted with 10 to 20 ml of 0.9% NaC1 can be stored at room temperature for 6 hours without agitation.

e. ABO, Rh and HLA-typed pluripotential mononuclear stem cells isolated from peripheral blood, treated to achieve 10% DMSO and the supernatant DMSO removed and the pluripotential mononuclear cells frozen and stored at -80 C for 1½ years, thawed, diluted with 10 to 20 ml of 0.9% NaCl can be stored at room temperature for 6 hours without agitation; and transfused into a compatible ABO, Rh and HLA recipient.

f. AB plasma obtained from donors stored at -80 C for at least 14 years, thawed can be stored at room temperature for 8 hours or at 4 C for 24 hours.

g. In vitro testing using the clot signature analyzer, platelet function analyzer and thromboelastogram does not correlate to the bleeding time;
h. The freezing of human and baboon platelets treated to achieve 6% DMSO, the supernatant DMSO removed, and stored at -80 C produce a bimodal population of platelets following thawing. One population of platelets that circulate is Gp1b normal and reduced annexin V binding and the other population of platelets Gp1b reduced and increased annexin V binding does not circulate but functions to reduce nonsurgical blood loss. The bimodal population of platelets and platelet microparticles are present in nonwashed platelets frozen after removal of the supernatant DMSO and in washed platelets frozen with the supernatant DMSO.
i. The book The Hypovolemic Anemia of Trauma: The Missing Blood Syndrome published by CRC Press, Boca Raton, FL in 1981 was written to report on our clinical experience during the years 1968 to 1974 when over 300 patients who sustained war injuries in South Vietnam were transferred to Chelsea Naval Hospital, Chelsea, MA. The patients were called to the attention of our research facility when the use of general anesthesia for routine debridement of wounds precipitated a life threatening state of hypotension. A series of tests established the presence of chronic hypovolemia.

When red blood cells were administered to the patients with this clinical disorder before their surgical procedure, hypotension during operations and post-operatively was eliminated. Determinations of blood volumes were made using 51 Cr labeled autologous red blood cells to measure red blood cell volume and ^{125}I iodinated cold agglutinin to measure the plasma volume. Cardiac output and the distribution of blood flow was measured in these patients to understand how the vital organs such as the brain, heart, and kidneys were protected in spite of 30-40% reduction in total blood volume.

To gain some understanding of the mechanism of the reduced blood volume, we studied the patient's red blood cell production as well as the destruction of the patient's own red blood cells and the donor red blood cells. Red blood cell production was impaired and both the patient and the donor RBC were being destroyed through an alternative pathway producing porphyrin-like substances instead of the usual urobilinogen breakdown product of the hemoglobin with the release of carbon monoxide. The degradation of red blood cell hemoglobin through the alternative pathway was thought to be responsible for the so-called missing blood syndrome.

The experience of Stinner and associates reported in Military Medicine, 175:1027-1029, 2010 is very different from our experience treating patients at the Chelsea Naval Hospital during the Vietnam War. They report that of 348 service members who sustained combat injuries in Afghanistan and Iraq, fifty-three (15.2%) required amputations 12 weeks to 5.5 years following their initial extremity injuries.

In our patients at the Chelsea Naval Hospital although these orthopedic patients had chronic hypovolemic anemia of trauma with reduction in both the RBC and plasma volume, normal hemoglobin and hematocrit values, and normal central blood volumes, they suffered from severely contracted peripheral blood volumes to their extremities. The aggressive transfusion of washed liquid preserved RBC and washed previously frozen deglycerolized RBC restored their peripheral blood volume and total blood volume. We studied 300 servicemen for at least 2 years during their hospitalization at Chelsea Naval Hospital. Chronic hypovolemic patients with traumatic injuries to their extremities were transfused RBC in order to restore their peripheral blood volume and to repair the injured extremities.

The observation of biphasic survival of compatible identifiable allogeneic RBC transfused to patients with traumatic injuries suggested the presence and removal of a toxic substance from the blood. The observation of biphasic survival curves suggested the removal from the patient's circulation of a toxic substance which had adversely affected the lifespan of donor RBC populations and of the patient's own RBC as well. Out hypothesis was that the presumed toxic substance had two adverse effects: it decreased the survival of the autologous and donor compatible and identifiable RBC and impaired erythropoiesis. The disappearance of the presumed toxic substance was accompanied by improvement in the survival of donor RBC populations as shown by the biphasic survival curves, and by improvement in the patient's RBC production as manifested by an increase in the reticulocyte count in peripheral blood. In patients with biphasic linear donor RBC survival curves, the lifespan in the initial phase ranged from 38 to 60 days, and in the second phase ranged from 78 to 113 days. This finding suggested that it took 30 to 40 days after transfusion for the toxic substance to be removed from the circulation and for the lifespan to increase (Valeri CR and Altschule MD, CRC Press, 1981, Boca Raton, FL).

Whose recommendations were correct regarding resuscitation of wounded casualties?

Colonel W. Crosby advocated correctly the optimum treatment of patients in hemorrhagic shock using compatible fresh whole blood from the walking blood bank.

Dr. EJ Cohn was not correct regarding the therapeutic effectiveness of human albumin. Human albumin with a MW of 68,000 distributes rapidly into the extravascular volume and should not be used to increase the plasma volume and the blood volume.

Dr. T. Shires was not correct advocating the use of Ringer's D,L lactate as the large volume resuscitation fluid. Ringer's D,L lactate is a toxic substance that produced the "Danang lung: and should not be used. Our studies support the need for "substrate resuscitation" using a Na-D-betahydroxybutyrate solution to maintain the metabolism and function of cells.

Current approaches to treat patients subjected to blood loss:

Use of hemostatic agents with external pressure to stop the blood loss by "buddy care";

Prevent hemodilution using large volumes of crystalloid or colloid solutions like hydroxyethylstarch (HES) (Myburgh JA et al, N Engl J Med 367:1901-1911, 2012). Hemodilution produces an anticoagulant effect by the dilution of RBC, platelets, plasma clotting proteins and increases blood loss.

Hypertonic saline should be avoided because it produces hyperchloremic acidosis.

Prevent hypothermia by warming the patients and by infusion of resuscitation fluids and blood products using warming devices to maintain a temperature of 40 C at the time of infusion. Hypothermia adversely affects hemostasis and increases blood loss.

Colonel Crosby observed in Korea and Vietnam that compatible fresh whole blood provided optimum treatment of patients subjected to hemorrhagic shock. What does fresh compatible whole blood free of infectious diseases and collected form screened blood donors contain? Fresh whole blood contains 95% viable RBC with normal function to provide oxygen to the brain, heart and kidneys and the fresh RBC exert a hemostatic effect to reduce nonsurgical blood loss.

In 1981, the testing of the fresh whole blood for infectious disease markers for AIDS, hepatitis and other infectious disease agents and the screening of the donor now require at least 24 to 48 hours before the fresh blood can be used.

Compatible donor red blood cells that are viable and functional provide oxygen to maintain high p02 tensions in brain, heart, and kidneys and exert a very important hemostatic effect to reduce nonsurgical blood loss. Viable and functional plasma clotting proteins are needed together with viable and functional platelets. The optimum ratio of RBC, plasma and platelets is now being investigated together with the optimum volume and composition of the resuscitation fluids that are used to treat the wounded casualties. The quality and quantity of the RBC, platelets, and plasma clotting proteins need to be defined when protocols to treat patients using ratios of RBC, plasma and platelets are recommended.

Small volume of 250 ml of 600 mM Na-D-betahydroxybutyrate solution needs to replace large volume resuscitation using crystalloid and colloid solutions. Therapeutic effects of Na-D-B-hydroxybutyrate to treat traumatic injury, insulin resistance, and head trauma are now being evaluated. Traumatic injury, infection and hemorrhage all cause insulin resistance. Insulin resistance results in inhibition of the activity of mitochondrial pyruvate dehydrogenase (PDH) and the conversion of pyruvate to acetyl CoA impairing cellular energy production. Na-D-B-hydroxybutyrate containing fluids increase metabolic efficiency and bypass metabolic blocks caused by insulin resistance and inhibition of pyruvate dehydrogenase (PDH) activity. Resuscitation should be undertaken in the field as soon as possible by infusion of 250 ml of ketone solution containing 600 mM Na-D-B-hydroxybutyrate to treat traumatic brain injury and to reduce the incidence of post-traumatic stress disorder. Na-D-B-hydroxybutyrate solutions should be used in the resuscitation following hemorrhage rather than hypertonic NaCl which produces hyperchloremic acidosis; large volumes of Ringer's lactate solutions which produce hemodilution and inhibit hemostasis; and albumin which is a non-effective colloid and does not increase the plasma volume and the blood volume.

CHAPTER 2 REFERENCES

1. Almond DV, Valeri CR. The in vivo effects of deglycerolized agglomerated erythrocytes transfused in multiple units to stable anemic patients. Transfusion 7:95-104, 1967.
2. Almond DV, Valeri CR. Relationship between lipid fractions of erythrocytes and their in vivo survival following preservation with glycerol and the slow freeze technic. Transfusion 7:10-16, 1967.
3. Alter HJ, Tabor E, Meryman HT, Hoofnagle JH, Kahn RA, Holland PV et al. Transmission of hepatitis B virus infection by transfusion of frozen-deglycerolize red blood cells, N Engl J Med 298:637-42, 1978.
4. Apstein CS, Dennis RC, Briggs L, Vogel WM, Frazer J, Valeri CR. Effect of erythrocyte storage and oxyhemoglobin affinity changes on cardiac function. Am J Physiol 248 (Heart Circ Physiol 17): H508-H515, 1985.
5. Asakura T, Reilly MP. Methods for the measurement of oxygen equilibrium: Curves of red cell suspensions and hemoglobin solutions: in Nicolau C (ed): Oxygen Transport in Red Blood Cells. Proceedings of the 12th Aharon Katzir Katchalsky Conference 1984. New York, Pergamon. Press. 1986, pp 57-75.
6. Axford TC Dearani JA, Ragno G, MacGregor H, Patel MA, Valeri CR, Khuri SF: Safety and therapeutic effectiveness of reinfused shed blood after open heart surgery. Ann Thorac Surg 57:615-622, 1994.
7. Biron PE, Howard I, Altschule MD, Valeri CR. Chronic deficits in red-cell mass in patients with orthopaedic injuries (stress anemia). J Bone Joint Surg 54-A:1001-1014, 1972.
8. Bellingham AJ, Huehns ER. Oxygen dissociation in red cells from patients with abnormal haemoglobins and pyruvate kinase deficiency. Forsvarsmedicin 5:207-211, 1969.
9. Benesch R, Benesch RE. The effect of organic phosphates from the human erythrocyte on the allosteric properties of hemoglobin. Biochem Biophys Res Commun 26:163-167, 1967.
10. Blevins FT, Shaw B, Valeri CR, Kasser I, Hall J, Reinfusion of shed blood after orthopaedic procedures in children and adolescents. J Bone & Joint Surg 75-A:363-371, 1993.

11. Brodine CE, Sell KW, Moss GS, Valeri CR. Navy surgical research programs. J Surg Res 7:545-548, 1967.
12. Carciero R, Valeri CR. Isolation of mononuclear leukocytes in a plastic bag system using Ficoll-Hypaque. Vox Sang 49:373-380, 1985.
13. Chanutin A, Curnish RF. Effect of organic and inorganic phosphates on the oxygen equilibrium of human erythrocytes. Arch Biochem Biophys 121:96-102, 1967.
14. Costa JL, Dobson CM, Kirk KL, Poulsen FM, Valeri CR, Vecchione JJ. Studies of human platelets by 19F and 31P NMR. FEBS Letters 99:141-146, 1979.
15. Cronkite EP. Measurement of the effectiveness of platelet transfusion. Transfusion 6:18-22, 1966.
16. Crowley JP, Valeri CR, Metzger J, Gray A, Schooneman F, Man NK, Merrill E: The estimation of whole blood viscosity by a porous bed method. Am J Clin Pathol 96:729- 737, 1991.
17. Danon D. A rapid micro method for recording red cell osmotic fragility by continuous decrease of salt concentration. J Clin Path 16:337-342, 1963.
18. Danon D, Markikovsky Y. Determination of density distribution of red cell population. J Lab Clin Med 64:668-674, 1964.
19. Dennis RC, Vito L, Weisel RD, Valeri CR, Berger RL, Hechtman HB. Improved myocardial performance following high 2,3 diphosphoglycerate red cell transfusions. Surgery 77:741-747, 1975.
20. Dennis RC, Hechtman HB, Berger RL, Vito L, Weise] PD, Valeri CR. Transfusion of 2,3 DPG-enriched red blood cells to improve cardiac function. Ann Thoracic Surg 26(1):17-26, 1978.
21. Dennis RC, Valeri CR. Measuring percent oxygen saturation of hemoglobin, percent carboxyhemoglobin and methemoglobin, and concentrations of total hemoglobin and oxygen in blood of man, dog, and baboon. Clin Chem 26:1304-1308, 1980.
22. Drabkin DL, Austin JR. Spectrophotometric studies. II. Preparations from washed red blood cells: nitric oxide hemoglobin and sulfhemoglobin. J Biol Chem 112:51-65, 1935.
23. D'Emmanuele di Villa Bianca R, Sorrentino R., Maffia P, Mirone V, Imbimbo C, Fusco F, De Palma R, Ignarro LI, Cirino G. Hydrogen sulfide as a mediator of human corpus

cavernous smooth-muscle relaxation. PNAS Early Edition: 1-6, 2008.

24. Farrugia A. Albumin usage in clinical medicine: tradition or therapeutic? Trans Med Rev 24:53-63, 2010.
25. Favuzza J, Hechtman HB. Hemostasis in the absence of clotting factors. J Trauma: 57:542-544, 2004.
26. Faris PM, Ritter MA, Keating EM, Valeri CR. Unwashed filtered shed blood collected after knee and hip arthroplasties: A source of autologous red blood cells. J Bone & Joint Surg 73A:1169-1178, 1991.
27. Festa RS, Asakura T. The use of an. oxygen dissociation curve analyzer in transfusion therapy. Transfusion 19:107-113, 1979.
28. Gabrio BW, Donohue DM, Finch CA. Erythrocytic preservation. V. Relationship between chemical changes and viability of stored blood treated with adenosine. J Clin Invest 34:1509-1512, 1955.
29. Gardos G. The mechanism of ion transport in human erythrocytes. I. The role of 2,3 diphosphoglyceric acid in the regulation of potassium transport. Acta Biochem et Biophysis Acad Sc Hung 1:139-148, 1966.
30. Handin RI, Lawler KC, Valeri CR. Automated platelet counting. Am J Clin Path 56:661-664, 1971.
31. Haynes LL, Tullis JL, Pyle HM, Sproul MT, Wallach S, Turville WC. Clinical use of glycerolized frozen blood. JAMA 1.73:1657-1663, 1960.
32. Haugen RK. Hepatitis after the transfusion of frozen red cells and washed red cells. N Engl J Med 301:393-395, 1979.
33. Healy WL, Wasilewski SA, Pfeifer BA, Kurtz SR, Hallack GN, Valerio M, Valeri CR. Methylmethacrylate monomer and fat content in shed blood after total joint arthroplasty. Clin Ortho Rel Res 286:15-17, 1993.
34. Healy WL, Pfeifer BA, Kurtz SR, Johnson C, Johnson W, Johnston R, Sanders D, Karpman R, Hallack ON, Valeri CR. Evaluation of autologous shed blood for autotransfusion after orthopaedic surgery. Clin Ortho Rel Res Number 99, pps 53-59, February 1994.
35. Huggins CE. Frozen blood: principles of practical preservation. Monogr Surg Sci 3:133-173, 1966.
36. Hunt SM, Lionetti FJ, Valeri CR, Callahan AB. Cryogenic

preservation of monocytes from human blood and plateletpheresis cellular residues. Blood 57:592-598, 1981.

37. Hunt SM, Lionetti FJ, Valeri CR. Isolation and cryopreservation of monocytes from plateletapheresis cellular residues. Transfusion 23:387-390, 1983.
38. Ikeda Y, Young YH, Vournakis JN, Leifer AM. Vascular effects of poly-N acetylglucosamine in isolated rat aortic rings. J Surg Res 102:215-220, 2002.
39. Kestin AS, Valeri CR, Khuri, SF, Loscalzo J, Ellis PA, MacGregor H, Birjiniuk V, Ouimet H, Pasche B, Nelson MI, Benoit SE, Rodino LI, Barnard MR, Michelson AD. The platelet function defect of cardiopulmonary bypass. Blood 82:107-117, 1993.
40. Lindner HB, Zhang A, Eldridge J, Demcheva M, Tsichilis P, Seth AA, Vournakis J, Muise-Helmericks, RC. Anti-bacterial effects of poly-N-acetyl. glucosamine nanofibers in cutaneous wound healing: requirement for AKTI. PLoS1 6: 18996, 2011.
41. Lionetti FJ, Valeri CR, Bond IC, Kivowitz C, Weinman E. Nucleotides in frozen glycerolized erythrocytes. Transfusion 6:116-123, 1966.
42. Kelechi TJ, Mueller M, Hankin CS, et al. A randomized investigator blinded controlled pilot study to evaluate the safety and efficacy of a poly-N-acetyl glucosamine-derived membrane material in patients with venous leg ulcers. J Am Assoc Derm May 20, 2011 DOI (dor; 10.1016 (jaad 2011 (OLO31) 2011.
43. Khuri SF, Wolfe JA, Josa M, Axford TC, Szymanski I, Assousa S, Ragno G, Patel M, Silverman A, Park M, Valeri CR. Hematologic changes during and after cardiopulmonary bypass and their relationship to the bleeding time and nonsurgical blood loss. J Thorac Cardiovasc Surg 104:94-107, 1992.
44. Melaragno AJ, Carciero R, Feingold H, Talarico L, Weintraub L, Valeri CR. Cryopreservation of human platelets using 6% dimethylsulfoxide and storage at -8O C. Effects of 2 years of frozen storage at -80C and transportation in dry ice. Vox Sang 49:245-258, 1985.
45. Merrill EW. Rheology of blood. Physiol Rev 49:863-888, 1969.
46. Merrill EW, Crowley JP, Valeri CR. Rapid and simple measurement of apparent whole blood viscosity. Lab Medica Int 10:19-24, 1993.

47. Michelson AD, MacGregor H, Barnard MR, Kestin AS, Rohrer MJ, Valeri CR. Reversible inhibition of human platelet activation by hypothermia in vivo and in vitro. J Thrombosis and. Haemostasis 5:633-640, 1994.
48. Moss GS. Massive transfusion of frozen preserved red cells in combat casualties. Report of three cases. Surgery 66:1108-1113, 1969.
49. Moss GS, Valeri CR, Brodine CE. Clinical experience with the use of frozen blood in combat casualties. NEJM 278:748-752, 1968.
50. Najjar SF, Healey NA, Healey CM, McGarry T, Khan B, Thatte HS, Khuri SF. Evaluation of poly-N-acetyl glucosamine as a hemostatic agent in patients undergoing cardiac catheterization: a double-blind, randomized study. J Trauma 57 Suppl:S38-S41, 2004.
51. O'Murchadha ET, Horland A, Neiman RS, Valeri CR. Morphology of cells grown in the CFU-GEMM tissue culture assay from mononuclear cells obtained from peripheral blood and bone marrow of normal volunteers. Exp Hematol 16:235-239, 1988.
52. Pietramaggiori G, Yang HJ, Scherer SS, Kaipainen A, Chan BK, Alperovich M, Newalder J, Demcheva M, Vournakis JN, Valeri CR, Hechtman HB, Orgill DP. Effects of poly-N-acetyl glucosamine (pGLcNAc) patch on wound healing in db/db mouse. J Trauma 64(3):803-808, 2008.
53. Prins ML. Cerebral metabolic adaptation and ketone metabolism after brain injury. J Cereb Blood Flow Metab 28:1-16, 2008.
54. Scherer SS, Pietramaggiori G, Matthews J, Perry S, Assmann A, Carothers A, Demcheva M, Muise-Helmericks RC, Seth A, Vournakis JN, Valeri CR, Fischer TH, Hechtman HB, Orgill DP. Poly-N-acetyl glucosamine (pGlcNAc) nano-fibers: a new bioactive material to enhance diabetic wound healing by cell migration and angiogenesis. Ann Surg 250:322-330, 2009.
55. Smith AU. Prevention of haemolysis during freezing and thawing of red blood cells. Lancet 2:910-911, 1950.
56. Smith AU, Polge C. Storage of bull spermatozoa at low temperature. Vet Rec 62:115- 116, 1950.
57. Stamler JS, Jaraki O, Osborne J, Simon DI, Keaney J, Vita J, Singel

D, Valeri CR, Loscalzo J. Nitric oxide circulates in mammalian plasma primarily as an S-nitroso adduct of serum albumin. Proc Natl Acad Sci USA 89:7674-7677, 1992.

58. Stinner DJ, Buns TC, Kirk KL, Scoville CR, Ficke JR, Hsu JR. Late Amputation Study Team (LAST). Prevalence of late amputations during the current conflicts in Afghanistan and Iraq. Mil Med 175:1027-1029, 2010.
59. Strumia MM, Colwell LS, Dugan A. The measure of erythropoiesis in anemias. I. The mixing time and the immediate posttransfusion disappearance of T1824 dye and of 51Cr tagged erythrocytes in relation to blood volume. Blood 13:128-145, 1958.
60. Strumia MM, Dugan A, Colwell LS. The measure of erythropoiesis in anemias. II. The immediate and subsequent loss of transfused erythrocytes in healthy subjects. Blood 19:115-122, 1962.
61. Strumia MM, Strumia PV, Dugan A. Significance of measurement of plasma volume and of indirect estimation of red cell volume. Transfusion 8:197-209, 1968.
62. Strumia MM, Strumia PV, Eusebi AJ. The preservation of blood for transfusion. VIII. Effect of adenine and inosine on the adenosine triphosphate and viability of red cells of stored blood at 1 C for 21 days. J Lab Clin Med 76:907-914, 1970.
63. Sweetman HE, Costa JL, Vecchione JJ, Valeri CR, Shepro D. Dense bodies and total calcium in human platelets following aspirin ingestion for a two-week period. Thrombosis Res 17:55-61, 1980.
64. Szymanski I0, Valeri CR, McCallum LE, Emerson CP, Rosenfield RE. Automated differential agglutination technic to measure red cell survival. I. Methodology. Transfusion 8:65-73, 1968.
65. Szymanski I0, Valeri CR. Automated differential agglutination technic to measure red cell survival. II. Survival in vivo of preserved red cells. Transfusion 8:74-83, 1968.
66. Szymanski IO, Valeri CR. Clinical evaluation of concentrated red cells. NEJM 280:281-287, 1969.
67. Thatte HS, Zagarins SE, Amiji M, Khuri SF. Poly-N-acetyl glucosamine-mediated red blood cell interactions. J Trauma 57 Suppl:S7-S12, 2004.
68. Thatte HS, Zagarins SE, Khuri SF, Fischer T. Mechanisms of

poly-N-acetyl glucosamine polymer-mediated hemostasis: platelet interactions. J Trauma 57 Suppl:S13-S21, 2004.

69. Tullis JL, Pennel RB, Melin M, Zemp JW, Henderson ME, Sproul MT, MacNeill A. Problems of albumin medium for deglycerolized red cell resuspension. Vox Sang 8:100-101, 1963.
70. Valeri CR, Henderson ME. Recent difficulties with frozen glycerolized blood. JAMA 188:1125-1131, 1964.
71. Valeri CR. Effect of resuspension medium on in vivo survival and supernatant hemoglobin of erythrocytes preserved with glycerol. Transfusion 5:25-35, 1965.
72. Valeri CR, Bond JC, Fowler K, Sobucki J. Quantitation of serum hemoglobin binding capacity using cellulose acetate membrane electrophoresis, Clin Chem 11:581-588, 1965.
73. Valeri CR, McCallum LE. Relationship between glutathione stability and in vivo survival of autologous, deglycerolized, resuspended red blood cells. Nature 205:561-563, 1965.
74. Valeri CR. Observations on recipient plasma hemoglobin concentration after transfusion with glycerolized frozen blood. Transfusion 5:36-53, 1965.
75. Valeri CR. The in vivo survival of Coombs positive autologous erythrocytes produced by agglomeration. Transfusion 6:247-253, 1966.
76. Valeri CR, Bond JC. Observations on the preservation of autologous human erythrocytes using glycerol, slow-freeze technic and agglomeration. Transfusion 6:254-262, 1966.
77. Valeri CR, Bond JC, McCallum LE. Relationships between metabolic state and (1) in vivo survival and (2) density distribution of previously frozen human erythrocytes. Transfusion 6:543-553, 1966.
78. Valeri CR, McCallum LE, Danon D. Relationships between in vivo survival and (1) density distribution, (2) osmotic fragility of previously frozen, autologous, agglomerated, deglycerolized erythrocytes. Transfusion 6:554-564, 1966.
79. Valeri CR, Runck AH, McCallum LE. Observations on autologous, previously frozen, deglycerolized, agglomerated, resuspended red cells. I. Effect of storage temperatures. II. Effect of adenine supplementation of glycerolized red cells prior to freezing. Transfusion 7(2) : 105-116, 1967.

80. Valeri CR, Brodine CE, Moss GS. Use of frozen blood in Vietnam. Bibl Haematol 29:735-8 1968.
81. Valeri CR, Runck AH. Long term frozen storage of human red blood cells: Studies in vivo and in vitro of autologous red blood cells preserved up to six years with high concentrations of glycerol. Transfusion 9:5-14, 1969.
82. Valeri CR, Fortier NL. Red-cell 2,3-diphosphoglycerate and creatine levels in patients with red-cell mass deficits or with cardiopulmonary insufficiency. NEJM 81:1452-1455, 1969.
83. Valeri CR, Szymanski IO, Runck AH. Therapeutic effectiveness of homologous erythrocyte transfusions following frozen storage at -80C for up to seven years. Transfusion 10:102-112, 1970.
84. Valeri CR, Zaroulis CG. Rejuvenation and freezing of outdated stored human red cells. NEJM 287:1307-1313, 1972.
85. Valeri CR, Zaroulis CG, Marchionni LD, Patti KJ. A simple method for measuring oxygen content in blood. J Lab Clin Med 79:1035-1040, 1972.
86. Valeri CR, Cooper AG, Pivacek LE. Limitations of measuring blood volume with iodinated I 125 serum albumin. Arch Int Med 132:534-538, 1973.
87. Valeri CR, Feingold, H, Marchionni LD. A simple method for freezing human platelets using 6% dimethylsulfoxide and storage at -80 C. Blood 43:131-136, 1974.
88. Valeri CR, Feingold H, Zaroulis CG, Sphar RL, Adams GM. Effects of hyperbaric exposure on human platelets. Aerospace Medicine 45:610-616, 1974.
89. Valeri CR. Simplification of the methods for adding and removing glycerol during freeze-preservation of human red blood cells with the high or low glycerol methods: Biochemical modification prior to freezing. Transfusion 5:195-218, 1975.
90. Valeri CR. Blood Banking and the Use of Frozen Blood Products. Chemical Rubber Company Press, Boca Raton, Florida, 1976.
91. Valeri CR, Weisel RD, Dennis RC, Mannick JA, Berger RL, Hechtman HB. Oxygen transport function of preserved red blood cells and myocardial performance. Prog Clin Biol Res 21:597-616, 1978.
92. Valeri CR, Yarnoz M, Vecchione JJ, Dennis RC, Anastasi J, Valeri DA, Pivacek LE, Hechtman HB, Emerson CP, Berger RL.

Improved oxygen delivery to the myocardium during hypothermia by perfusion with 2,3 DPG-enriched red blood cells. Ann Thoracic Surg 30:527-535, 1980.

93. Valeri CR, Valeri DA, Gray A, Melaragno AJ, Dennis RC, Emerson CP. Red blood cell concentrates stored at 4 C for 35 days in CPDA-1, CPDA-2, or CPDA-3 anticoagulant- preservative, biochemically modified, and frozen and stored in the polyvinyl chloride plastic primary collection bag with 40% W/V glycerol at -80 C. Transfusion 22:102-106, 1982.
94. Valeri CR, Valeri DA, Anastasi J, Vecchione JJ, Dennis RC, Emerson CP. Freezing in the primary polyvinylchloride plastic collection bag: A new system for preparing and freezing nonrejuvenated and rejuvenated red blood cells. Transfusion. 21:138-149, 1981.
95. Valeri CR, Altschule MD. Hypovolemic Anemia of Trauma: The Missing Blood Syndrome, Chemical Rubber Company Press, Boca Raton, Florida, 1981.
96. Valeri CR, Valeri DA, Gray A, Melaragno AJ, Dennis RC, Emerson CP. Red blood cell concentrates stored at 4 C for 35 days in CPDA-1, CPDA-2, or CPDA-3 anticoagulant-preservative, biochemically modified, and frozen and stored in the polyvinyl chloride plastic primary collection bag with 40% W/V glycerol at -80 C. Transfusion 22:102-106, 1982.
97. Valeri CR. Use of rejuvenation solutions in red blood cell preservation. CRC Crit Rev Clin Lab Sci 17:299-374, 1982.
98. Valeri CR. Measurement of viable ADSOL-preserved human red blood cells. NEJM 312:377-378, 1985.
99. Valeri CR, Landrock RD, Pivacek LE, Gray AD, Fink JG, Szymanski I0. Quantitative differential agglutination method using the Coulter Counter to measure survival of compatible but identifiable red blood cells. Vox Sang 49:195-205, 1985.
100. Valeri CR. Cryopreservation of human platelets and bone marrow and peripheral blood pluripotential mononuclear stem cells. Ann N Y Acad Sci 459:353-66, 1985.
101. Valeri CR, Melaragno AJ, Dittmer J, Roy AJ, Vecchione JJ, Cassidy GP, Gray AD, Carciero RE. Bone marrow reconstitution of lethally irradiated beagles by treatment with autologous previously frozen bone marrow or peripheral blood mononuclear cells obtained as a byproduct of plateletpheresis. **Technical Report 85-01**.

102. Valeri CR, Feingold H, Cassidy G, Ragno G, Khuri S, Altschule MD. Hypothermia induced reversible platelet dysfunction. Ann Surg 205:175-181, 1987.

103. Valeri CR, Pivacek LE, Palter M, Dennis RC, Yeston N, Emerson CP, Altschule MD. A clinical experience with ADSOL® preserved erythrocytes. Surg Gynec Obstet 166:33-46, 1988.

104. Valeri CR, Donahue K, Feingold HM, Cassidy GP, Altschule MD. Increase in plasma volume after the transfusion, of washed erythrocytes. Surg Gynec Obstet 162:30-36, 1986.

105. Valeri CR, Pivacek LE, Gray AD, Cassidy GP, Leavy ME, Dennis RC, Melaragno AJ, Niehoff J, Yeston N, Emerson CP, Altschule MD. The safety and therapeutic effectiveness of human red cells stored at -8O C for as long as 21 years. Transfusion 29:429-437, 1989.

106. Valeri CR. Blood and bone marrow products in the treatment of radiation injury. In Treatment of Radiation Injuries, Ed. D. Browne et al, Plenum Press, New York, NY, 1990, pps 19-27.

107. Valeri CR, Khabbaz K, Khuri SF, Marquardt C, Ragno G, Feingold H, Gray AD, Axford T. Effect of skin temperature on platelet function in patients undergoing extracorporeal bypass. J Thorac Cardiovasc Surgery 104:108-116, 1992.

108. Valeri CR, MacGregor H, Cassidy G, Tinney R, Pompei F. Effects of temperature on bleeding time and clotting time in normal male and female volunteers. Crit Care Med 23:698-704, 1995.

109. Valeri CR, Pivacek LE. Effects of the temperature, the duration of frozen storage, and the freezing container on in vitro measurements in human peripheral blood mononuclear cells. Transfusion 36:303-308, 1996.

110. Valeri CR, Crowley JP, Loscalzo J. The red cell transfusion trigger: has a sin of commission now become a sin of omission? Transfusion 38:602-610, 1998.

111. Valeri CR, Pivacek LE, Cassidy GP, Ragno G. Posttransfusion survival (24-hour) and hemolysis of previously frozen, deglycerolized RBCs after storage at 4 C for up to 14 days in sodium chloride alone or sodium chloride supplemented with additive solutions. Transfusion 40:1337-1340, 2000.

112. Valeri CR, Pivacek LE, Cassidy GP, Ragno G. The survival, function, and hemolysis of human RBCs stored at 4 C in additive

solution (AS-1, AS-3, or AS-5) for 42 days and then biochemically modified, frozen, thawed, washed, and stored. at 4 C in sodium chloride and glucose solution for 24 hours. Transfusion 40:1341-1345, 2000.

113. Valeri CR, Ragno G, Pivacek LE, Dennis RC, Hechtman HB, Khuri SF. Survival and function of baboon RBCs released from clotted blood and washed before autologous transfusion. Transfusion 41:1384-1389, 2001.
114. Valeri CR, Dennis RC, Ragno G, Pivacek LE, Hechtman HB, Khuri SF. Survival, function, and hemolysis of shed red blood cells processed as nonwashed blood and washed red blood cells. Ann Thoracic Surg 72:1598-1602, 2001.
115. Valeri CR, Cassidy G, Pivacek LE, Ragno G, Lieberthal W, Crowley JP, Khuri SF, Loscalzo J. Anemia-induced increase in the bleeding time: implications for treatment of nonsurgical blood loss. Transfusion 41:977-983, 2001.
116. Valeri CR, Pivacek LE, Cassidy GP, Ragno G. 24-hour 51Cr post-transfusion survival, 51Cr life span and haemolysis of red blood cells stored at 4 C for 56 days in AS-3. Vox Sang 80:48-50, 2001.
117. Valeri CR, Ragno G, Pivacek LE, O'Neill EM. In vivo survival of apheresis RBCs, frozen with 40-percent (wt/vol) glycerol, deglycerolized in the ACP 215, and stored at 4 C in AS-3 for up to 21 days. Transfusion 41:928-932, 2001.
118. Valeri CR, Ragno G, Pivacek LE, Srey R, Hess JR, Lippert LE, Mettille F, Fahie R, O'Neill EM, Szymanski I0. A multicenter study of in vitro and in vivo values in human RBCs frozen with 40-percent (wt/vol) glycerol and stored after deglycerolization for 15 days at 4 C in AS-3: assessment of RBC processing in the ACP 215. Transfusion 41:933-939, 2001.
119. Valeri CR, Ragno G, Srey R. Restoration of red blood cell volume following 2-unit red blood cell apheresis. Vox Sang 85(2):85-7, 2003.
120. Valeri CR, Srey R, Tilahun D, Ragno G. In vitro quality of red blood cells frozen with 40% w/v glycerol at -80 C for 14 years, deglycerolized with the Haemonetics ACP 215, and stored at 4C in additive solution-1 or additive solution-3 for up to 3-weeks. Transfusion 44:990-995, 2004.

121. Valeri CR, Srey R, Tilahun D, Ragno G. In vitro effects of polymerized N-acetyl glucosamine (NAG) on the activation of platelets in platelet rich plasma with and without red blood cells. J Trauma 57 (suppl):S22-S25, 2004.
122. Valeri CR, Ragno G, Veech RL. Effects of the resuscitation fluid and the HBOC excipient on the toxicity of the HBOC: Ringer's D, L-Lactate, Ringer's L-Lactate, and Ringer's ketone solutions. Art Cells, Blood Substitutes and Biotechnology 34(6): 601-606, 2006.
123. Valeri CR, Ragno G. The survival and function of baboon red blood cells, platelets, and plasma proteins: A review of the experience from 1972 to 2002 at the Naval Blood Research Laboratory, Boston, Massachusetts. Transfusion 46(8): 1-42, 2006.
124. Valeri CR, Morse DS, Ragno G, Dennis RC. Hemostatic defect in baboons infused non- treated and treated. autologous plasma. J Card Surg 21(6):565-571, 2006.
125. Valeri CR, Ragno G. Cryopreservation of human blood products. Trans Apher Sci 34:271-287, with an editorial on pages 267-269, 2006.
126. Valeri CR, MacGregor H, Ragno G, Healey N, Fonger J, Khuri SF. Effects of centrifugal and roller pumps on the survival of autologous red blood cells in cardiopulmonary bypass surgery. Perfusion 21:291-6, 2006.
127. Valeri CR, Dennis RC, Ragno G, MacGregor H, Menzoian JO, Khuri SF. Limitations of the hematocrit to assess the need for RBC transfusion in hypovolemic anemic patients. Transfusion 46:365-371, 2006.
128. Valeri CR, Khuri S, Ragno G. Non-surgical bleeding diathesis in anemic thrombocytopenic patients: Role of temperature, RBC, platelets, and plasma clotting proteins. Transfusion 47:206S-248S, 2007.
129. Valeri CR, Ragno G. The effects of preserved red blood cells on the severe adverse events observed in patients infused with hemoglobin based oxygen carriers. Artificial Cells, Blood Substitutes and Biotechnology 36 (1):3-18, 2008.
130. Valeri CR, Ragno G, Veech RL. Severe adverse events associated with hemoglobin based oxygen carriers: role of resuscitative fluids and liquid preserved RBC. Trans Apher Sci 39:205-211, 2008.

131. Valeri CR, Ragno G. An approach to prevent the severe adverse events associated with transfusion of FDA-approved blood products. Trans Aph Sci 42:223- 233, 2010.
132. Valeri CR, Vournakis JN. mRDH bandage for surgery and trauma: data summary and comparative review. J Trauma 71:S162-S166, 2011.
133. Valtis DJ, Kennedy AC. Defective gas transport function of stored red blood cells. Lancet 1:119-124, 1954.
134. Van Slyke DD, Phillips RA, Dole VP, Hamilton PB, Archibald RM, Flazin J. Calculation of hemoglobin from blood specific gravities, J Biol Chem 183:349-360, 1950.
135. Veech RL, Rogeness GA, Weil-Maiherbe H. Formation of protoporphyrin from hemoglobin in vitro. Biochem J 105:1209-1217, 1967.
136. Vournakis JN, Demcheva M, Whitson A, Guirca R, Pariser ER. Isolation, purification, and characterization of poly-N-acetyl glucosamine use as a hemostatic agent. J Trauma 57 (suppl) S2-S6, 2004.
137. The SAFE Study Investigators: A comparison of albumin and saline for fluid resuscitation in the intensive care unit. N Engl J Med 350:2247-2256, 2004.
138. Myburgh JA, Finfer S, Bellomo R, Billot L, Cass A, Gattas D, Glass P, Lipman J, Liu B, McArthur C, McGuinness S, Rajbhandari D, Taylor CB, Wenn SAR. Hydroxyethyl starch or saline for fluid resuscitation in intensive care. N Engl J Med 367:1901-1911, 2012.

CHAPTER 3

DECADES AND EVENTS: WHAT OCCURRED AND WHAT WAS LEARNED. PART II.

The initial attempt to produce a blood substitute was the isolation of human albumin from fresh whole blood using the Cohn Blood Fractionator which was designed by Edwin J. Cohn. A 25% concentrate of human albumin was used to resuscitate wounded casualties. Studies to evaluate bovine albumin instead of human albumin produced severe adverse events associated with anaphylactic shock in wounded servicemen in North Africa during World War II. Dr Edward Churchill, Professor of Surgery at the Harvard Medical School who was Chief of Surgery at the Massachusetts General Hospital, reported that blood and not human albumin or plasma should be used to resuscitate wounded casualties.

The Cohn Blood Fractionator was designed by Dr. Edwin J. Cohn and his associates to collect plasma from fresh whole blood to isolate albumin and the fresh red blood cells were discarded. Subsequently, the Cohn Blood Fractionator was used to salvage the fresh red blood cells by treating the RBC with 40% W/V glycerol, storage at -80C, thawing and deglycerolizing the RBC in the Cohn Blood Fractionator and storage at 4C in autologous plasma or in 5% albumin-glucose medium. The Cohn Blood Fractionator could not be used in a routine blood bank and was a research instrument. The deglycerolized red blood cells stored at 4 C in autologous plasma or in the 5% albumin-glucose medium were transfused to patients at the Chelsea Naval Hospital between 1957 and 1962.

The freezing of fresh human RBC collected in the Cohn Blood Fractionator used the glycerol procedure reported by Audrey Smith who initially successfully froze bull-spermatozoa and then human red blood cells with 40% W/V glycerol which were frozen and stored at -79C which is the temperature achieved by dry ice and alcohol.

Dr. James L. Tullis working with Captain Lewis Haynes, MC, USN who was the Commanding Officer of the Chelsea Naval Hospital utilized the Cohn Blood Fractionator to freeze the fresh red blood cells which were a byproduct of the procedures to isolate the plasma from fresh whole blood to prepare the 25% concentrate of human albumin. The Office of Naval Research (ONR) supported this project. Colonel William Crosby, MC, USA reported that the use of the Cohn Blood Fractionator was only a research instrument to freeze red blood cells and the U.S. Army was not interested in this approach to provide red blood cells to treat wounded servicemen. The U.S. Army wanted to use fresh whole blood obtained from the troops "as the walking blood bank" and the U.S. Army supported the use of fresh whole blood and procedures to preserve whole blood in the liquid state by storage at 4C.

The Office of Naval Research supported Dr. Edwin J. Cohn to provide albumin as the first blood substitute. Subsequently the immunoglobulins in the plasma were isolated for clinical use. The red blood cells that were initially discarded were frozen using the procedure reported by Audrey Smith using 40% W/V glycerol and frozen storage at -80 C in mechanical freezers.

Harris Refrigeration Company in the Boston area provided water cooled mechanical freezers to maintain the mean temperatures of-80 C with a range from -65 to -90 C and water cooled mechanical freezers to maintain the temperature of -120 C with a range from -110 to -135 C.

The preliminary studies reported that the RBC processed using the Cohn Blood Fractionator to glycerolize and deglycerolize the RBC produced safe and therapeutically effective RBC products at the Chelsea Naval Hospital. (Haynes LL et al JAMA 173:1657-1663, 1960) Unfortunately, severe adverse events (SAEs) were observed following the transfusion of previously frozen deglycerolized RBC processed using the Cohn Blood Fractionator. In July 1962 four (4) patients who received previously frozen deglycerolized RBC at the Chelsea Naval Hospital developed acute renal failure and required peritoneal and hemodialysis (Valeri CR and ME Henderson, JAMA 188:1125-1131, 1964). Studies were done to investigate the cause of these severe adverse events associated with previously frozen, deglycerolized RBC resuspended and stored at

4 C for 21 days. Studies were done to assess in vitro recovery following thawing (freeze-thaw recovery) and following washing (freeze-thaw-wash recovery), the 24-hour posttransfusion survival values, the level of hemolysis, the mechanism of removal of the nonviable autologous and allogeneic compatible RBC, and the sterility of the previously frozen deglycerolized RBC stored in autologous plasma and in the 5% albumin-glucose medium. The mechanism of removal of nonviable autologous and allogeneic compatible but identifiable red blood cells by intravascular hemolysis or by extravascular removal by the reticuloendothelial system was investigated. The 5% albumin-glucose medium was utilized to avoid plasma and the possible transmission of hepatitis.

These studies demonstrated that the deglycerolized RBC stored in autologous plasma at 4 C for 21 days were not associated with the severe adverse events. The studies demonstrated that deglycerolized RBC resuspended in the 5% outdated albumin- glucose medium for 21 days were associated with the severe adverse events. The severe adverse events were associated with the resuspension medium containing outdated albumin. The albumin utilized was from outdated albumin that was reprocessed and then added to glucose. Studies were performed by acrylamide gel electrophoresis, nephelometry, and ultracentrifugation procedures demonstrated aggregates of human albumin. The severe adverse events were associated with albumin aggregates that were transfused with the previously frozen, deglycerolized RBC resuspended in the outdated albumin-glucose medium which produced acute renal insufficiency in the 4 recipients. (Tullis JL, et al, Vox Sang 8:100-101, 1963).

From 1962 to 1965 the NBRL evaluated the freezing of human RBC using 40% W/V glycerol with a range from 35 to 45% W/V glycerol and storage in a polyvinylchloride plastic bag containing diethylhexylphthalate (DEHP) in a -80 C water-cooled mechanical freezer. RBC were glycerolized and deglycerolized in the Cohn Blood Fractionator and the previously frozen deglycerolized RBC were stored at 4 C in autologous plasma; the indated albumin-glucose medium, and the outdated albumin-glucose medium for 21 days.

Small aliquots of autologous RBC transfusions in normal volunteers were studied using the double radioisotope procedure which labeled the autologous preserved RBC with 51Cr and the recipient red blood cell volume was measured from the 125I albumin plasma volume and the total body hematocrit estimated from the peripheral venous hematocrit multiplied by 0.91. There was no evidence that the removal of the nonviable red blood cells was associated with intravascular hemolysis (Valeri CR, Transfusion 5:36-53, 1965, Valeri CR, et al Clin Chem 11:581-588, 1965, Lionetti FJ et al, J Lab Clin Med 64:519-528, 1964). The 51Cr 24-hour posttransfusion survival of the 10 ml aliquots of autologous previously frozen deglycerolized RBC were measured in normal male volunteers.

Autologous RBC glycerolized in the Cohn Blood Fractionator frozen with 40% W/V glycerol and stored at -80 C, deglycerolized in the Cohn Blood Fractionator using 4.0 liters of wash solutions and stored in autologous plasma for 21 days, stored in the indated albumin-glucose medium for 12 days, and stored in outdated albumin-glucose medium for only 6 days had 24 hour posttransfusion survival values of 70V% (Valeri CR, Transfusion 5:25-35, 1965). The 24-hour posttransfusion survival of allogeneic compatible but identifiable previously frozen deglycerolized RBC processed in the Cohn Blood Fractionator and resuspended in 5% indated albumin medium and stored at 4 C for 24 hours was measured in patients who required two or more units of red blood cells. The 24-hour posttransfusion survival and lifespan were measured by the automated differential agglutination procedure using the Technicon Autoanalyzer and 51Cr to measure the red blood cell volume of the recipient. The 24-hour posttransfusion survival value was 90% and the lifespan was 95 days for the deglycerolized RBC transfused to stable anemic patients who require two or more units of deglycerolized red blood cells resuspended in 5% indated albumin-glucose medium stored at 4 C for 24 hours. In vitro testing of the previously frozen deglycerolized RBC was done to assess the freeze-thaw (FT) recovery, the freeze-thaw-wash (F-T-W) recovery, percent hemolysis, RBC ATP level, glutathione level, glutathione stability, RBC osmotic fragility, RBC density distribution, RBC composition of lipids, RBC sodium and potassium levels, extracellular potassium and sodium levels, and supernatant osmolality (Valeri CR, Transfusion 5:25-35, 1965; Valeri CR, Transfusion 5:36-53, 1965; Valeri CR and McCallum LE, Nature 205:561-563, 1965; Valeri CR et al, Transfusion 5:267-272, 1965; Valeri CR, Transfusion 5:273-285, 1965; Lionetti FJ et al, Transfusion 6:116-123, 1966; Valeri CR, Transfusion 6:112-115, 1966; Valeri CR and McCallum LE, Transfusion 5:421-426, 1965; Valeri CR and Runck AH, Transfusion 9:5-14, 1969).

The replacement of the Cohn Blood Fractionator was accomplished by Allan"Jack" Latham former Senior Vice President at Arthur D. Little, Co.,

Cambridge, MA. Following his retirement, Allan "Jack" Latham made reusable stainless steel bowls and reusable polycarbonate boxes, and then disposable polycarbonate bowls to deglycerolize red blood cells frozen with 40% W/V glycerol and stored at -8O C in mechanical freezers.

The Haemonetics Corporation was founded by Allan "Jack" Latham and this company produced the Haemonetics Blood Processors 10, 15, 115 and 215 instruments to deglycerolize RBC frozen with 40% W/V glycerol and stored in -80 C mechanical freezers.

In 1965 Dr. Charles Huggins, a surgeon at the Mass General Hospital and son of the Nobel Prize Laureate, Charles Brenton Huggins, introduced a novel method to glycerolize and deglycerolize RBC using agglomeration which is the spontaneous sedimentation of RBC in low ionic solutions. This procedure was evaluated at the U.S. NBRL (Huggins CE, Monogr Surg Sci, 3:133-173, 1966).

The method utilized by Dr. Charles Huggins and Dr. Morton Grove Rasmussen at the Mass General Hospital transfused deglycerolized RBC to patients using the cross matching procedures which did not detect the Coombs positive RBC that were produced by the procedure. The addition of Na2EDTA to the glycerol solution prevented the accumulation of B_1C globulin on the red blood cells in the low ionic medium used during the original glycerolizing procedure. The volume of 250 ml of 0.9% NaC1 was used at the NBRL to disaggregate the agglomerated RBC to prevent the crenation of the deglycerolized resuspended RBC instead of the 75 to 100 ml of 0.9% NaC1 that was used by Dr. Charles Huggins. The phthalate esters provided by Dr. David Danon were utilized to document the increased red blood cell density, the increase in mean corpuscular hemoglobin concentration (MCHC), and the decrease in the mean corpuscular volume (MCV). The Danon Fragiligraph documented the decreased RBC osmotic fragility, and significant reduction in red blood cell potassium level was observed in the previously frozen RBC deglycerolized by agglomeration, and disaggregated with 75 to 100 ml of 0.9% NaC1. The disaggregation of the agglomerated RBC with 250 ml of 0.9% NaC1 restored towards normal the physical-structural parameters of the previously frozen deglycerolized agglomerated disaggregated RBC (Valeri CR, Transfusion 6:247-253, 1966; Valeri CR and Bond JC, Transfusion 6:254-262, 1966; Valeri CR et al, Transfusion 6:543-553, 1966; Valeri CR et al, Transfusion 6:554-564, 1966).

The safety and therapeutic effectiveness of the previously frozen deglycerolized RBC processed by the Huggins method were observed in the transfusion of multiple units to patients at the Chelsea Naval Hospital. The glycerolized RBC were washed by agglomeration, disaggregated with 250 ml of 0.9% NaC1, stored at 4 C for 24 hours, and concentrated by centrifugation to remove the supernatant solution containing hemolysis to adjust the hematocrit to 80% prior to transfusion (Almond DV and Valeri CR, Transfusion 7:10-16, 1967; Almond DV and Valeri CR, Transfusion 7:95-104, 1967).

A decision was made by the Naval Medical Research and Development Command of the U.S. Navy's Bureau of Medicine and Surgery to deploy a -80 C water cooled mechanical freezer to store RBC frozen with 40% W/V glycerol containing Na2EDTA at Danang, South Vietnam aboard the hospital ships U.S.S. Repose and U.S.S. Sanctuary; and at Clark Air Force Base in the Philippines. The frozen glycerolized RBC containing Na2EDTA were deglycerolized by agglomeration and disaggregated with 250 ml of 0.9% NaCl, stored at 4 C for 24 hours, centrifuged to remove the supernatant hemolysis, and the hematocrit adjusted to 80% prior to transfusion.

The agglomeration procedure introduced by Dr. Charles Huggins was to eliminate the need for centrifugation to deglycerolize the RBC. Studies at the NBRL demonstrated: a) the need to add Na2EDTA to the glycerol solution to prevent the Coombs positive RBC observed at the NBRL; b) the need to use a volume of 250 ml of 0.9% NaCl instead of 75 to 100 ml of 0.9% NaCl which was recommended by Dr. Huggins to disaggregate the agglomerated RBC in the low ionic solutions containing 8% glucose and 1% fructose; and c) the RBC frozen by the Huggins procedure, stored at -80 C, and washed by the Huggins procedure could be stored at -80 C for less than 2 years. RBC deglycerolized by agglomeration and disaggregated with 250 ml of 0.9% NaCl and stored at 4 C for 24 hours had to be centrifuged to remove the supernatant hemolysis to adjust the hematocrit to 80 V% prior to transfusion. The procedure to wash previously frozen glycerolized red blood cells by agglomeration was introduced to eliminate centrifugation. However, centrifuges were needed to concentrate the agglomerated-disaggregated RBC which contained high levels of hemolysis which had to be removed by centrifugation to increase the hematocrit to 80 V% prior to transfusion (Almond DV and Valeri CR, Transfusion 7:95-104, 1967; Valeri CR et al Transfusion 10:102-112, 1970).

Agglomeration of RBC and disaggregation was associated with

decreases in RBC potassium, decrease in mean corpuscular volume (MCV), and increase in mean corpuscular hemoglobin concentrations (MCHC) and increased RBC density.

Successful deployment of-80 C water-cooled mechanical freezers at Danang, South Vietnam, aboard the hospital ships U.S.S. Repose and the U.S.S. Sanctuary; and at the Clark Air Force Base, Philippines was demonstrated during the Vietnam War (Moss GS et al, N Engl J Med 278:748-752, 1968; Moss GS, Surgery 66:1108-1113, 1969).

The current -80 C mechanical freezers are air cooled using dual cascade compressors to maintain the mean temperature at -80 C with range of -65 to -90 C with tanks of liquid carbon dioxide attached to each mechanical freezer to be triggered to deliver the C02 when the temperature of the freezer decreases to -65 C.

The RBC frozen in low ionic medium with Na2 EDTA cannot be stored at -80 C for longer than 2 years and then washed by agglomeration. The freezer-thaw-wash (FTW) recovery was reduced, the in vitro hemolysis increased; and decreased 24-hour posttransfusion survival were observed for RBC frozen with glycerol in the low ionic medium with Na2 EDTA and stored at -80 C for longer than 2 years (Valeri CR et al, Transfusion 10:102-112, 1970).

The RBC frozen in low ionic medium with Na2 EDTA and stored at -80 C for up to 37 years have been washed successfully with electrolyte solutions using centrifugation in the Haemonetics Blood Processor 115 (Valeri CR et al, Transfusion 9:120-134, 1969; Valeri CR and Runck AH, Transfusion 9:5-14, 1969; Valeri CR et al, Transfusion 10:102-112, 1970; Valeri CR et al, Transfusion 29:429-437, 1989; Valeri CR et al, Vox Sang 79:168-174, 2000).

The experience at Danang, South Vietnam by the U.S. Navy demonstrated that the method used to deglycerolize red blood cells using the Huggins Cytoglomerator, refrigerated centrifuges, and -80 C water cooled mechanical freezer had in vitro recovery of about 70%; and excessive hemolysis which required removal of the supernatant solution after storage at 4 C for 24 hours prior to transfusion. The method to add and remove the glycerol were performed in open systems using the Huggins method. The safety and therapeutic effectiveness of frozen deglycerolized RBC processed by the Huggins procedure was demonstrated in the treatment of wounded casualties during the Vietnam War.

The use of frozen RBC preserved by the Huggins procedure was utilized by Dr. Gerald Moss when he returned from Danang and became Chief of Surgery at Cook County Hospital. in Chicago, IL.

Dr. Moss was concerned with the complexity of freezing red blood cells using glycerol, storage at -80 C, and need to deglycerolize the previously frozen RBC prior to transfusion. Dr. Moss decided to investigate the use of hemoglobin based oxygen carriers (HBOCs) in combination with Ringer's D,L lactate crystalloid solution to replace the need for RBC. Commercial and government funds have been allocated to study procedures to produce safe and therapeutically effective hemoglobin based oxygen carriers (HBOCs). Three commercial companies; Northfield Laboratories, Hemosol and Biopure have studied the safety and therapeutic effectiveness of HBOCs in patients. Severe adverse events have been observed with the HBOCs related to their vasoconstrictor activity and their generation of oxygen free radicals. As yet FDA has not approved a HBOC for clinical use.

Function of Preserved RBC and Platelets

The NBRL has performed extensive studies to assess the function of preserved RBC to deliver oxygen at high tissue oxygen tension to the brain, heart and kidneys. The increased affinity for oxygen was reported by Valtis, DJ and Kennedy AC in 1954 for red blood cells stored at 4 C in acid-citrate-dextrose anticoagulant for 7 days. The increase in affinity for oxygen by the preserved RBC was reported in 1967 to be related to the decrease in red blood cell 2,3 DPG level by Benesch R and Benesch RE and Chanutin A and Curnish RF. (Valtis DJ and AC Kennedy, Lancet 1:119-124, 1954; Benesch R and Benesch RE, Biochem Biophys Res Commun 26:163-267, 1967; Chanutin A and Curnish RF. Arch Biochem Biophys 121:96-102, 1967).

Studies were done at the NBRL between 1968 to 1974 to document that patients with chronic hypovolemic anemia following traumatic injuries and patients with cardiopulmonary insufficiency had red blood cells with increased levels of RBC 2,3 DPG and RBC creatine.

Two significant serendipitous observations were made at the NBRL. Tissue hypoxia in patients was associated with decrease in red blood cell volume and in patients with cardiopulmonary insufficiency produced red blood cell with increased 2,3 DPG and creatine levels. The release of muscle creatine occurred from the hypoperfused hypoxic muscles in these patients. Two red blood cell indicators of tissue hypoxia; increases in RBC 2,3 DPG and RBC creatine levels were observed in patients hospitalized at the Chelsea Naval Hospital. These observations were published by Valeri CR and NL Fortier in the New England Journal of Medicine 81:1452-1455, 1969. Studies performed at the NBRL were unable to demonstrate that creatine could enter the RBC under the various in vitro conditions used in our investigations. However, in vivo studies showed that the increased RBC creatine levels observed in patients with chronic hypovolemic anemia of trauma following RBC transfusions to increase the red blood cell volume and the peripheral red blood cell volume to normal were associated with a decrease in the red blood cell creatine levels. In vivo studies suggested that the red blood cell creatine level in these patients with chronic hypovolemic anemia of trauma correlated to the hypoxia in their muscles. The decrease in the peripheral red blood cell volume reduced the perfusion of the muscle of the extremities. Restoration of the peripheral blood volume by the red blood cell transfusions restored perfusion to the hypoxic muscles and the reduction in red blood cell creatine level occurred. (Valeri CR, Altschule MD. Chemical Rubber Co. Press, Boca Raton, FL, 1981).

In the hypovolemic anemic patients the decrease in red blood cell volume stimulated an increase in red blood cell 2,3 DPG level. Our subsequent studies were done to ensure that the transfused RBC not only circulated but functioned to provide oxygen to the brain, heart and kidneys which require high tissue oxygen tension. In addition, RBC transfusions need to increase the peripheral red blood cell volume to perfuse the muscle of the extremities and prevent the loss of muscle creatine which increased the RBC creatine level. Previous clinical studies have reported that patients with congestive heart failure had increased level of creatine in their urine. Reduction in urinary creatine was associated with treatment of the congestive heart failure.

Studies between 1968 to 1974 at the NBRL documented that both red blood cell 2,3 DPG level and RBC creatine level were useful measurements to assess the presence of tissue hypoxia. The level of RBC 2,3 DPG level was affected by the patient's blood pH and the level of plasma inorganic phosphorus.

The initial studies performed at the NBRL to assess the quality of the preservation of the RBC were to measure the 24-hour posttransfusion survival using the double radioisotope procedure using 1251/51Cr method and the RBC 51Cr lifespan. In addition, the index of therapeutic effectiveness reported the in vitro recovery of the RBC and the 24 hour posttransfusion survival of the preserved red blood cells with a potential for normal long-term lifespan was related to the compatibility of the donor RBC and the state of health of the recipient. However, the index of therapeutic effectiveness (ITE) did not consider the functions of the compatible RBC to carry and release oxygen at high tissue oxygen tension to brain, heart and kidneys.

Although Valtis DJ and Kennedy AC reported in 1954 that blood collected in acid citrate dextrose (ACD) anticoagulant and stored at 4 C for one week had significantly increased affinity for oxygen, the function of the preserved red blood cells has never been established as a criterion for the safety and therapeutic effectiveness of a RBC product. In 1967 the role of RBC organic phosphates 2,3 DPG and ATP were reported to affect the RBC loading of oxygen in the lungs and oxygen delivery to the tissues.

Since 1968 the NBRL has studied not only the in vivo survival of preserved RBC but their function to provide oxygen at high tensions to the brain, heart and kidneys. Studies performed at the NBRL located at Chelsea Naval Hospital demonstrated that wounded servicemen who returned from South Vietnam with wounds of the extremities had chronic hypovolemic anemia of trauma. The hematocrit values did not reflect the red blood cell volume, plasma volume, and total blood volume deficits in these patients. The direct measurement of the red blood cell volume with the radioisotope 51Cr was routinely done to measure the red blood cell volume. In addition, the plasma volume was measured using 125I human albumin. The use of the radiolabeled albumin demonstrated rapid loss of the 125I labeled albumin into the extravascular volume and overestimation of the plasma volume was observed in these patients. The simultaneous measurement of the plasma volume using 125I radiolabeled human 7S albumin with a molecular weight of 68,000 and 131I radiolabeled human 19S cold agglutinin with a molecular weight of 1,000,000 demonstrated that human albumin cannot be used to measure the plasma volume in patients and should not be used as a resuscitation solution to increase the plasma volume and blood volume in patients subjected to hemorrhagic shock. (Farrugia A. Trans Med Rev 24:53-63, 2010). The data demonstrated that iodinated human albumin and Evans blue dye which binds to human albumin should not be used to measure the plasma volume in patients.

However, in normal volunteers 125I human albumin can estimate the plasma volume and with the total body hematocrit (peripheral venous hematocrit multiplied by 0.89) to estimate the red blood cell volume and the total blood volume but not in patients (Valeri CR, et al, Arch It Med 132:534-538, 1973).

Studies in wounded servicemen who returned from Vietnam with primarily orthopedic injuries of the extremities had chronic hypovolemia of trauma with significant decreases in red blood cell volume, plasma volume, peripheral blood volume, and total blood volume. The peripheral venous hematocrit did not reflect the hypovolemic state and the significant reduction in the red blood cell volume, peripheral blood volume, and the plasma volume. The central blood volume assessed by arterial blood pressure, heart rate, arterial blood gases, cardiac output, pulmonary artery wedge pressure, and renal function was normal. The deficit in red blood cell volume was documented by the measured 51Cr red blood cell volume, the deficit in the plasma volume by its measurement using ^{125}I labeled cold agglutinin. The deficit in peripheral blood volume was demonstrated by the transfusion of multiple units of compatible but identifiable red blood cells that were washed liquid preserved RBC and washed previously frozen deglycerolized red blood cells. The aggressive transfusion of multiple units of washed compatible liquid preserved and previously frozen deglycerolized red blood cells increased the red blood cell volume, the peripheral red blood cell volume, and the plasma volume with only a slight increase in the hematocrit values and hemoglobin concentrations in the peripheral venous blood. The central red blood cell volume increased slightly following the transfusion but within 4 hours returned to the baseline level (Valeri CR et al, Transfusion 38:602-610, 1998).

The effect of the transfusion of washed red blood cells to restore the red blood cell volume, peripheral red cell volume and the plasma volume was associated with improvement in blood flow to the gastrointestinal tract and increase in appetite, improvement in libido, improvement in general health and decrease in the edema which was present in extremity wounds. The increase in red blood cell volume restored perfusion to the extremities with only rare amputation of extremities by the aggressive transfusion to increase the peripheral red blood cell volume. The increase in the plasma volume associated with the transfusion of washed red blood cells was associated with decrease in the edematous fluid in the wounds of the extremities. (Valeri CR, Altschule MD, Hypovolemic Anemia of

Trauma: The Missing Blood Syndrome, Chemical Rubber Co. Press, Boca Raton, Florida, 1981).

The therapeutic effectiveness of the transfusion of washed viable red blood cells produced a significant reduction in both the elevated RBC 2,3 DPG and creatine levels. The elevated red blood cell creatine level could not be explained by an increased in red blood cell production which was reduced in these patients. The increased red blood cell creatine in these patients with chronic hypovolemia of trauma was not due to an increase in erythropoiesis.

The study of the patients wounded in Vietnam revealed that their wounds required debridement under anesthesia but bleeding of these wounds was minimal. The reduction in nonsurgical blood loss observed with debridement of these wounds of the extremities was related to the normal or slightly increased hematocrit of these patients. This serendipitous clinical observation stimulated the NBRL to study the effect of hematocrit on the bleeding time and nonsurgical blood loss. Viable and functional RBC exert a significant effect on the bleeding time and nonsurgical blood. In 1910 Duke WW reported that fresh RBC had an important effect on the bleeding time in three (3) patients with anemia and thrombocytopenia. More recently Turitto VT and HJ Weiss and Marcus AJ have demonstrated that RBC survival and function are critical to restore hemostasis by reduction in the bleeding time and nonsurgical blood loss. (Duke WW, JAMA 60:1185-1192, 1910; reprinted 250:1201-1209, 1983; Turitto VT, Weiss HJ, Science 207:541-543, 1980; Marcus AJ, Blood 76:1903-1907, 1990; Valles J et al, Blood 78:154-162,1991; Santos MT et al, J Clin Invest 87:571-580, 1991).

The focus of adequate preservation of RBC has been to maintain RBC ATP level, RBC morphology; RBC rheology; hemolysis of 1% or less; reduced levels of extracellular potassium; sterility; and presence of nontoxic additives to maintain the metabolic and physical-structural integrity of the preserved red blood cells. Gabrio BW and associates and Strumia MM and associates have reported that adenosine, adenine, and inosine maintain the metabolic state of RBC stored at 4 C. (Gabrio BW et al J Clin Invest 34:1509-1512, 1955; Strumia MM et al, J lab Clin Med 76:907-914, 1970). The preservation of whole blood at 4 C was done to maintain the sterility and the quality of the red blood cells, platelets, and plasma proteins. Blood is an excellent medium to support the growth of bacterial infectious agents. Dr. Max Strumia in 1965 was the Chairman of the National Research Council and he stressed the need to collect blood in anticoagulant-preservative solution immediately with agitation on wet ice at 4 C to prevent bacterial contamination. The major risk with the collection and preservation of blood and blood products is the potential for bacterial contamination. Dr. Max Strumia was an advocate for the safety and therapeutic effectiveness of preserved blood and blood products. Dr. Max Strumia stressed that the anticoagulant used to collect the blood must prevent clotting and must preserve the blood components, i.e. red blood cells, platelets, and plasma proteins. The acid-citrate- dextrose (ACD) and citrate-phosphate-dextrose (CPD) anticoagulants contain citrate to chelate the calcium and prevent the clotting of the blood. The volume of the ACD and CPD anticoagulants contains 50% more citrate to chelate the calcium to prevent the clotting. Dr. Max Strumia stressed the volume of ACD and CPD in excess of that needed to chelate the calcium was essential for the preservation of the RBC, platelets and plasma proteins in the whole blood (Valeri CR, Physiology of Blood Transfusion; Little, Brown & Co., In Surgical Intensive Care, ed PS Barie and GT Shires, 1993),

The U.S. Army under the direction of Colonel William Crosby made an important clinical decision to provide compatible fresh whole blood to treat wounded servicemen in the Korean War and in the Vietnam War using "the walking blood bank". Surgeons have always recognized that compatible fresh whole blood produced the best clinical results in the treatment of patients subjected to traumatic injuries. The testing of blood for potential transmission of infectious diseases like hepatitis and syphilis were the major risks associated with the transfusion of ABO and Rh compatible RBC during the Korean and Vietnam Wars.

The preservation of whole blood in acid-citrate-dextrose (ACD) anticoagulant in glass bottles was done to provide liquid preserved whole blood stored at 4 C for 21 days to treat patients when fresh whole blood was not available. Methods to preserve blood were significantly changed with the introduction of plastic bags composed of polyvinylchloride plastic bags containing diethylhexylphthalate (DEHP) introduced by Dr. Carl W. Walter. The use of the multiple plastic bag collection systems permitted the separation of whole blood into its components: red blood cells, platelets, and plasma and each component was preserved: red blood cell concentrates at 4C, platelets at 4C and then at 22C, and plasma and cryoprecipitate at -20 C.

Freezing of whole blood has been investigated using extracellular additives like hydroxyethyl starch (HES) and pentastarch and freezing using liquid nitrogen and storage in the gas phase of liquid nitrogen at -150 C or in the liquid phase of liquid nitrogen at -197 C. Following thawing the whole blood frozen with the extracellular additives the safety and therapeutic effectiveness of the thawed whole blood containing the extracellular additives have not been demonstrated. Instead platelets and pluripotential mononuclear adult stem cells treated with cryoprotective agents and freezing at 1 C per minute using controlled rate freezing instruments with liquid nitrogen and storage in the gas phase or liquid phase of liquid nitrogen are now utilized.

Freezing red blood cells with glycerol, platelets with DMSO and pluripotential mononuclear adult stem cells with DMSO and other substances require post-thaw processing to remove the glycerol and DMSO prior to transfusion. Platelets and pluripotential mononuclear adult stem cells treated with DMSO and other substances can be frozen at a controlled rate of freezing at 1 C per minute using controlled rate freezing instruments with liquid nitrogen and storage in the gas phase of liquid nitrogen or in the liquid phase of liquid nitrogen.

Alternatively the red blood cells treated with glycerol and platelets and pluripotential mononuclear adult stem cells treated with DMSO can be frozen and stored in a mechanical freezer maintained at a mean temperature of -80 C with a range of -65 to -90 C.

The freezing of red blood cells, platelets, and pluripotential mononuclear adult stem cells with intracellular additives and storage at -80 C is simpler and less expensive than freezing using a controlled rate freezer with liquid nitrogen and storage in the liquid nitrogen. The NBRL has pioneered the use of the -80 C mechanical freezer to freeze group O Rh positive and Group O Rh negative RBC with 40% W/V glycerol with the removal of the supernatant glycerol prior to freezing for at least 10 years; freeze single donor leukoreduced group O platelets with 6% DMSO with the removal of the supernatant DMSO prior to freezing for at least 2 years; freeze pluripotential ABO, Rh, and HLA mononuclear adult stem cells isolated from peripheral blood with 10% DMSO with the removal of the supernatant DMSO prior to freezing for 1 ½ years; and freeze AB plasma for at least 10 years. The RBC, platelets, pluripotential mononuclear adult stem cells, and plasma are all frozen and stored in a mechanical freezer maintained at a mean temperature of -80C with a range from -65C to -90C.

The composition of the containers used to freeze the blood products using polyvinylchloride (PVC) plastic bags and ethylvinyl acetate (EVA) plastic bags; the breakage of the container associated with freezing; the maintenance of a functionally closed system to add and remove the cryoprotective agents; post-thaw processing of the frozen blood product; the resuspension medium used to store the thawed previously frozen blood products in the liquid state; the temperatures and the length of storage at room temperature and at 4 C; and the sterility of the blood product prior to transfusion have been investigated at the NBRL.

The major concern with the preservation of blood and blood products is the potential for bacterial contamination. The FDA has approved the storage of blood products processed in an open system for storage at 4 C for only 24 hours and for storage at room temperature for only 8 hours. The FDA has approved the storage of single donor and pooled platelets at room temperature (22 $\pm$ 2C) for 5 days. A major breakthrough in the processing of blood products was the introduction of the Haemonetics Blood Processor ACP2I5, a functionally closed instrument to add glycerol, freeze the red blood cells, deglycerolize the red blood cells and storage of the red blood cells at 4 C in the additive solution AS-3 (Nutricel) for 2 weeks. The FDA licensure of the Haemonetics Blood Processor ACP215 permits for the first time a technology to add substances to blood products and then wash the blood products which can then be stored safely at 4C for at least 2 weeks without contamination.

The testing of the red blood cells preserved in the liquid state and the frozen state includes the in vitro recovery of the red blood cells; in vivo survival of the preserved red blood cells including the 24-hour posttransfusion survival and RBC lifespan; the in vitro hemolysis of the red blood cells assessed by the release of hemoglobin and potassium from the preserved red blood cells, the function of the preserved red blood cells to release oxygen at high tissue oxygen tension to the brain, heart, and kidneys and to restore hemostasis by the reduction in the bleeding time and nonsurgical blood loss; the quantity of nonviable red blood cells and mechanism of the in vivo removal of the nonviable compatible red blood cells by hemolysis and increase in plasma hemoglobin level or by extravascular removal by the reticuloendothelial system in the recipient; the presence of red blood cell microvesicles in the blood product; the sterility; and the potential toxicity of the preservative solutions. (Valeri CR, Ragno G, Trans Aph Sci 34:271- 287, 2006).

The testing of platelets preserved in the liquid state and the frozen state includes the measurement of the in vitro recovery of platelets, in vivo recovery of the platelets one to two hours following infusion and the lifespan of the platelets; the function of the preserved platelets to reduce the increased bleeding time in stable thrombocytopenic patients and to correct an aspirin induced prolonged bleeding time in normal volunteers by the reduction in the bleeding time and the increase in shed blood thromboxane A2 level at the bleeding time site; the in vitro production of thromboxane by the platelets following stimulation with arachidonic acid (AA) and adenosine diphosphate (ADP); the quantity of platelet microparticles; the sterility and the potential toxicity of solutions used to preserve the platelets. Plasma is tested to assess the clotting proteins Factors V, VIII, von Willebrand's factor and fibrinogen levels and the opsonic protein - fibronectin and the oncotic protein albumin. (Valeri CR et al, Transfusion 47:206S-248S, 2007).

The NBRL demonstrated that autologous peripheral blood pluripotential mononuclear adult stem cells frozen with 10% DMSO and stored at -80 C in a mechanical freezer, thawed, washed, and reinfused repopulated the bone marrow in beagle dogs subjected to radiation injury. (Valeri CR, et al, Technical Report 85-01). Studies were done to demonstrate that human peripheral blood mononuclear cells (PBMC) adult stem cells can be frozen with 10% DMSO and stored at -80 C for 1 ½ years and following thawing and washing were sterile and had in vitro recovery that was 90% that of fresh PBMCs, viability that was 90% and growth in CFU-GEMM (granulocyte-erythroid-monocyte-megakaryocytic) tissue culture assay, that was similar to that of fresh PBMC. The measurements of CD34 antigens, CFU GEMM (granulocyte-erythroid-monocyte-megakaryocytic) tissue culture assay, and 7 aminoactinomycin D (7 AAD) test were used to assess the viability and function of fresh and preserved pluripotential mononuclear adult stem cells isolated from peripheral blood. (Valeri CR Ragno G, Trans Aph Sci 34:271-287, 2006).

The current severe adverse events associated with mortality and morbidity following transfusion of compatible blood products tested for infectious disease markers reflect the poor quality of the FDA approved red blood cells and platelets and the presence of plasma and nonplasma substances that produce the reported severe adverse events related to the liquid preserved red blood cells, platelets and plasma that are infused (Valeri CR, Ragno G, Trans Aph Sci 42:223-233, 2010).

Platelet survival does not mean they function to exert a hemostatic effect. The NBRL has reported the use of aspirin treatment of normal volunteers and baboons to increase the bleeding time and then evaluated the ability of the autologous preserved platelets in normal volunteers and baboons to correct the aspirin prolonged bleeding time. This protocol has been utilized by Pineda AA and associates to assess the survival and functional integrity of washed autologous platelets in Transfusion 29:524-527, 1989 and by Brecher ME and associates to assess the survival and functional integrity of filtered autologous platelets in Transfusion 30:718-721, 1990 (Pineda AA et al, Transfusion 29:524-527, 1989; Brecher ME et al, Transfusion 30:718-721, 1990).

Khuri and associates have studied the survival and function of liquid preserved allogeneic platelets and previously frozen allogeneic DMSO platelets washed prior to transfusion in patients following cardiopulmonary bypass surgery. Patients transfused with single donor allogeneic previously frozen DMSO platelets, washed and stored in autologous plasma at room temperature without agitation for 5 hours were more hemostatically effective than liquid preserved allogeneic single donor and pooled platelets stored at room temperature for a mean of 3.4 days with agitation. Reduction in nonsurgical blood loss and reduction in need for allogeneic red blood cells and allogeneic fresh frozen plasma were observed in patients who received the frozen washed platelets compared to patients who received liquid preserved platelets stored at room temperature on for a mean of 3.4 days. (Khuri SF et al, J Thorac Cardiovasc Surg 117:172-184, 1999; Valeri CR et al, Transfusion 47:206S-248S, 2007; Valeri CR, Ragno G, Trans Aph Sci 42:223-233, 2010).

CHAPTER 3 REFERENCES

1. Almond DV, Valeri CR. Relationship between lipid fractions of erythrocytes and their in vivo survival following preservation with glycerol and the slow freeze technic. Transfusion 7:10-16, 1967.
2. Almond DV, Valeri CR. The in vivo effects of deglycerolized agglomerated erythrocytestransfused in multiple units to stable anemic patients. Transfusion 7:95-104, 1967,
3. Benesch R, Benesch RE. The effect of organic phosphates from the human erythrocyte on the allosteric properties of hemoglobin. Biochem Biophys Res Commun 26:162-167, 1967.
4. Brecher ME, Pineda AA, Zylstra-Halling VW, Chowdhury S, Forstrum LA. In vivo viability and functional integrity of filtered platelets. Transfusion 30:71.8-721, 1990.
5. Chanutin A. Curnish RF. Effect of organic and inorganic phosphates on the oxygen equilibrium of human erythrocytes. Arch Biochem Biophys 121:96-102, 1967.
6. Duke WW. The relation of blood platelets to hemorrhagic disease. JAMA 60:1185-1192, 1910; reprinted JAMA 250:1201-1209, 1983.
7. Farrugia A. Albumin usage in clinical medicine: tradition or therapeutic? Trans Med Rev 24:53-63, 2010.
8. Gabrio BW, Donohue DM, Finch CA. Erythrocytic preservation. V. Relationship between chemical changes and viability of stored blood treated with adenosine. J Clin Invest 34:1509-1512, 1955.
9. Haynes LL, Tullis IL, Pyle NM, Sproul MT, Wallach S, Turville WC. Clinical use of glycerolized frozen blood. JAMA 173:1657-1663, 1960.
10. Huggins CE. Frozen blood: principles of practical preservation. Monogr Surg Sci 3:133-173, 1966.
11. Khuri SF, Healey N, MacGregor H, Barnard MR., Szymanski I0, Birjiniuk V, Michelson AD, Gagnon DR, Valeri CR. Comparison of the effects of transfusions of cryopreserved and liquid preserved platelets on hemostasis and blood loss after cardiopulmonary bypass. J Thorac Cardiovasc Surg 117:172-184, 1999.
12. Lionetti FJ, Valeri CR, Bond JC, Fortier NL. Measurement of

hemoglobin binding capacity of human serum or plasma by means of dextran gels. J Lab Clin. Med 64 :519-528, 1964.

13. Lionetti FJ, Valeri CR, Bond JC, Kivowitz C, Weinman E. Nucleotides in frozen glycerolized erythrocytes. Transfusion 6:116-123, 1966.
14. Marcus AJ. Thrombosis and inflammation as multicellular processes: pathophysiologic significance of transcellular metabolisms. Blood 76 :1903-1907, 1990.
15. Moss GS, Valeri CR, Brodine CE. Clinical experience with. the use of frozen blood in combat casualties. NEJM 278:748-752, 1968.
16. Moss GS. Massive transfusion of frozen preserved red cells in combat casualties. Report of three cases. Surgery 66:1108-1113, 1969.
17. Pineda AA, Zylstra VW, Clare DE, Dewanjee MK, Forstrom LA. Viability and functional integrity of washed platelets. Transfusion 29:524-527, 1989.
18. Santos MT, Valles J, Marcus AJ, Safier LB, Broekman MJ, Islam N, Ullman HL, Eiroa AM, Aznar J. Enhancement of platelet reactivity and modulation of eicosanoid production by intact erythrocytes: A new approach to platelet activation and recruitment. J Clin Invest 87:571-580, 1991.
19. Strumia MM, Strumia PV, Eusebi AJ. The preservation of blood for transfusion. VIII. Effect of adenine and inosine on the adenosine triphosphate and viability of red cells of stored blood at 1 C for 21 days. J Lab Clin Med 76:907-914, 1970.
20. Tullis JL, Pennel RB, Melin M, Zemp JW, Henderson ME, Sproul MT, MacNeill A. Problems of albumin medium for deglycerolized red cell resuspension. Vox Sang 8:100-101, 1963.
21. Turitto VT, Weiss HJ. Red blood cells: their dual role in thrombus formation. Science 207:541-543, 1980.
22. Valeri CR, Henderson. ME. Recent difficulties with frozen glycerolized blood. JAMA 188:1125-1131, 1964.
23. Valeri CR. Effect of resuspension medium on in vivo survival and supernatant hemoglobin of erythrocytes preserved with glycerol. Transfusion 5:25-35, 1965.
24. Valeri CR. Observations on recipient plasma hemoglobin concentration after transfusion with glycerolized frozen blood. Transfusion 5:36-53, 1965.

25. Valeri CR, McCallum LE. Relationship between glutathione stability and in vivo survival of autologous, deglycerolized, resuspended red blood cells. Nature 205:561-563, 1965.
26. Valeri. CR, Mercado-Lugo R, Danon D. Relationship between osmotic fragility and in vivo survival of autologous deglycerolized resuspended red blood cells. Transfusion 5:267-272, 1965.
27. Valeri CR. The in vivo survival, mode of removal of the non-viable cells, and the total amount of supernatant hemoglobin in deglycerolized, resuspended erythrocytes. I. The effect of the period of storage in ACD at 4C prior to glycerolization. II. The effect of washing deglycerolized, resuspended erythrocytes after a period of storage at 4C. Transfusion 5:273-285, 1965.
28. Valeri CR, McCallum LE. The age of human erythrocytes lost during freezing and thawing with glycerol using the Cohn Fractionator. Transfusion 5:421-426, 1965.
29. Valeri CR, Bond JC, Fowler K, Sobucki J. Quantitation of serum hemoglobin binding capacity using cellulose acetate membrane electrophoresis. Clin Chem 11:581-588, 1965.
30. Valeri CR. In vivo survival and supernatant hemoglobin of autologous, deglycerolized, resuspended erythrocytes processed using centrifugation. III. The effect of the length of storage at-80 C. Transfusion 6:112-115, 1966.
31. Valeri CR. The in vivo survival of Coombs positive autologous erythrocytes produced by agglomeration. Transfusion 6:247-253, 1966.
32. Valeri CR, Bond JC. Observations on the preservation of autologous human erythrocytes using glycerol, slow-freeze technic and agglomeration. Transfusion 6:254-262, 1966.
33. Valeri CR, Bond JC, McCallum LE. Relationships between metabolic state and (1) in vivo survival and (2) density distribution of previously frozen human erythrocytes. Transfusion 6:543-553, 1966.
34. Valeri CR, McCallum LE, Danon D. Relationships between in vivo survival and (1) density distribution, (2) osmotic fragility of previously frozen, autologous, agglomerated, deglycerolized erythrocytes. Transfusion 6:554-564, 1966.
35. Valeri CR, Fortier NL. Red-cell 2,3-diphosphoglycerate and

creatine levels in patients with red-cell mass deficits or with cardiopulmonary insufficiency. NEJM 81:1452-1455, 1969.

36. Valeri CR, Runck AM. Long term frozen storage of human red blood cells: Studies in vivo and in vitro of autologous red blood, cells preserved up to six years with high concentrations of glycerol. Transfusion 9:5-14, 1969.
37. Valeri CR, Runck AM, Sampson WT. Effects of agglomeration on human red blood cells. Transfusion 9:120-134, 1969.
38. Valeri CR, Szymanski I0, Runck AM. Therapeutic effectiveness of homologous erythrocyte transfusions following frozen storage at -80C for up to seven years. Transfusion 10:102-112, 1970.
39. Valeri CR, Cooper AG, Pivacek LE. Limitations of measuring blood volume with iodinated 1125 serum albumin. Arch Int Med 132:534-538, 1973.
40. Valeri CR. Blood Banking and the Use of Frozen Blood Products. Chemical Rubber Company Press, Boca Raton, Florida, 1976.
41. Valeri CR, Altschule MD. Hypovolemic Anemia of Trauma: The Missing Blood Syndrome, Chemical Rubber Company Press, Boca Raton, Florida, 1981.
42. Valeri CR, Melaragno AJ, Dittmer J, Roy At, Vecchione H, Cassidy GP, Gray AD and Carciero RE: Bone marrow reconstitution of lethally irradiated beagles by treatment with autologous previously frozen bone marrow or peripheral blood mononuclear cells obtained as a byproduct of plateletpheresis. **Technical Report 85-01**.
43. Valeri CR, Pivacek LE, Gray AD, Cassidy GP, Leavy ME, Dennis RC, Melaragno AJ, Niehoff J, Yeston N, Emerson CP, Altschule MD. The safety and therapeutic effectiveness of human red cells stored at -80C for as long as 21 years. Transfusion 29:429-437, 1989.
44. Valeri CR. Physiology of Blood Transfusion. Little, Brown & Co. In Surgical Intensive Care, Ed PS Barie and GT Shires, 1993.
45. Valeri CR, Crowley JP, Loscalzo J. The red cell transfusion trigger: has a sin of commission now become a sin of omission? Transfusion 38:602-610, 1998.
46. Valeri CR, Ragno G, Pivacek LE, Cassidy GP, Srey R, Hansson-Wicher M, Leavy ME. An experiment with glycerol-frozen red

blood cells stored at -80 C for up to 37 years. Vox Sang 79:168-174, 2000.

47. Valeri CR, Ragno G. Cryopreservation of human blood products. Trans Apher Sci 34:271-287, with an editorial on pages 267-269, 2006.
48. Valeri CR, Khuri SF, Ragno G. Non-surgical bleeding diathesis in anemic thrombocytopenic patients: Role of temperature, RBC, platelets, and plasma clotting proteins. Transfusion 47:206S-248S, 2007.
49. Valeri CR, Ragno G. An approach to prevent the severe adverse events associated with transfusion of FDA-approved blood products. Trans Aph Sci 42:223-233, 2010.
50. Valles J, Santos MT, Aznar J, Marcus AJ, Martinez-Sales V, Portoles M, Broekman MJ, Safier LB. Erythrocytes metabolically enhance collagen -induced platelet responsiveness via increased thromboxane production, adenosine diphosphate release, and recruitment. Blood 78:154-162, 1991.
51. Valtis DJ, Kennedy AC. Defective gas transport function of stored red blood cells. Lancet:119-124, 1954.

CHAPTER 4

DECADES AND EVENTS: WHAT OCCURRED AND WHAT WAS LEARNED. PART III.

1. Renal shutdown in four (4) patients at the Chelsea Naval Hospital in July 1962 produced by the outdated aggregated albumin in the resuspension medium to store deglycerolized red blood cells at 4 C for 21 days (Valeri CR, Henderson ME, JAMA 188:1125-1131, 1964).
2. Eternal Red Blood Cells reported in 1965 by Dr. Max Strumia, Chairman of National Research Council, at a meeting attended by Dr. Scott Swisher, Dr. Ernest Simon, Dr. Ernest Beutler, and Dr. Joseph Bove. Single label and double label procedures were discussed to assess the in vivo 24-hour posttransfusion survival of preserved RBC with no requirement that the preserved red blood cells function. Valtis DJ, Kennedy AC in 1954 reported that RBC stored in acid citrate dextrose (ACD) anticoagulant at 4 C for 7 days had increased affinity for oxygen. Benesch R, Benesch RE and Chanutin A, Curnish RF in 1967 reported that the decrease in RBC DPG produced the increased in RBC affinity for oxygen in the liquid preserved RBC (Valtis DJ, Kennedy AC, Lancet 1:119-124 1954; Benesch R, Benesch RE, Biochem Biophys Res Comm 26:162-167, 1967; Chanutin A, Curnish RF, Arch Biochem Biophys, 121:96-102, 1967).
3. Evans blue dye was used to measure the plasma volume and with the total body hematocrit to estimate the red blood cell volume and the blood volume in normal volunteers and in patients. Evans blue dye binds to human albumin. Human albumin labeled with 125I or 1311 was used to measure plasma volume in normal volunteers and in patients. In 1960 the Volemetron was introduced to measure the plasma volume using 125I

human albumin to estimate the red blood cell volume and the blood volume from the total body hematocrit estimated from the peripheral venous hematocrit and a factor ranging from 0.89 to 0.91 for an individual with a normal size spleen (Strumia MM et al, Transfusion 8:197-209 1968; Valeri CR, et al, Arch Int Med 132:534-538, 1973; Valeri CR, Altschule MD, CRC Press Boca Raton, FL, 1981).

4. Studies by Dr. Max Strumia and associates demonstrated that Evans Blue dye and 125I albumin could not be used to measure the plasma volume, red blood cell volume, and blood volume in patients. At the AABB meeting in 1965 Dr. Max Strumia reported that he had published data that were not correct. Dr. Strumia published that Evans Blue dye which binds to albumin accurately measured the plasma volume in patients. This statement was not correct because the Evans Blue dye distributed rapidly into the extravascular volume and overestimated the plasma volume, red blood cell volume and the blood volume in patients.

5. Studies at the Chelsea Naval Hospital, Chelsea, MA documented that radiolabeled human albumin did not accurately measure the plasma volume in patients. Rapid distribution, of radiolabeled iodinated albumin with a molecular weight of 68,000 into the extravascular volume in patients was reported by the NBRL (Valeri CR, et al Arch Int Med 132:534-538, 1973).

6. The cold agglutinin 19S macroglobulin with a molecular weight of 1,000,000 was isolated, purified, and labeled with 125I or 131I radioisotope from a patient with cold agglutinin disease by Dr. Amiel Cooper who worked with Sir John Dacie in London, England prior to his working at the NBRL (Valeri CR et al, Arch Int Med, 132:534-538. 1973).

7. Measurement of osmotic fragility of fresh and preserved RBC using the Danon fragiligraph was performed at the NBRL. Dr. Paul Marks at Columbia University was promised the fragiligraph which was left at the NBRL by Dr. David Danon for evaluation to assess the relationship between the osmotic fragility of preserved red blood cells and the 24 hour posttransfusion survival (Danon D, J Clin Path 16:337-342, 1963).

8. Density distribution of fresh and preserved RBC was measured using the phthalate esters provided the NBRL by Dr. David Danon (Danon D, Markikovsky Y, J Lab Clin Med 64:668-674, 1964).

9. Huggins Cytoglomerator was evaluated at the NBRL to deglycerolize RBC using low ionic solutions by the agglomeration procedure. Coombs positive autologous RBC were observed at the NBRL. Na2EDTA was

added to the glycerol solution to prevent Coombs positive RBC which was due to the accumulation of B_1C globulin onto the RBC in the low ionic glycerol solution that was used by Dr. Charles Huggins (Valeri CR, Transfusion 6:247-253 1966).

10. The volume of 0.9% NaCl used to disaggregate the agglomerated RBC was critical. A volume of 250 ml of 0.9% NaCl was needed to restore the physical-structural parameters of the agglomerated and disaggregated RBC and not the 75 to 100 ml of 0.9% NaCl which was recommended by Dr. Charles Huggins (Valeri CR, Bond JC, Transfusion 6:254-262 1966; Valeri CR et al, Transfusion, 6:543-553, 1966; Valeri CR. et al, Transfusion 6:554-564, 1966).

11. The Ashby procedure to measure the 24-hour posttransfusion survival, and lifespan of compatible but identifiable RBC by the manual differential agglutination procedure was automated at the NBRL. Dr. Richard Rosenfield recommended the use of the Technicon Autoanalyzer using Anti-A, Anti-B, and Anti-M antibodies to quantitate compatible but identifiable allogeneic RBC. At the NBRL Dr. I.0. Szymanski and associates automated the differential agglutination procedure using the Technicon Autoanalyzer and subsequently the NBRL personnel utilized the Coulter Counter to automate the manual Ashby differential agglutination procedure (Szymanski IO et al, Brit J Hematol 13:50-53,1967; Szymanski I0, et al, Transfusion 8:65-73, 1968; Szymanski IO, Valeri CR, Transfusion 8:74-83, 1968; Valeri CR et al, Vox Sang 49:195-205, 1985).

12. Wounded servicemen from Vietnam with chronic hypovolemic anemia of trauma had hematocrit values which did not assess the red blood cell volume deficit and whether the patients were hypovolemic or normovolemic (Valeri CR, Altschule MD, CRC Press, Boca Raton, FL. 1981; Biron PE et al, J Bone J Surg, 54A, 1001-1014, 1972).

13. 51Cr labeling of autologous RBC in patients was required to measure the red blood cell volumes to document the red blood cell volume deficits. (Valeri CR, Altschule MD, CRC Press, Boca Raton, FL, 1981).

14. Transfusion of multiple units of allogeneic compatible identifiable red blood cells increased the peripheral red blood cell volume and increased the plasma volume in patients with chronic hypovolemic anemia of trauma. (Valeri CR, Altschule MD, CRC Press, Boca Raton, FL, 1981; Valeri CR et al., Transfusion 38:602-610, 1998).

15. Washed autologous dog RBC transfusions increased both the red blood cell volume and the plasma volume in dogs subjected to acute hypovolemic

anemia. (Valeri CR, et al, Surg Gynecol Obstet 162:30-36, 1986; Valeri CR et al, Transfusion, 38:602- 610, 1998).

16. The normal central blood volume and the decreased peripheral blood volume were observed in patients with chronic hypovolemic anemia of trauma (Valeri CR, Altschule MD, CRC Press, Boca Raton, FL, 1981).

17. The limitations of the peripheral venous hematocrit to determine the transfusion trigger for red blood cells were observed in hypovolemic anemic patients subjected to cardiac and vascular surgical procedures (Valeri CR, et al., Transfusion 46:365-371, 2006).

18. The limitations of the platelet count, platelet size, and platelet hemostatic function to determine the transfusion trigger for platelets were observed in patients following cardiac surgery. (Valeri CR, et al, Transfusion 47:206S-248S, 2007).

19. Viability and function of RBC:
    a. Maintain high p02 tensions in brain, heart and kidneys
    b. Restore hemostasis and reduce the bleeding time and reduce nonsurgical blood loss
    c. Maintain blood pH and RBCATP, DPG and p50 levels.
    d. Increase the red blood cell volume, plasma volume, peripheral blood volume, and the blood volume in hypovolemic patients with traumatic injuries to their extremities.
    e. Transfusion of compatible identifiable viable red blood cells increase the peripheral blood volume to the gastrointestinal tract and to the bone, muscle and skin of the extremities in hypovolemic patients, (Valeri CR ,Altschule MD, CRC Press, Boca Raton, FL, 1981; Valeri CR et al, Transfusion 38:602-610, 1998).
    f. The RBC volume was measured using the radioisotopes 51Cr, 99mTc, and 111In oxine and biotin-X-NHS hydroxysuccinimide and fluorescent streptavidin in the flow cytometer in the baboon (Valeri CR, et al, Transfusion 42:343-348, 2002; Valeri CR et al, Transfusion 43:1366-1373, 2003).
    g. Increase in bleeding time by the reduction in the peripheral venous hematocrit by the removal of 360 ml of red blood cells and infusion of one liter of 0.9% NaCl in healthy male and female blood donors. (Valeri CR et al, Transfusion, 41:977-983, 2001).

20. Viability and Function of Platelets
    a. 51Cr, 111 In oxine, and biotin X-N-hydroxysuccinimide (NHS) and fluorescent streptavidin to measure baboon platelet in vivo

recovery and lifespan. (Valeri CR et al, Trans Aph Sci 32:275-281, 2005).

b. Aspirin induced increase in bleeding and reduction in shed blood thromboxane level at the bleeding time site in normal volunteers and in baboons (Handin RI and Valeri CR, NEJM, 285:538-543, 1971; Valeri CR, NEJM 290:353-358, 1974; Valeri CR et al, Blood 43:131-136, 1974; Valeri CR, Transfusion 16:20-23, 1976; Valeri CR et al, Transfusion 42:1206-1216, 2002).

c. Platelet aggregation response in vitro to a combination of arachidonic acid (AA) and adenosine diphosphate (ADP) and platelet production of thromboxane in vitro following stimulation with a combination of AA and ADP in human platelets to assess the function of the fresh and preserved platelets, (Valeri CR et al, Transfusion 45:596-603, 2005).

d. Platelet function analyzer (PFA) to assess sodium citrated blood response to a combination of ADP and collagen and to a combination of ADP and epinephrine; the thromboelastogram; and the clot signature analyzer were used to assess in vitro the hemostatic mechanism (Valeri CR, Ragno G, Trans Aph Sc. 35:33-41, 2006).

e. Testing of blood in the platelet function analyzer; thromboelastogram; and clot signature analyzer did not correlate to the bleeding time (Valeri CR, Ragno G, Trans Aph Sci 35:33-41.2006).

f. Dr. Nathaniel Shulman produced an increase in. bleeding time by the reduction in the healthy normal male volunteer platelet count by removal of platelets from whole blood and the reinfusion of platelet poor plasma and red blood cells. (Shulman NR et al, Trans Assoc Phys 81:302-313, 1968).

21. Gibson's Lesion of Collection. Our observation in baboons transfused with autologous RBC labeled with 51Cr and biotin X-NHS demonstrated that the lesion of collection reported by Gibson and associates was due to 51Cr elution from the labeled RBCs and not to the removal of nonviable 51Cr labeled RBCs. When the pH value of the medium in which the RBCs were labeled with 51Cr was 6.0 or lower elution of 51Cr from the RBCs occurred. (Valeri CR et al., Transfusion 42:343-348. 2002; Valeri CR et al, Transfusion 43:1366-1373, 2003).

22. Studies at the NBRL were performed in the baboon to assess the circulation and function of thrombin treated degranulated autologous

PLTs. The data showed that circulating degranulated PLTs rapidly lose surface P-selectin to the plasma pool but continue to circulate and function in a normal manner. In addition, a novel three-color whole blood flow cytometric method for tracking PLTs and measurement of PLT function in vivo were developed in collaboration with Dr. A. Michelson and Marc Barnard (Michelson AD, et al, Proc. Natl Acad Sci USA. 93:11877-11882, 1996).

The NBRL study demonstrated in vivo that circulating degranulated PLTs rapidly lose surface P-selectin to the plasma pool, thereby enabling the continued circulation and function of degranulated PLTs. Circulating degranulated PLTs rapidly lose surface P-selectin to the plasma pool, which indicates that PLT surface P-selectin is not an ideal marker for the detection of circulating degranulated PLTs. The NBRL study suggested that assay of the plasma concentration of soluble P-selectin could be used as a marker of PLT activation in clinical settings. An increase in the plasma concentration of soluble P-selectin in clinical setting, however, may also reflect the release of P-selectin from activated and/or damaged endothelial cells. (Michelson AD, et al, Proc Natl Acad Sci USA 93:11877-11882, 1996).

23. Studies were performed at the NBRL to assess the sensitivity of PLT surface P-selectin and PLT-rnonocyte heterotypic aggregates to detect PLT activation in vivo in the baboon. PLT surface P-selectin is considered the gold standard marker of PLT activation. Degranulated P-selectin,-positive PLTs, however, aggregate with WBCs and rapidly lose surface P-selectin in vivo. Flow cytometric tracking of autologous biotinylated PLTs in baboons enabled us to directly demonstrate for the first time in vivo that transfused degranulated PLTs rapidly form circulating aggregates with monocytes and neutrophils. (Michelson AD, et al, Proc Natl Acad Sci USA 93:11 877-11882, 1996).

PLT surface P-selectin is generally considered to be the gold standard marker of PLT activation. In this study, however, we demonstrated by in vivo tracking of activated PLTs in baboons, in patients subjected to percutaneous coronary catheterization, and in patients with acute myocardial infarction (AMI) that circulating PLT monocyte aggregates are a more sensitive indicator of in vivo PLT activation than PLT surface P-selectin (Furman MI et al, J Am Coll Cardiol 31:352-358, 1998; Peyton BD et al, J Vasc Surg, 27:1109-1116, 1998; Michelson AD et al, Circulation 104:1533-1537, 2001).

24. The Merrill porous bed viscometer measures the blood viscosity at

shear rate of 19 inverse seconds. The effects of red blood cells, platelets, white blood cells, and plasma on blood viscosity were evaluated. The whole blood viscosity at a shear rate of 19 inverse seconds measured in the Merrill porous bed viscometer was affected by the cube of the hematocrit and the square of the fibrinogen in the blood. (Merrill EW, Physiol Rev 49:863-888, 1969; Merrill EW et al, Lab Medica Inst 10:19-24, 1993; Crowley JP et al, Am J Clin Path 96:729-737, 1991; Crowley JP et al, Ann Clin Lab Sci 24:533-541, 1994).

25. Temperature affects the function of RBC, platelets, and plasma clotting proteins and affects hemostasis. Hypothermia produces a reversible platelet dysfunction by an increase in the bleeding time, a reduction in thromboxane level at the bleeding time site, and an increase in nonsurgical blood loss at the bleeding time site. Normothermia restores the function of the platelets, increases the thromboxane level at the bleeding time site, reduces the bleeding time, and reduces nonsurgical blood loss at the bleeding time site. Frank Pompei provided the infra red laser technology to measure the skin temperature at the bleeding time site. (Valeri CR et al, Ann Surg 205:175-181, 1987; Valeri CR, et al, Crit Care Med 23:698-704. 1995; Valeri et al, J Thorac Cardiovasc Surg 104:108-116, 1992; Michelson AD et al, J Thromb Hemost 5:633-640, 1994).

26. Volume loading of the hearts of patients following cardiopulmonary bypass surgery at University Hospital at Boston University Medical Center was assessed by the measurement of cardiac output and cardiac work by Dr. Herbert Hechtman, Dr. Robert Berger, Dr. Neil Yeston, Dr. Richard Dennis and Dr. Richard Weisel using the IL cooximeter to measure the percent (%) oxygen saturation. (HBO2); percent (%) methemoglobin; percent (%) carboxyhemoglobin; (HbCO); indocyanine green dye and thermal dilution to measure cardiac output, and using pulmonary artery wedge pressure to measure the filling pressure of the heart. (Dennis RC et al, Surgery 77:741 -747, 1975; Dennis RC et al, Ann Thorac Surg 26:17-26, 1978; Valeri CR et al, Prog Clin Bio Res 21:597-616, 1978; Valeri CR, Chemical Rubber Company Press, Boca Raton, FL, 1976).

27. Dr. Shukri Khuri and associates at West Roxbury VA Hospital studied the effect of membrane and bubble oxygenators and roller pump and centrifugal pump on the red blood cells in patients subjected to cardiopulmonary bypass surgery. No difference was observed whether the membrane oxygenator or bubble oxygenator was used (Khuri S et

al, J Thorac Cardiovasc Surg 104:94-107, 1992). Likewise, no difference was observed whether the roller pump or the centrifugal pump was used during cardiopulmonary bypass surgery (Valeri CR et al, Perfusion 21:291-296, 2006). In addition, Dr. Khuri and associates studied the effects of temperature, heparin, hemodilution, hematocrit, platelet count, platelet size, platelet function, and plasma clotting proteins on nonsurgical blood loss in patients subjected to cardiopulmonary bypass surgery. (Valeri CR et al, Transfusion 47:206S-248S, 2007).

a. Hobson RW II and associates studied the ischemic-reperfusion injury in isolated canine gracilis muscle: the effects of the rate of reperfusion blood flow, hypothermia, heparin, and urokinase.
b. Wright JG et al, J Surg Res 44:754-763, 1988
c. Wright JG et al, J Trauma 28:1026-1031, 1988
d. Wright JG et at, Current Surgery 45:25-27, 1988
e, Wright JG et al, Arch Surg 123:470-472, 1988
f. Belkin M et al, J Vasc Surg 9:161-168. 1989
g. Callow AD and associates studied small caliber grafts in baboons to assess platelet-arterial synthetic graft interaction and its modification to prevent thrombosis of the graft.
h. Callow AD et al, Ann Surg 191:362-366, 1980
i. Callow AD et al, Arch Surg 117:1447-1455, 1982
j. Mackey WC et al. Ann Surg 200:93-99, 1.984
k. Gembarowicz R et al, Surg Forum 33:466-468, 1982
l. Shoenfeld NA et al, J Vasc Surg 5:76-82, 1987
m. Shoenfeld NA et al, Surg Gynecol Obstet 166:454-457, 1988.
n. Eldrup-Jorgenson J et al, Arch Surg 121:778-781, 1986
o. Shoenfeld NA et al, J Vasc Surg 8:49-54, 1988

28. Dr. Alan Michelson, Mare Barnard and associates studied platelet function in patients subjected to cardiopulmonary bypass surgery and the effect of hypothermia on platelet function in vitro and in vivo (Kestin AS et al, Blood 82:107-117, 1993; Michelson AD et at, J Thromb and Haemost 5:633-640, 1994; Michelson AD et al, Br J Hematol 104:64-68, 1999). Dr. L. Michael Snyder and associates studied the effect of hydrogen peroxide on the survival of autologous red blood cells in the baboon. (McKenney J, et al Blood 76:206-211, 1990).

29. Studies were performed in baboons to evaluate the survival and function, of autologous and allogeneic baboon red blood cells, autologous baboon platelets, and autologous baboon plasma proteins subjected to

preservation by liquid and freezing procedures (Valeri CR, Ragno G, Transfusion 46:1S-42S, 2006).

30. Dr. Herbert Hechtman and associates studied the effects of ischemia to the lower extremities, the gastrointestinal tract, and skeletal muscle of animals. In these studies the role of granulocytes, prostacyclin (PGI2), thromboxane, leukotrienes, tumor necrosis factor (TNF), interleukin 2 (IL-2), complement, soluble P-selectin, oxygen free radicals, and neutrophil elastase were evaluated in the pathogenesis of acute lung injury associated with reduction in blood flow (Hechtman HB et al, Surg Clin NA 63:263-283, 1983; Welbourn R et al, Ann Surg 212:728-733, 1990; Goldman G et al, Surgery 17:83-89, 1995; Goldman G et al, Surgery 107:428-433, 1990; Weiser MR, et al, Surgical Forum 45:389-391, 1994; Woodcock SA et at, Shock 14:610-615, 2000; Gibbs SAL et al, Surgery 119:652-656, 1996; Kyriakides C ct al, Am J Physiol Cell Physiol. 279:C520-C528, 2000; Weiser MR et at, Shock 5:402-407, 1996).

31. Dr. Francis Moore, Jr. and associates studied the effect of hypothermia, hemodilution, and heparin to inhibit the activation of complement in patients subjected to cardiopulmonary bypass surgery (Moore FD, Jr.. et al, Ann Surg 208:95-103, 1988).

32. Large volume parenteral solutions were prepared from potable water using membrane technology without heat sterilization using the Millipore Sterimatics ST 30 at the NBRL. The Sterimatics ST 30 was deployed, aboard the LHA USS Saipan and at Fort Hood, TX to prepare resuscitation solutions and. solutions to deglycerolize red blood cells.

33. Sawka and associates at U.S. Army Natick Environmental Laboratory studied physical performance and cognitive function in healthy servicemen during hyperthermia and following transfusions of autologous red blood cells at normal and at low p02 environments (Muza SR et al, Aviat Space Environ Med 58:1001-1004, 1987; Sawka MN et al, J Appl Physiol 62:912-918, 1987; Sawka MN et al, Am J Physiol 255:R456-463, 1988; Sawka M, et al, Am J Physiol 257:R311 -R316, 1989; Young AS et at J App Physiol 81:252-259, 1996; Sawka MN et al, J Appl Physiol 81:636-642, 1996; Pandolf KB et al, Europ J App Occup Physiol 79:1-6, 1998).

34. Red blood cell p50 (partial pressure at which 50% of the hemoglobin is saturated) was measured using the Lex 02 Con fuel cell and the Van Slyke instrument; IL cooximeter; Bellingham-Huehns tonometer attached to a cuvette in the Unicam spectrophotometer; the Hemoscan instrument; and Hemoxanalyzer instrument (Dennis RC, Valeri CR, Clin Chem 26:1304-

1308, 1980; Bellingham AJ, Huehns ER, Forsvarsmedicin 5:207-211, 1969; Asakura T, Reilly MP, In Nicolau G, editor, Oxygen transport in red blood cells, NewYork, Pergamon Press, 57-75, 1986; Festa RS, Asakura T, Transfusion 19:107-113, 1979).

35. Reduction in endothelial nitric oxide (NO) by oxidation to perioxynitrite, nitrite and nitrate released endothelin from the endothelial cells. Dr. A.M. Lefer; Dr. W. Lieberthal, and Dr. Hechtman reported on the effect of endothelin to vasoconstrict blood vessels. Like nitric oxide (NO), hydrogen sulfide is a vasodilator and binds to hemoglobin to produce sulfhemoglobin whereas nitric oxide inhibits platelet function and binds to the heme (NOHb) and the globin (SNOHb) of hemoglobin (Ikeda Y et al, J Surg Res 102:215-220, 2002; Favuzza J, Hechtman HB, J Trauma Suppl 57: S42-S44, 2004; Thompson A et al, Am J Physiol 269:H743-H748, 1995; Drabkin DL, Austin JH, J Biol Chem 112:51-65, 1935; D'Emmanuele di Villa Bianca R et al, PNAS Early Edition 1-6, 2008).

36. Large volumes of resuscitation solutions containing Ringer's DL lactate, Ringer's L lactate, Ringer's ketone solutions have been used to treat animals in hemorrhagic shock. Substrate mediated small volume resuscitation is now being investigated using the ketone-Na D betahydroxybutyrate. Large volume resuscitation with crystalloid and colloid solutions produce hemodilution and adversely affect hemostasis. DL lactate is a toxic solution and should not be used. Dr. Richard Veech and Dr. Kieran Clarke have reported perfusion of isolated rat hearts in vitro with Ringer's DL lactate solution and with Ringer's ketone solution. Improved myocardial function was observed in rat hearts perfused with Ringer's ketone solution associated with decreased oxygen consumption compared to isolated perfused rat hearts perfused with Ringer's DL lactate solution. Dr. Richard L. Veech has reported that Na D B hydroxybutyrate is the ideal substrate to treat patients with traumatic injuries, head trauma and insulin resistance (Koustova E et al, Surgery 134:267-274, 2003; Ayuste EC, et al, J Trauma 60:52-63, 2006; Jaskille A, et al, J Am Coll Surg, 202:25-35, 2006; Valeri CR et al, Art Cells, Blood Subst and Biotech 34:601-606, 2006; Valeri CR et al, Trans Aph Sci 39:205-211, 2008).

37. In 1980, Dr. Carl Apstein and Dr Mark Vogel studied in vitro isolated rabbit hearts perfused with human red blood cells with decreased affinity for oxygen and increased RBC DPG levels and human red blood cells with increased affinity for oxygen and decreased RBC DPG levels on

myocardial function at normothermic and hypothermic temperatures. Human red blood cells with decreased affinity for oxygen containing elevated levels of RBC DPG improved myocardial function at normothermic and hypothermic temperatures (Apstein CS et al, Am J Physiol 248:H508-H515, 1985).

38. In 1982, Dr. Carl Apstein and Dr. Mark Vogel reported that isolated rabbit hearts perfused in vitro with human hemoglobin based oxygen carriers (HBOC) produced severe vasoconstriction of the coronary circulation and markedly decreased myocardial function (Vogel WM et al, Am J Physiol 251:H413-H420, 1986; Vogel WM et al, Life Science 41:89-93. 1987).

39. In 1978, Dr. M. Yarnoz and associates studied in vitro perfusion of isolated dog hearts with human red blood cells with increased affinity for oxygen and decreased RBC DPG levels and human RBC with reduced affinity for oxygen and increased RBC DPG levels on myocardial function at hypothermic and. normothermic temperatures. Human red blood cells with increased DPG levels and decreased affinity for oxygen improved myocardial function of the isolated dog hearts at normothermic and hypothermic temperatures (Valeri CR et al., Ann Thor Surg 30:527-535, 1980).

40. In 1980, Dr. Y. Castany and associates at the Retina Foundation, Boston, MA studied in vitro perfusion of bovine eyeballs with human red blood cells with increased affinity for oxygen and decreased DPG levels and with human RBC with decreased affinity for oxygen and increased DPG levels on flicker fusion response. Bovine eyeballs perfused with human red blood cells with decreased affinity for oxygen and increased DPG levels had improved flicker fusion response compared to bovine eyeballs perfused in vitro with human red blood cells with increased affinity for oxygen and decreased DPG levels.

41. The approach to support combat casualties using fresh whole blood was advocated by Colonel William Crosby, MC, USA and the U.S. Army during the Korean War between 1952 to 1954; Vietnam War between 1965 to 1974; and the Persian Gulf War between 1980 to 1981. Prescreened personnel donated fresh whole blood not tested for FDA mandated infectious disease markers prior to transfusion are now used to treat wounded military and civilian casualties in Afghanistan and Iraq by the U.S. Army.

42. The quality of the liquid preserved whole blood and liquid preserved

RBC concentrates is assessed by sterility, 24-hour posttransfusion survival values, less than 1% hemolysis and the non-toxic additives used to preserve the RBC.

43. Since 1981 following the observation that fresh frozen plasma and cryoprecipitate transfused to patients with. hemophilia produced autoimmune deficiency syndrome (AIDS). Testing for AIDS together with hepatitis and other infectious disease markers is now mandated by FDA prior to transfusion. In addition, the screening of the donor with regards where the donor resides and screening the dietary history of the donor are now required by the FDA to eliminate high risk blood donors. Today extensive testing of the blood and screening of the blood donor are required, so that fresh whole blood is no longer available. The testing for all the infectious disease markers usually takes 24 to 48 hours. AIDS, hepatitis, Leishmaniasis, malaria, shistosomiasis, brucellosis, tuberculosis are diseases that the U.S. military are exposed to in the Middle East countries during their deployment in combat zones.

44. The recent experience by the U.S. Armed Services Blood Program in the Middle East has demonstrated that fresh whole blood produces significantly better clinical outcome compared to liquid preserved red blood cells stored at 4 C in the additive solution AS-5 (Optisol) for 42 days when used to resuscitate casualties in the Middle East conflicts by the U.S. Military.

45. Hemoglobin based oxygen carriers (HBOCs) provided by commercial companies like Northfield, Hemosol, and Biopure have produced severe adverse events and are not FDA approved.

46. The beneficial effects have been observed, using ABO compatible fresh whole blood which contain viable and functional RBC, platelets, and plasma clotting proteins.

47. In collaboration with the NBRL, Saba and his associates have reported that the storage at - 20 C and at -80 C for one year reduced the function of the opsonic protein fibronectin. (Blumenstock FA et al Vox Sang 54:129-137, 1988).

48. In collaboration with the NBRL, Ichikura T and associates have reported increased metastatic growth of Lewis lung carcinoma following the transfusion of liquid preserved syngeneic mice blood compared to the transfusion of fresh syngeneic mice blood. Liquid preserved syngeneic mice blood increased metastatic growth of tumor cells in mice.(Ichikura T et al, J Nippon Geka Gakkai Zasshi 92:734-739, 1991).

49. In collaboration with the NBRL, Crowley and associates have reported no adverse effects of autologous nonviable dog red blood cells and human hemoglobin based oxygen carriers (HBOC) on the clearance of E. coli bacteria in dogs. In addition, no difference was observed in the clearance of E. coli in dogs treated with normal dog plasma and hyperimmune dog plasma. (Crowley JP et al. Cir Shock 26:287-295. 1988; Crowley JP et al, Cir Shock 36:31-37, 1992; Crowley JP et al, Cir Shock 41: 144-149 1993).

50. The studies in mice indicate that tumor cell metastases were accelerated by liquid preserved syngeneic mice blood compared to fresh syngeneic mice blood whereas the clearance of E. coli in dogs was not affected by autologous nonviable dog red blood cells, human hemoglobin based oxygen carriers (HBOC) and hyperimmune dog plasma.

51. Optimum Treatment of Combat Casualties

a. Use of the hemostatic agents N acetyl glucosamine nanofibers to prevent blood loss, stimulates wound healing and exerts an antimicrobial effect. (Valeri CR, J Trauma 61:240-241, 2006; Valeri CR et al, J Trauma 57:S22-S25, 2004; Pietromaggiori G et al, J Trauma 64;803-808, 2008; Fisher TH et al, J Trauma 71:S176-S182, 2011; Scherer SS et al, Ann Surg 250:322-330, 2009; Linder HB et al PLoSI 6:e18996, 2011).

b. Prevent hypothermia — rewarm patients and infuse resuscitation solutions and blood products at 40 C using devices to warm the resuscitation solutions and blood products.

c. Use of substrate resuscitation fluids to provide 250 ml volumes of a hypertonic ketone solution containing 600 mM of Na D B hydroxybutyrate to treat patients with traumatic injuries, head trauma, and insulin resistance.

d. Transfusion of universal donor group O Rh positive and group O Rh negative RBC that circulate and function; transfusion of group AB plasma clotting protein from male donors that circulate and function; and transfusion of single donor leukoreduced group O platelets that circulate and. function.

e. Optimum volume of hypertonic ketone containing resuscitation solution and optimum ratio of RBC, plasma and platelets need to be determined, Quality and quantity of RBC, platelets and plasma need to be defined in addition to the compatibility and possible transmission of infectious diseases. Small volumes of resuscitation solutions need to replace large volumes of resuscitation solutions to prevent hemodilution and nonsurgical blood loss.

52. Deployment of -80 C air cooled mechanical freezers maintained at a mean temperature of -80 C with a range of -65 C to -90 C with a tank of liquid CO2 to be delivered into the mechanical freezer when the temperature is less than -65 C. Transportation of frozen blood products in insulated containers with dry ice will provide safe and therapeutically effective blood products.

53. Current requirements for testing of blood products for infectious disease markers and the donor history now supports the use of universal donor frozen RBC, platelets, and plasma which can be tested and quarantined for six months to retest the blood donor for infectious disease markers to ensure the safety of the frozen blood products. The quarantine of frozen blood products and retesting of the donors after 6 months may eliminate the need to disinfect the blood products that are now being investigated (Horowitz B et al Blood Cells 18:141-150, 1992; Valeri CR et al, Photochem Photobiol 65:446-450, 1997; BenHur E et al, Dev Biol 102:149-155, 2000; Purmal A, et al, Transfusion 42:139-145, 2002).

54. Adverse effects of hypothermia and the need to warm the operating room and wounded individuals, use of blood warmers to maintain the temperatures of the resuscitation fluids and blood products at 40 C at the time of transfusion to ensure optimum function of these products in the recipient

55. Need to provide resuscitation solutions, irrigation fluids, and solutions to process frozen RBC from potable water and concentrates at the site of need. Successful deployment of the Millipore Sterimatics (ST) 30 instrument was demonstrated to prepare resuscitation solutions, irrigation fluids, and solutions to deglycerolize the RBC frozen with 40% W/V glycerol and stored at -80 C aboard the USS Saipan LHA ship and at Fort Hood, TX by the NBRL.

56. Hemostatic agents prepared by Marine Polymer Technologies, Inc. provide the polymerized N-acetylglucosamine nanofiber referred to as the long NAG (lNAG) which must be removed from the bleeding site. Gamma radiation of lNAG produces the short NAG (sNAG) nanofiber which allows the sNAG to be left in the wounded area to exert its hemostatic effect, stimulate wound healing, and exert an antimicrobial effect. The glycocalyx on which the endothelial cells adhere may produce derivatives of the N acetylglucosamine which activate the platelets and red blood cells

to produce a hemostatic effect, stimulate wound healing, and exert an antimicrobial effect (Weinbaum S et al, Annals Rev Biomed Eng 9:121-167, 2007).

57. Blood products frozen and stored in -80 C mechanical freezers provide universal donor group O Rh positive and group O Rh negative RBC, single donor leukoreduced group O platelets and AB plasma clotting proteins that circulate and function immediately or shortly after transfusion.

58. Universal donor Group O Rh positive and Group O Rh negative RBC treated with 40% W/V glycerol, the supernatant glycerol removed prior to freezing and stored at -8O C for at least 10 years, thawed, washed using the Haemonetics Blood Processor ACP215, and stored at 4 C in the additive solution AS-3 (Nutricel) for 2 weeks have been FDA approved. (Valeri CR, et al, Transfusion 41:933-939,2001.; Valeri CR et al, Transfusion 41:928-932, 2001).

59. Universal donor AB plasma collected from donors stored at -80 C for at least 14 years maintain the function of the clotting protein factors V and VIII and fibrinogen (Valeri CR and Ragno G, Transfusion 45:1829-1830, 2005).

60. Single donor group O leukoreduced platelets treated with 6% DMSO, the supernatant DMSO removed, and DMSO treated platelets frozen and stored at -80 C, thawed, diluted with 10 to 20 ml of 0.9% NaC1 and stored at room temperature without agitation for 4 hours have in vitro recovery of 90% of the platelets; in vivo recovery 1 to 2 hours following autologous transfusion of 25-30% and a lifespan of 7 days (Valeri CR et al, Transfusion 45:1890-1898, 2005).

61. There is no correlation between the BT and surgical blood loss, but a correlation does exist between the BT and nonsurgical blood loss. Bleeding time is affected by several factors such as the local and systemic temperatures, hematocrit, platelet count, platelet size, platelet function and plasma clotting proteins (Valeri CR et al Transfusion. 47:206S-248S, 2007).

62. Transfusion of red blood cells to increase the hematocrit to 35 volume percent was observed to reduce nonsurgical blood loss in patients after cardiopulmonary bypass surgery. (Khuri SF et al, J Thorac Cardiovasc Surg 104:94-107, 1992).

63. Red blood cells administered to anemic thrombocytopenic patients with a nonsurgical bleeding diathesis need to survive and function immediately or shortly after transfusion. Liquid preserved red blood cells should be

stored at 4 C in CPDA1 and additive solutions CPD/AS-1, CP2D/AS-3 and CPD/AS-5 for no more than 2 weeks. (Valeri CR Trans Aph Sci 42:223-233, 2010).

64. The survival and function of preserved platelets are necessary to restore the hemostatic defect in patients with reduced platelet counts and/or with platelet dysfunction provided the hematocrit of the patient is 35 V% (Valeri CR, Ragno G, Trans Aph Sci 42:223-233, 2010).

65. The in vivo interaction of platelets and RBC provide hemostasis by the platelet producing thromboxane and RBCs scavenging of endothelial nitric oxide which vasodilates blood vessels and inhibits platelet function. When the endothelial cell nitric oxide is reduced platelet function is restored and endothelin, a potent vasoconstrictor substance, is released from the endothelial cells to vasoconstrict the bleeding site. Activation of the platelets and RBCs at the BT site produces thrombin which converts fibrinogen to fibrin (Valeri. CR et al, Transfusion 47:206S-248S, 2007).

66. RBCs in platelet rich plasma treated with poly N acetylglucosamine long nanofiber reduced the fibrinogen levels demonstrating that poly N acetylglucosamine long nanofibers activate the RBC to produce thrombin (Valeri CR et al, J Trauma Supplement 57:S22-S25, 2004).

67. Red blood cells interact with poly N acetylglucosamine long nanofiber to assume a stomatocytic morphology and generates a prothrombotic membrane to produce thrombin (Fischer TH et al, Biomed Mater 3:1-9, 2008).

68. Like platelets, RBCs at the BT site provide both the adenosine diphosphate (ADP) and arachidonic acid (AA) used by platelets to produce thromboxane. Two (2) vasoconstrictor substances are produced at the bleeding site, thromboxane by the platelets and endothelin released from the endothelial cells to vasoconstrict the nonsurgical bleeding site. The activation of the platelets and red blood cells generate thrombin. (Valeri CR, Khuri S, Ragno G, Transfusion 47:206S-248S, 2007).

69. Viable and functional. RBCs are more effective than viable and functional platelets in reducing the bleeding time and the volume of shed blood at the bleeding time site. The transfusion of preserved RBC that survive and function to achieve a hematocrit level of 35 volumes percent in anemic thrombocytopenic patients may reduce or eliminate the need for prophylactic leukoreduced platelet transfusions to restore hemostasis and reduce or prevent nonsurgical blood loss. Repeated transfusions of leukoreduced platelets in anemic thrombocytopenic patients may

alloimmunize the recipient and produce transfusion related acute lung injury (TRALI) (Valeri CR et al, Transfusion 47: 206S- 248S, 2007).

70. Anemic thrombocytopenic patients with nonsurgical blood loss have both reduced numbers of platelets and dysfunctional platelets. Transfusion of viable and functional RBCs to achieve a hematocrit of 35 volumes percent will restore platelet function in anemic thrombocytopenic patients before prophylactic platelet transfusions (Valeri CR et al, Transfusion 47:206S-248S, 2007).

71. The hemostatic effect of viable and functional RBCs is due to their scavenging of endothelial cell nitric oxide by oxidation and by nitric oxide binding to the hemoglobin. Platelet function is restored to normal by the reduction in nitric oxide and the decrease in platelet cyclic guanosine monophosphate (cGMP) level. RBCs, like platelets, may provide both ADP and AA for the platelet production of thromboxane at the bleeding site. RBCs increase blood viscosity and shear stress which may release ADP to aggregate platelets and may release endothelin, a potent vasoconstrictor substance from the endothelin cells (Valeri CR et al, Transfusion 47:206S-248S, 2007).

72. The endothelial cells are attached to a layer of subendothelial glycocalyx which may produce derivatives of the N acetylglucosamine which is made by the microalgae used by the Marine Polymer Technologies, Inc. to produce the polymerized N acetylglucosamine nanofiber.

73. Combat casualty care general approach to optimum treatment should utilize low volume resuscitation fluid containing the ketone substrate Na D betahydroxybutyrate instead of the large volume crystalloid or colloid solutions. Toxicity of Ringer's D,L lactate solution has been documented and the adverse effects of hemodilution on hemostasis produced by large volume resuscitation fluids increases nonsurgical blood loss. Albumin is a poor oncotic protein and should not be used to resuscitate hypovolemic patients. The use of large volumes of crystalloid and colloid solutions should be avoided in the treatment of patients in hemorrhagic shock.

74. General approach to optimum treatment of combat casualties.

   a. Control surgical blood loss and use of hemostatic agents with external pressure
   b. Resuscitation solutions using low volume ketone containing solutions to avoid large volume crystalloid or colloid resuscitation solutions. Substrate resuscitation requires that the ketone (NaD BHB) should be utilized to replace lactate, glucose, and pyruvate

to maintain cellular metabolism and function in patients with traumatic injuries, insulin resistance and head trauma.

c. Nonsurgical bleeding diathesis is associated with hypothermia, hemodilution, anemia,thrombocytopenia, and reduction in plasma clotting proteins.
d. Quality of blood products must ensure the optimum survival and function of RBC, platelets, and plasma clotting proteins that are transfused together with the resuscitation solutions containing NaD BHB. The ketone NaD BHB is needed to treat patients with head trauma to prevent the posttraumatic stress syndrome.

75. Safety and therapeutic effectiveness of the polymerized N acetylglucosamine (NAG) hemostatic bandages with external pressure have been demonstrated in the following clinical situations:

a. Maintain patency of the catheter used to perform hemodialysis (Unlap A et al, Am J Med Sci 338(3):178-184, 2009).
b. Restoration of hemostasis and accelerates wound healing at the arterial site used to catheterize patients to perform diagnostic and therapeutic interventions (Najjar SF et al, J Trauma 57:S38-S41, 2004).
c. Acceleration of wound healing in patients with chronic venous leg ulcers. (Kelechi TJ et al, J Amer Acad Derm 2011 (DOI:10.1016/j.jaad.2011.01.031).
d. Exert an antimicrobial effect by stimulating endothelial cells to produce alpha and beta defensin peptides (Lindner HB et al, PLoSJ 6:e18996, 2011).

76. Resuscitation:

a. Low volume substrate resuscitation using hypertonic NaD BHB ketone solution instead of large volumes of crystalloid and colloid which produce hemodilution and an anticoagulant effect with an increase in nonsurgical blood loss. Human albumin does not increase the plasma volume and the blood volume in patients subjected to hemorrhagic shock.
b. Hemodilution. and hypothermia produce adverse effects on hemostasis and increase nonsurgical blood loss.
c. Nonsurgical blood loss treated with hemostatic agents and external pressure.
d. Use of parenteral hypertonic ketone solution to treat traumatic injuries, insulin resistance and head trauma.

e. Production of parenteral solutions for resuscitation, irrigation solutions, and solutions to process blood products at site of need from potable water and concentrates. The Millipore Sterimatic Instrument (ST-30) using membrane technology produces 30 liters per hour of sterile, pyrogen free water for injection without heat sterilization. Applied Research Associates instrument uses membrane technology and heat sterilization to produce sterile, pyrogen free water for injection from potable water.

77. To ensure the optimum survival and function liquid preserved RBC should be stored in CPDA-1 and in the additive solutions CPD/AS-1, CP2D/AS-3 and CPD/AS-5 at 4 C for only 2 weeks.

78. To ensure the optimum survival and hemostatic function, platelets should be stored at room temperature with agitation for only 2 days.

79. To ensure the optimum function of clotting proteins fresh frozen plasma and cryoprecipitate should be stored at-20 C for one year, thawed and stored at 4 C for 24 hours (Valeri CR, Ragno G, Trans Aph Sci 42:223-233, 2010).

80. Fresh whole blood was advocated by Colonel William Crosby and the U.S. Army and Department of Defense to treat wounded casualties in Korea, Vietnam, Persian Gulf War and now in the Middle East combat zones. In 1981 the need to test fresh whole blood for autoimmune deficiency syndrome (AIDS), hepatitis, and other infectious disease agents and need to obtain the history of the blood donor related to where the donor lives and what the donor eats have limited the availability of fresh whole blood, which now requires 24 to 48 hours to test the blood products prior to transfusions.

81. An approach to prevent the severe adverse events associated with FDA approved blood products now requires the optimum use of liquid methods and freezing procedures to provide safe and therapeutically effective blood products to treat combat casualties. The experience by the Netherlands military under the direction of Dr. Charles Lelkens, Dr. Femke Noorman, and Dr. John. Badloe have demonstrated that frozen RBC, frozen platelets, frozen plasma all stored at -80 C in mechanical freezers provided safe and therapeutically effective blood products to treat wounded casualties without the need for fresh whole blood in the combat zones (Valeri CR, Ragno G, Trans Aph Sci 42:223.-233, 2010).

82. Current requirements to provide optimum treatment of casualties in the war zone:

a. Hemostatic agents need to be stockpiled
b. Technology to warm wounded casualties, the operating room, and the use of warming devices to maintain the temperature of resuscitation fluid and blood products at 40 C at the time of transfusion
c. Production of a hypertonic ketone (NaD BHB) solution to provide 250 ml volumes prior to transfusion of viable and functional RBC, platelets and plasma clotting proteins Na D betahydroxybutyrate (NaD BHB) ketone substrate should be utilized to treat patients with traumatic injuries who have insulin resistance and patients with head injury. The quality and quantity of the RBC, platelets, and plasma need to be defined. The optimum ratio of red blood cells, plasma and platelets has to be determined.
d. Technology to prepare resuscitation fluids, irrigation solutions, and solutions to deglycerolize frozen red blood cells at the site of need using membrane technology.
e. Nonrejuvenated and indated rejuvenated group O Rh positive and group O Rh negative RBC stored at 4 C for 3 to 6 days and outdated rejuvenated liquid preserved RBC stored at 4 C in additive solutions (CPD/AS-l, CP2D/AS-3, and CPD/AS-5) for 42 days using pyruvate, inosine, phosphate and adenine (PIPA) solution and incubation at 37 C for one hour to increase the RBC ATP, DPG and p50 levels to 1 ½ to 2 times normal and then treated with 40% W/V glycerol and removal of supernatant glycerol prior to freezing and storage at -80 C for at least 10 years, thawed, deglycerolized using the Haemonetics ACP215 instrument can be stored at 4 C in AS-3 for two weeks for nonrejuvenated and indated rejuvenated RBC and for at least 24 hours for outdated rejuvenated RBC (Valeri CR, Ragno G, Trans Aph Sci 42:223-233, 2010).
f. Salvaging outdated group O Rh positive and group O Rh negative RBC by biochemical treatment with PIPA solution prior to treatment with 40% W/V glycerol, removal of the supernatant glycerol prior to freezing and storage at -80 C will eliminate the outdating of universal donor group O Rh positive and group O Rh negative red blood cells and provide previously frozen deglycerolized RBC using the Haemonetics Blood Processor ACP215 instrument, stored at 4 C in AS-3 for at least 24 hours

that circulate and function shortly following transfusion. (Valeri CR, Ragno G, Trans Aph Sci 42:223-233, 2010).

g. Deployment of -8O C mechanical freezers to store frozen universal donor group O Rh positive and group O Rh negative RBC processed as nonrejuvenated, indated rejuvenated and outdated rejuvenated RBC treated with 40% W/V glycerol with the removal of the supernatant glycerol prior to freezing for at least 10 years at - 80 C in mechanical freezers, thawed, washed using the Haemonetics Blood Processor ACP215 instrument and stored at 4 C in AS-3 for at least 24 hours that circulate and function immediately or shortly after transfusion; single donor leukoreduced group O platelets treated with 6% DMSO, the supernatant DMSO removed, the DMSO treated platelets frozen and stored at -80 C for at least 2 years, thawed, diluted with 10 to 20 ml of 0.9% NaCl or with a unit of AB plasma and stored at room temperature without agitation for 6 hours that circulate and function shortly after transfusion; and group AB plasma collected from male donors, frozen at -80 C for at least 14 years thawed and stored at 4 C for 24 hours to provide functional clotting proteins factors V and VIII and fibrinogen (Valeri CR, Ragno G, Trans Aph Sci 42:223-233, 2010).

CHAPTER 4 REFERENCES

1. Almond DV, Valeri CR. The in vivo effects ofdeglycerolized agglomerated erythrocytes transfused in multiple units to stable anemic patients. Transfusion 7:95-104, 1967.
2. Almond DV, Valeri, CR. Relationship between lipid fractions of erythrocytes and their in vivo survival following preservation with glycerol and the slow freeze technic. Transfusion 7:10-16, 1967.
3. Alter NJ, Tabor E, Meryman HT, Hoofnagle JH, Kahn BA, Holland PV et al. Transmission of hepatitis B virus infection by transfusion. of frozen-deglycerolized red blood cells. N Engl J Med 298:637-642, 1978.
4. Apstein CS, Dennis RC, Briggs L, Vogel WM., Frazer J, Valeri CR. Effect of erythrocyte storage and oxyhemoglobin affinity changes on cardiac function. Am J Physiol 248 (Heart Circ Physiol 17): H.508-H515, 1985.
5. Asakura T, Reilly MP. Methods for the measurement of oxygen equilibrium: Curves of red cell suspensions and hemoglobin solutions: in Nicolau C (ed): Oxygen Transport in Red Blood Cells. Proceedings of the 12th Aharon Katzir Katchalsky Conference 1984. New York, Pergamon Press. 1986, pp 57-75.
6. Axford TC, Dearani JA, Ragno G, MacGregor H, Patel MA, Valeri CR, Khuri SF. Safety and therapeutic effectiveness of reinfused shed blood after open heart surgery. Ann Thorac Surg 57:615-622, 1994.
7. Ayuste EC, Chen H, Koustova E, Rhee P, Ahuja N, Chen Z, Valeri CR, Spaniolas K, Mehrani T, Alam HB. Hepatic and pulmonary apoptosis after hemorrhagic shock in swine can he reduced through modifications of conventional Ringer's solution. J Trauma 60(1):52-63. 2006.
8. Belkin M, Valeri CR, Hobson. RW II. Intraarterial urokinase increases skeletal muscle viability after acute ischemia. J Vasc Surg 9:161-168. 1989.
9. Bellingham AS and Huehns ER. Oxygen dissociation in red cells from patients with abnormal haemoglobins and pyruvate kinase deficiency. Forsvarsmedicin 5:207-211, 1969.
10. Benesch R, Benesch RE. The effect of organic phosphates from

the human erythrocyte on the allosteric properties of hemoglobin. Biochem Biophys Res Commun 26:162-167, 1967.

11. Ben-Hur E, Chan WS, Yim Z, Zuk MM, Dayal V, Roth N, Heldman E, Lazo A, Valeri CR, Horowitz B. Photochemical decontamination of red blood cell concentrates with the silicon phthalocyanine PC 4 and red light. Dev Biol 102:149-155, 2000.
12. Biron PE, Howard J, Altschule MD, Valeri CR, Chronic deficits in red cell mass in patients with orthopaedic injuries (stress anemia). J Bone Joint Surg 54-A:1001-1014, 1972.
13. Blevins FT, Shaw B, Valeri CR, Kasser J, Hall J. Reinfusion of shed blood after orthopaedic procedures in children and adolescents. J Bone & Joint Surg 75-A:363-371, 1993.
14. Blumenstock FA, Valeri CR, Saba TM, Cho F, Melaragno A, Gray A, Lcwis M. Progressive loss of fibronectin-mediated opsonic activity in plasma cryoprecipitate with storage: Role of fibronectin fragmentation. Vox Sang 54:129-137, 1.988.
15. Brodine CE, Sell KW, Moss GS, Valeri CR Navy surgical research programs. J Surg Res 7:545-548, 1967.
16. Callow AD, Ledig CB, O'Donnell TF, Kelly JJ, Rosenthal D, Korwin S, Hotte C, Kahn PC, Vecchione JJ, Valeri CR. A primate model for the study of the interaction of ^{111}In-labeled baboon platelets with dacron® arterial prostheses. Ann Surg 191:362-366, 1980.
17. Callow AD, Connolly R, O'Donnell TF Jr, Gembarowicz R, Keough E, Ramberg-Laskaris K, Valeri CR. Platelet-arterial synthetic graft interaction and its modification. Arch Sur 117:1447-1455, 1982.
18. Carciero R, Valeri CR. Isolation of mononuclear leukocytes in a plastic bag system using Ficoll-Hypaque. Vox Sang 49:373-380, 1985.
19. Chanutin A, Curnish RF. Effect of organic and inorganic phosphates on the oxygen equilibrium of human erythrocytes. Arch Biochem Biophys 121:96-102, 1967.
20. Costa JL, Dobson CM, Kirk KL, Poulsen FM, Valeri CR, Vecchione JJ. Studies of human platelets by ^{19}F and ^{31}P NMR. FEBS Letters 99:141-146, 1979.
21. Cronkite EP. Measurement of the effectiveness of platelet transfusion. Transfusion 6:18-22, 1966.

22. Crowley JP, Metzger J, Pivacek L, Dennis RC, Valeri CR. Effects of plasma administration on gram negative shock in granulocytopenic dogs. Circulatory Shock 26:287-295, 1988.
23. Crowley JP, Valeri CR, Metzger J, Gray A, Schooneman F, Man NK, Merrill E: The estimation of whole blood viscosity by a porous bed method. Am J Clin Pathol 96:729-737, 1991.
24. Crowley JP, Metzger J, Pivacek, L, Valeri CR. The effect of viable and nonviable autologous red blood cell transfusions on experimental bacteremia. Circ Shock 36:31-37, 1992.
25. Crowley JP, Metzger J, Gray A, Pivacek LE, Cassidy G, Valeri CR. Infusion of stroma free cross-linked hemoglobin, during acute gram-negative bacteremia. Circ Shock 41:144-149, 1993.
26. Crowley JP, Metzger J, Assaf A. Carleton RC, Merrill E, Valeri CR. Low density lipoprotein cholesterol and whole blood viscosity. Ann Clin Lab Sci 24:533-541, 1994.
27. Danon D. A rapid micro method for recording red cell osmotic fragility by continuous decrease of salt concentration. J Clin Path 16:337-342, 1963.
28. Danon D, Markikovsky Y. Determination of density distribution of red cell population. J Lab Clin Med 64:668-674, 1964.
29. Dennis RC, Vito L, Weisel RD, Valeri CR, Berger RL, Hechtman HB. Improved myocardial performance following high 2,3 diphosphoglycerate red cell transfusions. Surgery 77:741 -747, 1975.
30. Dennis RC, Hechtman HB, Berger RL, Vito L, Weisel, RD, Valeri CR. Transfusion of 2,3 DPG-enriched red blood cells to improve cardiac function. Ann Thoracic Surg 26(1):17-26. 1978.
31. Dennis RC, Valeri CR. Measuring percent oxygen saturation of hemoglobin, percent carboxyhemoglobin and methemoglobin, and concentrations of total hemoglobin and oxygen in blood of man, dog, and baboon. Clin Chem 26:1304-1308, 1980.
32. D'Emmanuele di Villa Bianca R, Sorrentino R, Maffia P, Mirone V, Imbimbo C, Fusco P, De Palma R, Ignarro LJ, Cirino G. Hydrogen sulfide as a mediator of human corpus cavernous smooth-muscle relaxation. PNAS Early Edition: 1-6, 2008.
33. Drabkin DL, Austin JH. Spectrophotometric studies. II. Preparations from washed red blood cells: nitric oxide hemoglobin and sulfhemoglobin. J Biol Chem 112:51-65, 1935.

34. Eldrup-Jorgensen J, Connelly RJ, Mackey WC, Ramberg K, O'Donnell TF, Valeri CR, Callow AD. Antiplatelet therapy and vascular grafts: Studies in a baboon ex vivo shunt. Arch Surg 121:778-781, 1986,
35. Faris PM, Ritter MA, Keating EM, Valeri CR. Unwashed filtered shed blood collected after knee and hip arthroplasties: A source of autologous red blood cells. J Bone & Joint Surg 73A:1169-1178, 1991.
36. Favuzza J, Hechtman HB. Hemostasis in the absence of clotting factors. J Trauma: 57:S42-S44. 2004.
37. Festa RS, Asakura T. The use of an oxygen dissociation, curve analyzer in transfusion therapy. Transfusion 19:107-113, 1979.
38. Fischer TH, Hays WE, Valeri CR. Poly-N-acetyl glucosamine fibers accelerate hemostasis in patients treated with antiplatelet drugs. J Trauma 71:S176-S182, 2011.
39. Fischer TH, Valeri CR, Smith CJ, Scull CM, Merricks EP, Nichols TC, Demcheva M, Vournakis JN. Non classical processes in surface hemostasis: mechanisms for the poly-N-acetyl glucosamine induced alteration of red blood cell morphology and surface prothrombogenicity. Biomed Mater 3(1):1-19, 2008.
40. Furman MI, Benoit SE, Barnard MR, Valeri CR, Borbone ML, Becker RC, Hechtman HB, Michelson AD. Increased platelet reactivity and circulating monocyte-platelet aggregates in patients with stable coronary artery disease. J Am Coll Cardiol 31:352-358, 1998.
41. Gabrio BW, Donohue DM, Finch CA. Erythrocytic preservation. V. Relationship between chemical changes and viability of stored blood, treated with adenosine. J Clin Invest 34:1509-1512, 1955.
42. Gardos G. The mechanism of ion transport in human erythrocytes. I. The role of 2,3 diphosphoglyceric acid in the regulation of potassium transport. Acta Biochem et Biophysis Acad Sc Hung 1:139-148, 1966
43. Gembarowicz R, Connolly R, Callow AD, O'Donnell TF, Keough E, Schultz M, Ramberg-Laskaris K, Melaragno AJ, Valeri CR. Effect of PGI2 on the interaction of platelets with small, caliber dacron grafts. Surg Forum. 33:466-468, 1982.
44. Gibbs SAL, Weiser MR, Kobzik L, Valeri CR, Shepro D, Hechtman HB. P-selectin mediates intestinal ischemic injury

by enhancing complement deposition. Surgery 119:652-656, 1996.

45. Goldman G, Welbourn R, Paterson IS, Klausner JM, Kobzik L, Valeri CR, Shepro D, Hechtman HR. Ischemia-induced neutrophil activation and diapedesis is lipoxygenase dependent. Surgery 107:428-433, 1990.
46. Goldman G, Welbourn R, Kobzik L, Valeri CR, Shepro D, Hechtman NB. Neutrophil adhesion receptor CD 18 mediates remote but not localized acid aspiration injury. Surgery 17:83-89, 1995.
47. Handin RI, Lawler KC, Valeri CR. Automated platelet counting. Am. J Clin Path 56:661-664, 1971.
48. Handin RI, Valeri CR. Hemostatic effectiveness of platelets stored at 22 C. NEJM 285:538-543. 1971.
49. Haugen RK. Hepatitis after the transfusion of frozen, red cells and washed red cells. N Engl J Med 301:393-395, 1979.
50. Haynes LL, Tullis JL, Pyle HM, Sproul MT. Wallach S, Turville WC. Clinical use of glycerolized frozen blood. JAMA 173:1657-1663, 1960.
51. Healy WL, Wasilewski SA, Pfeifer BA, Kurtz SR, Hallack GN, Valerio M, Valeri CR. Methylmethacrylate monomer and fat content in shed blood after total joint arthroplasty. Clin Ortho Rel Res 286:15-17, 1993.
52. Healy WL, Pfeifer BA, Kurtz SR. Johnson C, Johnson W, Johnston R, Sanders D, Karpman R, Hallack GN, Valeri CR. Evaluation of autologous shed blood fox autotransfusion after orthopaedic surgery. Clin Ortho Rel Res Number 99, pps 53-59, February 1994.
53. Hechtman HB, Huval WV, Mathieson MA, Stemp LI, Valeri CR, Shepro D. Prostaglandin and thromboxane mediation of cardiopulmonary failure. Surg Clin NA 63:263-283, 1983.
54. Horowitz B, Rywkin S, Margolis-Nunno H, Williams B, Geacintov N, Prince AM, Pascual D, Ragno G, Valeri CR, Huima-Byron T. Inactivation of viruses in red cell and platelet concentrates with aluminum phthalocyanine (AlPc) sulfonates. Blood Cells 18:141-150, 1992.
55. Huggins CE. Frozen blood: principles of practical preservation. Monogr Surg Sci 3:133-173, 1966.

56. Ichikura T, Tamakuma S, Ito H, Tomimatsu. S, Valeri CR. [Effects of syngeneic preserved blood cells on metastatic growth of the Lewis lung carcinoma.] Nippon Geka Gakkai Zasshi 92(6):734-9, 1991 (Japanese).
57. Ikeda Y, Young YH, Vournakis JN, Leifer AM. Vascular effects of poly-N-acetylglucosamine in isolated rat aortic rings. J Surg Res 102:215-220, 2002.
58. Jaskille A, Koustova E, Rhee P, Britten-Web J, Chen H, Valeri CR, Kirkpatrick JR, Alam HB. Hepatic apoptosis following hemorrhagic shock in rats can be reduced through modifications of conventional Ringer's solution. J Am Coll Surg 202(l):25-35, 2006.
59. Kelechi TJ, Mueller M, Hankin CS, et al. A randomized investigator blinded controlled pilot study to evaluate the safety and efficacy of a poly-N-acetyl glucosamine-derived membrane material in patients with venous leg ulcers. J Am Assoc Derm (doi:10.1016/j.jaad.2011.01.031).
60. Kestin AS, Valeri CR, Khuri SF, Loscalzo J, Ellis PA, MacGregor H, Birjiniuk V, Ouimet H, Pasche B, Nelson MJ, Benoit SE, Rodino LJ, Barnard MR, Michelson AD. The platelet function defect of cardiopulmonary bypass. Blood 82:107-117, 1993.
61. Khuri SF, Wolfe LA, Josa M, Axford TC, Szymanski I, Assousa S, Ragno G, Patel M, Silverman A, Park M, Valeri CR. Hematologic changes during and after cardiopulmonary bypass and their relationship to the bleeding time and nonsurgical blood loss. J Thorac Cardiovasc Surg 104:94-107, 1992.
62. Koustova E, Rhee P, Hancock T, Chen H, Inocencio R, Valeri CR, Alam HB. Ketone and pyruvate Ringer's solutions decrease pulmonary apoptosis in a rat model of severe hemorrhagic shock and resuscitation. Surgery 134(2):267-274, 2003.
63. Kyriakides C, Woodcock SA, Wang Y, Favuzza J, Austen Jr WG, Kobzik L, Moore FD, Valeri CR, Shepro D, Hechtman HB. Soluble p-selectin moderates complement-dependent reperfusion injury of ischemic, skeletal muscle. Am J Physiol Cell Physiol 279:C520-C528, 2000.
64. Lindner HB, Zhang A, Eldridge J, Demcheva M, Tsichilis P, Seth A, Vournakis J, Muise Helmericks RC. Anti-bacterial effects of poly-N-acetylglucosamine nanofibers in cutaneous wound healing: requirement for AKTI. PLoSI 6:el8996, 2011.

65. Lionetti FJ, Valeri CR, Bond JC, Kivowitz C, Weinman E. Nucleotides in frozen glycerolized erythrocytes. Transfusion 6:116-123, 1966.
66. Mackey WC, Connolly RJ, Callow AD, Keough EM, Ramberg-Laskaris K, McCullough JL, O'Donnell TF, Melaragno AJ, Valeri CR, Weiblen BJ. Aspirin decreases platelet uptake on Dacron vascular grafts in baboons. Ann Surg 200:93-99, 1984.
67. McKenney J. Valeri CR, Mohandas N, Fortier N, Giorgio A, Snyder L. Decreased in vivo survival of hydrogen peroxide-damaged baboon red blood cells. Blood 76:206-211, 1990.
68. Melaragno AJ, Carciero R, Feingold H, Talarico L, Weintraub L, Valeri CR. Cryopreservation of human platelets using 6% dimethylsulfoxide and storage at -80C. Effects of 2 years of frozen storage at – 80 C and transportation in dry ice. Vox Sang 49:245-258, 1985.
69. Merrill EW. Rheology of blood. Physiol Rev 49:863-888. 1969.
70. Merrill EW, Crowley JP, Valeri CR. Rapid and simple measurement of apparent whole blood viscosity. Lab Medica Int 10:19-24, 1993.
71. Michelson AD, MacGregor H, Barnard MR, Kestin AS, Rohrer MJ, Valeri CR. Reversible inhibition, of human platelet activation by hypothermia in vivo and in vitro. J Thrombosis and Haemostasis 5:633-640, 1994.
72. Michelson AD, Barnard MR. Hechtman HB, MacGregor H, Connolly RJ, Loscalzo J, Valeri CR. In vivo tracking of platelets: circulating degranulated platelets rapidly lose surface P-selectin but continue to circulate and function. Proc Natl Acad Sci USA 93:11877-11882, 1996.
73. Michelson AD, Barnard MR, Khuri SF, Rohrer MJ, MacGregor H, Valeri CR. The effects of aspirin and hypothermia on platelet function in vivo. Br J Haematol 104:64-68, 1999.
74. Michelson AD, Barnard MR., Krueger LA, Valeri CR, Furman I. Circulating monocyte platelet aggregates are a more sensitive marker of in vivo platelet activation than platelet/surface P selectin: Studies in baboons, human coronary intervention, and human acute myocardial infarction. Circulation 104(3):1533-1537, 2001.
75. Moore FD Jr, Warner KG, Assousa S, Valeri CR, Khuri SF. The effects of complement activation during cardiopulmonary bypass.

Attenuation by hypothermia heparin and hemodilution. Ann Surg 208:95-103, 1988.

76. Moss GS, Valeri CR, Brodine CE. Clinical experience with the use of frozen blood in combat casualties. NEJM 278:748-752, 1968.
77. Moss GS. Massive transfusion of frozen preserved red cells in combat casualties. Report of three cases. Surgery 66:1108-1113, 1969.
78. Muza SR, Sawka MN, Young AJ, Dennis RC, Gonzalez RR, Martin JW, Pandolf KB, Valeri CR. Elite special forces: Physiological description and ergogenic influence of blood reinfusion. Aviat Space Environ Med 58:1001-1004, 1987.
79. Nader RG, Garcia JC, Drushal K, Pesek T. Clinical evaluation of the SyvekPatch® in consecutive patients undergoing interventional, EPS and diagnostic cardiac catheterization procedures. J Inves Cardiol 14:305-307, 2002.
80. Najjar SF, Healey NA, Healey CM. McGarry T, Khan B, Thatte HS, Khuri SF. Evaluation of poly-N-acetyl glucosamine as a hemostatic agent in patients undergoing cardiac catheterization: a double-blind, randomized study. J Trauma 57Supp S38-S41, 2004.
81. O'Murchada ET, Horland A, Neiman RS, Valeri CR. Morphology of cells grown in the CFU-GEMM tissue culture assay from mononuclear cells obtained from peripheral blood and bone marrow of normal volunteers. Exp Hematol 16:235-239, 1988.
82. Pandolf KB, Young AJ, Sawka MN, Kenney JL, Sharp MW, Cote RR, Freund BJ, Valeri CR. Does erythrocyte infusion improve 3.2-km run performance at high altitude? Europ J Appl Physiol Occup Physiol 79:1-6, 1998.
83. Peyton BD, Rohrer MJ, Furman MI, Barnard MR, Rodino LJ, Benoit S, Hechtman HB, Valeri CR, Michelson AD. Patients with venous stasis ulceration have increased monocyte-platelet aggregation. J Vasc Surg 27:1109-1116, 1998.
84. Pietramaggiori G, Yang HJ, Scherer SS, Kaipainen A, Chan RK, Alperovich M, Newalder J, Demcheva M, Vournakis JN, Valeri CR, Hechtman HB, Orgill DP. Effects of poly-N-acetyl glucosamine (pGLcNAc) patch on wound healing in a db/db mouse. J Trauma 64(3):803-808, 2008.

85. Purmal A, Valeri CR, Dzik W, Pivacek L, Ragno G, Lazo A, Chapman J. Process for the preparation of pathogen-inactivated RBC concentrates using PEN110 chemistry: Preclinical studies. Transfusion 42:139-145. 2002.

86. Sawka MN, Dennis RC, Gonzalez RR, Young AJ, Muza SR, Martin JW, Wenger CB, Francesconi RP, Pandolf KB, Valeri CR. Influence of polycythemia on blood volume and thermoregulation during exercise-heat stress. J Appl Physiol 62:912-918, 1987.

87. Sawka MN, Gonzalez RR, Young AJ, Muza SR, Pandolf KB, Latzka WA, Dennis RC, Valeri CR. Polycythemia and hydration: effects on thermoregulation and blood volume during exercise-heat stress, Am J Physiol 255 (Regulatory Integrative Comp Physiol 24):R456-463, 1988.

88. Sawka MN, Gonzalez RR, Young AJ, Dennis RC, Valeri CR, Pandolf KB. Control of thermoregulatory sweating during exercise in the heat. Am J Physiol 257 (Regulatory Integrative Comp Physiol 26):R311-R316, 1989.

89. Sawka MN, Young AJ, Rock PB, Lyons TP, Boushel R, Freund BJ, Muza SR, Cymerman A, Dennis .RC, Pandolf KB, Valeri. CR. Altitude acclimatization and blood volume: Effects of exogenous erythrocyte volume expansion. J Appl Physiol 81:636-642, 1996.

90. Scherer SS, Pietramaggiori G, Matthews J, Perry S, Assmann A, Carothers A, Demcheva M, Muise-Helmericks RC, Seth A, Vournakis JN, Valeri CR, Fischer TH, Hechtman HB, Orgill DP. Poly-N-acetyl glucosamine (pGlcNAc) nano-fibers: a new bioactive material to enhance diabetic wound healing by cell migration and angiogenesis. Ann Surg 250:322-330, 2009.

91. Shoenfeld NA, Eldrup-Jorgensen J. Connolly R, Callow AD, Valeri CR, Ramberg K, Mackey WC, O'Donnell TF. The effect of low molecular weight dextran on platelet deposition onto prosthetic materials. J Vasc Surg 5:76-82, 1987.

92. Shoenfeld NA, Yeager A, Connolly R, Ramberg K, Forgione L, Giorgio A, Valeri CR, Callow AD. A new primate model for the study of intravenous thrombotic potential and its modification. J Vasc Surg 8:49-54, 1988.

93. Shoenfeld NA Connolly R, Ramberg K, Valeri CR, Eldrup-

Jorgensen J, Callow AD. The systemic activation of platelets by Dacron® grafts. Surg Gynec Obstet 166:454-457, 1988.

94. Shulman NR, Wattkins SP Jr, Itscoitz SB, Students AB. Evidence that the spleen retains the youngest and hemostatically most effective platelets. Trans Assoc Am Phys 81:302-313, 1968.
95. Smith AU. Prevention of haemolysis during freezing and thawing of red blood cells. Lancet 2:910-911, 1950.
96. Smith AU, Pulge C. Storage of bull spermatozoa at low temperature. Vet Rec 62:115-116, 1950.
97. Stamler JS, Jaraki O, Osborne J, Simon DI, Keaney J, Vita J, Singel D, Valeri CR, Loscalzo J. Nitric oxide circulates in mammalian plasma primarily as an S-nitroso adduct of serum albumin. Proc Natl Acad Sci USA 89:7674-7677, 1992.
98. Strumia MM, Colwell LS, Dugan A. The measure of erythropoiesis in anemias. I. The mixing time and the immediate posttransfusion disappearance of T1824 dye and of 51Cr tagged erythrocytes in relation to blood volume. Blood 8:128-145, 1958.
99. Strumia MM, Dugan A, Colwell LS. The measure of erythropoiesis in anemias. II. The immediate and subsequent loss of transfused erythrocytes in healthy subjects. Blood 19:115-122, 1962.
100. Strumia MM, Strumia PV, Duga A. Significance of measurement of plasma volume and of indirect estimation of red cell volume. Transfusion 8:1.97-209, 1968.
101. Sweetman HE, Costa JL, Vecchione JJ, Valeri CR, Shepro D. Dense bodies and total calcium in human platelets fo1lowing aspirin ingestion for a two-week period. Thrombosis Res 17:55-61, 1980.
102. Szymanski IO, Valeri CR, Almond DV, Emerson CP, Rosenfield RE. Automated differential agglutination for measurement of red cell survival. Brit J Haemat 13 (Supp:50-53. 1967.
103. Szymanski IO, Valeri CR, McCallum LE, Emerson CP, Rosenfield RE. Automated differential agglutination technic to measure red cell survival. I. Methodology. Transfusion 8:65-73, 1968.
104. Szymanski IO, Valeri CR. Automated differential agglutination technic to measure red cell survival. II. Survival in vivo of preserved red cells. Transfusion 8:74-83, 1968.
105. Szymanski IO, Valeri CR. Clinical evaluation of concentrated red cells. NEJM 280:281- 287, 1969.

106. Thatte MS, Zagarins SE, Amiji M, Khuri SF. Poly-N-acetyl glucosamine mediated red blood cell interactions. J Trauma 57Suppl:S7-S12, 2004.

107. Thatte HS, Zagarins S, Khuri SF, Fischer T. Mechanisms of poly-N-acetyl glucosamine polymer-mediated hemostasis: platelet interactions. J Trauma 57Suppl:S13-S21, 2004.

108. Thompson A, Valeri CR, Lieberthal W. Endothelin receptor A blockade alters hemodynamic response to nitric oxide inhibition in rats. Am J Physiol 269 (Heart and Circ Physiol 38):H743-H748, 1995.

109. Tullis JL, Pennel RB, Melin M, Zemp JW, Henderson ME, Sproul MT, MacNeill A. Problems of albumin medium for deglycerolized red cell resuspension. Vox Sang 8:100-101, 1963.

110. Unlap A, Ploth DW, Colvin R, Counts C, Shepp PH, Van Natta MI, Meinert CL. Does poly-N-acetyl glucosamine patch use reduce arteriovenous fistula and graft failure rates in hemodialysis patients with end state renal disease? Am J Med Sci 338(3):178-184, 2009.

111. Valeri CR, Henderson ME. Recent difficulties with frozen glycerolized blood. JAMA 188:1125-1131. 1964.

112. Valeri CR. McCallum LE. Relationship between glutathione stability and in vivo survival of autologous. deglycerolized, resuspended red blood cells. Nature 205:561-563, 1965.

113. Valeri CR, Bond JC, Fowler K, Sobucki J. Quantitation of serum hemoglobin binding capacity using cellulose acetate membrane electrophoresis. Clin Chem 11:581-588, 1965.

114. Valeri CR. Observations on recipient plasma hemoglobin concentration after transfusion with glycerolized frozen blood. Transfusion 5:36-53, 1965.

115. Valeri CR. Effect of resuspension medium on in vivo survival and supernatant hemoglobin of erythrocytes preserved with glycerol. Transfusion 5:25-35 1965.

116. Valeri CR. The in vivo survival of Coombs positive autologous erythrocytes produced by agglomeration. Transfusion 6:247-253, 1966.

117. Valeri CR, McCallum LE, Danon D. Relationships between in vivo survival and (1) density distribution, (2) osmotic fragility

of previously frozen, autologous, agglomerated, deglycerolized erythrocytes. Transfusion 6:554-564, 1966.

118. Valeri CR, Bond JC, McCallum LE. Relationships between metabolic state and (1) in vivo survival and (2) density distribution of previously frozen human erythrocytes. Transfusion 6:543-553, 1966.

119. Valeri CR, Bond JC. Observations on the preservation of autologous human erythrocytes using glycerol, slow-freeze technic and agglomeration. Transfusion 6:254-262, 1966.

120. Valeri CR, Runck AH, McCallum LE. Observations on autologous, previously frozen, deglycerolized, agglomerated, resuspended red cells. I. Effect of storage temperatures. II. Effect of adenine supplementation of glycerolized red cells prior to freezing. Transfusion 7(2): 105-116, 1967.

121. Valeri CR, Brodine CE, Moss GE. Use of frozen blood in Vietnam. Bibl Haematol 29:735-738, 1968.

122. Valeri CR, Fortier NL. Red-cell 2,3-diphosphoglycerate and creatine levels in patients with red cell mass deficits or with cardiopulmonary insufficiency. NEJM 81:1452-1455, 1969.

123. Valeri CR, Runck AH. Long term, frozen storage of human red blood cells: Studies in vivo and in vitro of autologous red blood cells preserved up to six years with high concentrations of glycerol. Transfusion 9:5-14, 1969.

124. Valeri CR, Szymanski IO, Runck AH. Therapeutic effectiveness of homologous erythrocyte transfusions following frozen storage at -80 C for up to seven years. Transfusion 10:102-112, 1970.

125. Valeri CR, Zaroulis CG. Rejuvenation and freezing of outdated stored human red cells. NEJM 287:1307-1313, 1972.

126. Valeri CR, Zaroulis CG, Marchionni LD, Patti KJ. A simple method for measuring oxygen content in blood. J Lab Clin Med 79:1035-1040, 1972.

127. Valeri CR, Cooper AG, Pivacek LE. Limitations of measuring blood volume with iodinated I 125 serum albumin. Arch lnt Med 132:534-538, 1973.

128. Valeri CR, Feingold H, Marchionni LD. A simple method for freezing human platelets using 6% dimethylsulfoxide and storage at -80 C. Blood 43:131-136, 1974.

129. Valeri CR. Hemostatic effectiveness of liquid preserved and previously frozen human platelets. NEJM 290:353-358, 1974.
130. Valeri CR, Feingold H, Marchionni LD. The relation between response to hypotonic stress and the 51Cr recovery in vivo of preserved platelets. Transfusion 14:331-337, 1974.
131. Valeri CR. Simplification of the methods for adding and removing glycerol during freeze-preservation of human red blood cells with the high or low glycerol methods: Biochemical modification prior to freezing. Transfusion 5:195-218, 1975.
132. Valeri CR. Blood Banking and the Use of Frozen Blood Products. Chemical Rubber Company Press, Boca Raton, Florida, 1976.
133. Valeri C. Circulation and hemostatic effectiveness of platelets stored at 4 C or 22 C: studies in aspirin-treated normal volunteers. Transfusion 16:20-23, 1976.
134. Valeri CR, Weisel RD, Dennis RC, Mannick JA, Berger RL, Hechtman HB. Oxygen transport function of preserved red blood cells and myocardial performance. Prog Clin Biol Res 21:597-616. 1978.
135. Valeri CR, Yarnoz M, Vecchione JJ, Dennis RC, Anastasi J, Valeri DA, Pivacek LE, Hechtman HB, Emerson CP, Berger RL. Improved oxygen delivery to the myocardium during hypothermia by perfusion with 2,3 DPG-enriched red blood cells. Ann Thoracic Surg 30:527-535, 1980.
136. Valeri CR and Altschule MD. Hypovolemic Anemia of Trauma: The Missing Blood Syndrome, Chemical Rubber Company Press, Boca Raton, Florida, 1981.
137. Valeri CR, Valeri DA, Anastasi J, Vecchione JJ, Dennis RC, Emerson CP. Freezing in the primary polyvinylchloride plastic collection bag. A new system for preparing and freezing nonrejuvenated and rejuvenated red blood cells. Transfusion 21:138-149, 1981.
138. Valeri CR. Use of rejuvenation solutions in red blood cell preservation. CRC Crit Rev Clin Lab Sci 17:299-374, 1982.
139. Valeri CR. Measurement of viable ADSOL-preserved human red blood cells. NEJM 312:377-378, 1985.
140. Valeri CR, Landrock RD, Pivacek LE, Gray AD, Fink JG, Szymanski IO. Quantitative differential agglutination method using the Coulter Counter to measure survival of compatible but identifiable red blood cells. Vox Sang 49:195-205, 1985.

141. Valeri CR. Cryopreservation of human platelets and bone marrow and peripheral blond totipotential mononuclear stem cells, Ann NY Acad Sci 459:353-366, 1985.

142. Valeri CR, Donahue K, Feingold HM, Cassidy GP, Altschule MD. Increase in plasma volume after the transfusion of washed erythrocytes. Surg Gynecol Obstet 162:30-36 1986.

143. Valeri CR, Feingold H, Cassidy G, Ragno F, Khuri S, Altschule MD. Hypothermia-induced reversible platelet dysfunction. Ann Surg 205:175-181, 1987.

144. Valeri CR, Pivacek LE, Palter M, Dennis RC, Yeston N, Emerson CP, Altschule MD. A clinical experience with ADSOL® preserved erythrocytes. Surg Gynec Obstet 166:33-46, 1988.

145. Valeri CR, Pivacek LE, Gray AD, Cassidy GP, Leavy M, Dennis RC, Melaragno AJ, Niehoff J, Yeston N, Emerson CP, Altschule MD. The safety and therapeutic effectiveness of human red cells stored at -80 C for as long as 21 years. Transfusion 29:429-437, 1989.

146. Valeri CR. Blood and bone marrow products in the treatment of radiation injury. In Treatment of Radiation Injuries, Ed. D. Browne et al, Plenum Press, New York, NY, 1990, pps 19-27.

147. Valeri CR, Khabbaz K, Khuri SF, Marquardt C, Ragno G, Feingold H, Gray AD, Axford T. Effect of skin temperature on platelet function in patients undergoing extracorporeal bypass. J Thorac Cardiovasc Surgery 104:108-116, 1992.

148. Valeri CR, MacGregor H, Cassidy G, Tinney R. Pompei F. Effects of temperature on bleeding time and clotting time in normal male and female volunteers. Crit Care Med 23:693-704, 1995.

149. Valeri CR, Ragno G, MacGregor H, Pivacek LE. The effect of disinfection on viability and function of baboon red blood cells. Photochem Photobiol 65:446-450, 1997.

150. Valeri CR, Crowley JP, Loscalzo. The red cell transfusion trigger: has a sin of commission now become a sin of omission? Transfusion 38:602-610, 1998.

151. Valeri CR, Pivacek LE, Cassidy GP, Ragno G. The survival, function, and hemolysis of human RBCs stored at 4 C in additive solution (AS-1, AS-3, or .AS-5) for 42 days and then biochemically modified, frozen, thawed, washed, and stored at 4 C in sodium chloride and glucose solution for 24 hours. Transfusion 40:1341-1345, 2000.

152. Valeri CR, Dennis RC, Ragno G, Pivacek LE, Hechtman HB, Khuri SF. Survival, function, and hemolysis of shed red blood cells processed as nonwashed blood and washed red blood cells. Ann Thoracic Surg 72:1598-1602, 2001.

153. Valeri, CR, Ragno G, Pivacek LE, Dennis RC, Hechtman HB, Khuri SF. Survival and function of baboon RBCs released from clotted blood and washed before autologous transfusion. Transfusion 41:1384-1389, 2001.

154. Valeri CR, Ragno G, Pivacek LE, O'Neill EM. In. vivo survival of apheresis RBCs, frozen with 40-percent (wt/vol) glycerol, deglycerolized in the ACP 215, and stored at 4 C in AS-3 for up to 21 days. Transfusion 41:928-932, 2001.

155. Valeri CR, Ragno G, Pivacek LE, Srey R, Hess JR, Lippert LE, Mettille F, Fahie R, O'Neill EM, Szymanski IO. A multicenter study of in vitro and in vivo values in human RBCs frozen with 40-percent (wt/vol) glycerol and stored after deglycerolization for 15 days at 4 C in AS-3: assessment of RBC processing in the ACP 215. Transfusion 41:933-939, 2001.

156. Valeri CR, Pivacek LE, Cassidy GP, Ragno G. 24-hour 51Cr post-transfusion survival, 51Cr life span and haemolysis of red blood cells stored at 4 C for 56 days in AS-3. Vox Sang 80:48-50, 2001.

157. Valeri CR, Cassidy G, Pivacek LE, Ragno G, Lieberthal W, Crowley JP, Khuri SF, Loscalzo J. Anemia-induced increase in the bleeding time: implications for treatment of nonsurgical blood loss. Transfusion 41:977-983, 2001.

158. Valeri CR, Pivacek LE, Cassidy GP, Ragno G. Volume of RBCs, 24- and 48-hour posttransfusion survivals, and the lifespan of (51)Cr and biotin-X-N-hydroxysuccinimide (NHS)-labeled autologous baboons RBCs: effect of the anticoagulant and blood pH on (51)Cr and biotin-X-NHS elution in vivo. Transfusion 42:343-8, 2002.

159. Valeri CR, MacGregor H, Giorgio A, Ragno G. Circulation and hemostatic function of autologous fresh, liquid preserved, and cryopreserved baboon platelets transfused to correct an aspirin-induced thrombocytopathy. Transfusion 42:1206-1216, 2002.

160. Valeri CR, MacGregor H, Giorgio A, Srey R, Ragno G. Comparison of radioisotope methods and a non-radioisotope

method to measure the RBC volume and RBC survival in the baboon. Transfusion 43(10):1366-1373, 2003.

161. Valeri CR, Ragno G, Srey R. Restoration of red blood cell volume following 2-unit red blood cell apheresis. Vox Sang 85(2):85-87, 2003.
162. Valeri CR, Srey R, Tilahun D, Ragno G. In vitro quality of red blood cells frozen with 40% w/v glycerol at -80 C for 14 years, deglycerolized with the Haemonetics ACP 215, and stored at 4C in additive solution-I or additive solution-3 for up to 3-weeks. Transfusion 44:990-995, 2004.
163. Valeri CR, Srey R, Tilahun D, Ragno G. In vitro effects of polymerized N-acetyl glucosamine (NAG) on the activation of platelets in platelet rich plasma with and without of red blood cells. J Trauma 57 (suppl):S22-S25, 2004.
164. Valeri CR, MacGregor H, Ragno G. Correlation between in vitro aggregation and thromboxane A2 production in fresh, liquid-preserved, and cryopreserved human platelets: Effects of agonists, pH, and plasma and saline resuspension. Transfusion 45:596-603, 2005.
165. Valeri CR, MacGregor H, Giorgio A, Ragno G: Comparison of radioisotope methods and a non-radioisotope method to measure platelet survival in the baboon. Trans Aph Sci 32(3):275-281, 2005.
166. Valeri CR, Ragno G, van Houten P, Rose L, Rose M, Egozy Y, Popovsky M. Automation of the glycerolization of the RBC using the high separation bowl in the Haemonetics ACP2I 5 instrument. Transfusion 45(10):162l-I627, 2005.
167. Valeri CR, Ragno G, Khuri S. Freezing human platelets using 6% DMSO with removal of the supernatant solution prior to freezing and storage at -80 C without post-thaw processing. Transfusion 45(1):1890-1898, 2005.
168. Valeri CR. Ragno G. The effect of storage of fresh frozen plasma at -80 C for as long as 14 years on plasma clotting proteins. Transfusion 45(11):1829-1830, 2005.
169. Valeri CR, Ragno G, Veech RL. Effects of the resuscitation fluid and the HBOC excipient on the toxicity of the HBOC: Ringer's D, L-Lactate, Ringer's L-Lactate, and Ringer's ketone solutions. Art Cells, Blood Subst and Biotech 34(6) 601-606, 2006.

170. Valeri CR, Ragno G. The survival and function of baboon red blood cells, platelets, and plasma proteins: A review of the experience from 1972 to 2002 at the Naval Blood Research Laboratory, Boston, Massachusetts. Transfusion 46(8): 1-42, 2006.

171. Valeri. CR, Dennis RC, Ragno G, MacGregor H, Menzoian JO, Khuri SF. Limitations of the hematocrit to assess the need for RBC transfusion in hypovolemic anemic patients. Transfusion 46:365-371, 2006.

172. Valeri CR, Ragno G. Cryopreservation of human blood products. Trans Aph Sci 34:271-287, with an editorial on pages 267-269, 2006.

173. Valeri CR, MacGregor H, Ragno G, Healey N, Fonger J, Khuri SF. Effects of centrifugal and roller pumps on the survival of autologous red blood cells in cardiopulmonary bypass surgery. Perfusion 21:291-296, 2006.

174. Valeri CR, Ragno G. In vitro testing of platelets using the thromboelastogram, platelet function analyzer, and the clot signature analyzer to predict the bleeding time. Trans Aph Sci 35(1):33-41, 2006.

175. Valeri CR, Morse DS, Ragno G, Dennis RC. Hemostatic defect in baboons infused non- treated and treated autologous plasma. J Card Surg 21(6):565-571, 2006.

176. Valeri CR. "Making sense of the preclinical literature on advanced hemostatic products". Letter to the Editor. J Trauma 61:240-241, 2006.

177. Valeri CR, Khuri, S, Ragno G. Non-surgical bleeding diathesis in anemic thrombocytopenic patients: Role of temperature, RBC, platelets, and plasma clotting proteins. Transfusion 47:206S-248S, 2007.

178. Valeri. CR, Ragno G. The effects of preserved red blood cells on the severe adverse events observed in patients infused with hemoglobin based oxygen carriers. Art Cells Blood Subst Biotech 36 (1):3-18, 2008.

179. Valeri CR, Ragno G, Veech RL. Severe adverse events associated with hemoglobin based oxygen carriers: role of resuscitative fluids and liquid preserved RBC. Trans Aph Sci 39:205-211, 2008.

180. Valeri CR, Ragno G. An approach to prevent the severe adverse

events associated with transfusion of FDA-approved blood products. Trans Aph Sci 42:223- 233, 2010.

181. Valtis DJ, Kennedy AC. Defective gas transport function of stored red blood cells. Lancet 1:119-124, 1954.
182. Van Slyke DD, Phillips RA, Dole VP, Hamilton PB, Archibald RM, Plazin J. Calculation of hemoglobin from blood specific gravities. J Biol Chem 183:349-360, 1950.
183. Veech RL, Rogeness GA, Weil-Malherbe H. Formation of protoporphyrin from hemoglobin in vitro. Biochem J 105:1209-1217, 1967.
184. Vogel WM, Dennis RC, Cassidy GP, Apstein CS, Valeri CR. Coronary constrictor effect of stoma-free hemoglobin solutions. Am J Physiol (Heart Circ Physiol 20) 251:H413-H420, 1986.
185. Vogel. WM, Hsia JC, Briggs LL, Er SS, Cassidy G, Apstein CS, Valeri CR. Reduced coronary vasoconstrictor activity of hemoglobin solutions purified by ATP-agarose affinity chromatography. Life Sciences 41:89-93, 1987.
186. Vournakis JN, Demcheva M, Whitson A, Guirea R, Pariser ER. Isolation, purification, and characterization of poly-N-acetyl glucosamine use as a hemostatic agent. J Trauma 57(suppl) S2-S6, 2004.
187. Weinbaum S, Tarhell JM, Damiano ER. The structure and function of the endothelial glycocalyx layer. Ann Rev Biomed Eng 9:121-167, 2007.
188. Weiser MR, Gibbs SAL, Kobzik L, Valeri CR, Shepro D, Hechtman HB. P-selectin mediates local reperfusion injury after lower torso ischemia. Surgical Forum 45:389-394, 1994.
189. Weiser MR, Gibbs SAL, Valeri CR, Shepro D, Hechtman HB. Anti-selectin therapy modifies skeletal muscle ischemia and reperfusion injury. Shock 5:402-407, 1996.
190. Welbourn R, Goldman G, Kobzik L, Valeri CR, Shepro D, Hechtman HB. Involvement of thromboxane and neutrophils in multiple-system organ edema with interleukin-2. Ann Surg 212:728-733. 1990.
191. Woodcock SA, Kyriakides C, Wang Y, Austen WG Jr, Moore ED Jr, Valeri CR, Hartwell D, Hechtman HB. Soluble p-selectin moderates complement dependent injury. Shock 14:610-615, 2000.

192. Wright JG, Kerr JC, Valeri CR, Hobson RW II. Endothelial permeability to iodine-125- labeled albumin predicts skeletal muscle injury after ischemia reperfusion. Current Surgery 45:25-27, 1988.

193. Wright JG, Kerr JC, Valeri CR, Hobson RW II. Heparin decreases ischemia-reperfusion injury in isolated canine gracilis model. Arch Surg 123:470-472, 1988.

194. Wright JG, Kerr JC, Valeri CR, Hobson RW II. Regional hypothermia protects against ischemia-reperfusion injury insulated canine gracilis muscle. J Trauma 28:1026-1031, 1988.

195. Wright JG, Fox D, Kerr JC, Valeri CR, Hobson RW II. Rate of reperfusion blood modulates reperfusion injury in skeletal muscle. J Surg Res 44:754-763, 1988.

196. Young AJ, Sawka MN, Muza SR, Boushel R, Lyons T, Rock PB, Freund BJ, Waters R, Cymerman A, Pandolf KB, Valeri CR. Effects of erythrocyte infusion on VO2max at high altitude. J Appl Physiol 81:252-259, 1996

CHAPTER 5

DECADES AND EVENTS: WHAT OCCURRED AND WHAT WAS LEARNED. PART IV

Studies have been performed at the Naval Blood Research Laboratory to assess survival and function of fresh and preserved RBC and platelets. Studies have been performed to assess the function of fresh and preserved RBC to transport oxygen to tissues, exert a hemostatic effect to prevent nonsurgical blood loss, and to regulate the plasma volume. (Valeri CR, NEJM 284:81-88, 1971; Valeri CR and Zaroulis CG, NEJM 287:1307-1313, 1972; Valeri CR et al, Surg Gynec Obstet 162:30-36, 1986; Valeri CR et al, Transfusion 41:977-983, 2001; Khuri SF et al, J Thorac Cardiovasc Surg 117:172-184, 1999; Valeri CR et al, Transfusion 42:1206-1212, 2002; Valeri CR, J Med 5:278-291, 1974; Valeri CR, NEJM 290:353-358, 1974; Valeri CR et al, Transfusion 47:206S-248S, 2007; Valeri CR, Ragno G, Trans Aph Sci 42:223-233, 2010; Valeri CR et al, Transfusion 38:602-610, 1998).

Studies were performed to quantitate the percent of nonviable preserved RBC using a double radioisotope procedure in normal volunteers, in patients, and in baboons (Szymanski IO, Valeri CR, Transfusion 8:74-83, 1968; Szymanski IO, Valeri CR, Vox Sang 15:287-292, 1968; Valeri CR et al Transfusion 24:105-108, 1984; Valeri CR et al, Surg Gynec Obstet 166:33-46, 1988; Valeri CR , Ragno G, Transfusion 46(8):1S-42S, 2006).

The double radioisotope procedure is needed to quantitate the percentage of irreversibly damaged RBC removed during the mixing of the preserved RBC within the blood volume. Comparable results are observed using the single and double labeling procedures when the irreversibly damaged preserved RBC are not removed during the mixing of the preserved RBC within the blood volume of the recipient. The 24-hour posttransfusion survival of preserved autologous RBC in normal volunteers was measured by 51Cr labeling of the preserved RBC and the RBC volume of the healthy volunteer was estimated from the 1251-albumin plasma volume and the total body hematocrit (peripheral venous hematocrit multiplied, by 0.89). In patients the survival of allogeneic preserved, compatible identifiable RBC was measured using two (2) differential agglutination procedures (Szymanski IO and Valeri CR, Transfusion 8:74-83, 1968; Valeri CR et al, Vox Sang 49:195-205, 1985). The Technicon autoanalyzer and the Coulter counter were used to measure the compatible but identifiable allogeneic preserved RBC and the RB C volume of the recipient was measured using 51 Cr labeled autologous RBC to measure the 24-hour posttransfusion survival value and the lifespan of the preserved allogeneic RBC. In vitro testing of RBC morphology, RBC filterability, and RBC deformability did not correlate to the 24-hour posttransfusion survival value which quantitates the irreversibly damaged RBC that are removed at an accelerated rate during the 24-hour posttransfusion period. In vitro testing of RBC ATP did not correlate to the 24-hour posttransfusion survival value. The oxygen transport function of preserved RBC can be assessed by the measurement of RBC ATP, DPG and. P50 values.

The vasoconstrictor activity of hemoglobin based oxygen carriers was reported in studies performed at the Naval Blood Research Laboratory in collaboration with Dr. Carl Apstein and Dr. Mark Vogel in the perfusion of isolated rabbit hearts with hemoglobin based oxygen carriers (HBOCs) compared to the human red blood cells with increased and decreased affinity for oxygen (Apstein CS et al, Am J Physiol 248:H508-H515, 1985; Vogel WM et al, 251:H413-H420, 1986). Human RBCs with increased levels of RBC DPG and decreased affinity for oxygen were observed in servicemen wounded in Vietnam who were hypovolemic and anemic (Valeri CR and Fortier NL,NEJM 81:1452-1455, 1969; Fortier NL et al, Forsvarsmedicin 5:250-257, 1969). The reduction in the number of RBC in these chronic hypovolemic anemic patients was compensated for by the increase in the DPG level of RBCs in the circulation of these patients. These clinical observations on wounded servicemen demonstrated that the compensatory response to reduction in red blood cell volume was to increase the RBC DPG in the circulating RBC and to facilitate oxygen delivery to tissues in patients with a reduced volume of RBC. RBC were treated with solutions containing pyruvate, inosine, phosphate and adenine to increase the RBC ATP and DPG to levels observed in these hypovolemic anemic patients prior to freezing with glycerol, an intracellular cryoprotectant. The washing of the glycerolized RBC to reduce the concentration of glycerol also reduced the levels of substrates used to biochemically modify the RBC prior to freezing.

In the study of wounded servicemen during the Vietnam War who were hospitalized at the Chelsea Naval Hospital, the limitation of the peripheral venous hematocrit to detect patients with significant reductions in red blood cell volume was documented (Valeri CR et al, Arch Int Med 132:534-538, 1973; Valeri CR and Altschule MD, Hypovolemic Anemia of Trauma: The Missing Blood Syndrome, CRC Press, Boca Raton, FL, 1981).

The studies performed at the Chelsea Naval. Hospital from 1968 to 1974 demonstrated that the peripheral venous hematocrit did not estimate the red blood cell volume deficit and the hypovolemic state in the wounded servicemen that were injured in Vietnam. 51Cr labeled autologous RBC were used to measure the RBC volume in these injured servicemen (Biron PE et al, J Bone Joint Surg 54-A:1001-1014, 1972). The plasma volume was not accurately measured using 1251-albumin with a molecular weight of 68,000 whereas the plasma volume was accurately measured using 131I-labeled cold agglutinin with a molecular weight of 1,000,000 in these patients (Valeri CR, Altschule MD, Hypovolemic Anemia of Trauma: The Missing Blood Syndrome, CRC Press, Boca Raton, FL, 1981). The reduction in the 51Cr labeled RBC volume and the reduction in the plasma volume using ^{125}1 labeled cold agglutinin documented the chronic hypovolemic anemic state of these wounded servicemen with normal peripheral venous hematocrit values with "stress anemia". The reduction in both the red blood cell volume and the plasma volume documented the chronic hypovolemic. anemic state in which the peripheral venous hematocrit did not detect the red blood cell volume deficit and the hypovolemic state of these patients with traumatic injuries.

The rapid loss of the 125I-albumin with a molecular weight of 68,000 into the extravascular volume was similar to the rapid loss of the hemoglobin based oxygen carrier with molecular weights of 32,000 and 64,000 into the extravascular volume (Valeri CR et al, Arch Int Med 132:524-538, 1973).

The rapid loss of the non-modified stroma-free hemoglobin into the extravascular volume was associated with vasoconstriction which was due to either oxidation of nitric oxide by oxygen free radicals released from the stroma free hemoglobin or the binding of nitric oxide by hemoglobin. Polymerization of the stroma-free hemoglobin with glutaraldehyde reduced the vasoconstrictor activity observed in the original NBRL studies performed in the early 1980's in isolated perfused rabbit hearts. (Vogel WM et al, Am J Physiol (Heart Circ Physiol 20) 251:H413-H420, 1986; Vogel WM et al, Life Sciences 41:89-93, 1987).

The major progress in the development of safe and therapeutically effective hemoglobin based oxygen carriers has been the reduction in vasoconstrictor activity associated with the elimination of 32,000 and 64,000 molecular weight molecules from the hemoglobin based oxygen carriers.

Studies performed at the Chelsea Naval Hospital on wounded servicemen from Vietnam revealed several important clinical findings reported in the book Hypovolemic Anemia of Trauma: The Missing Blood Syndrome by Valeri CR, Altschule MD published by CRC Press, Boca Raton, Florida 1981.

a. The limitation of the peripheral venous hematocrit to detect significant reductions in RBC volume to assess the need for RBC transfusions;
b. The increase in RBC DPG and decrease in RBC affinity for oxygen to compensate for the reduction in RBC volume in the wounded servicemen;
c. The need to measure the RBC volume with. autologous RBC labeled with 51Cr to determine the need for RBC transfusions;
d. Limitations of 1251 albumin to measure the plasma volume in patients;
e. Limitation of human albumin as a blood substitute to increase the plasma volume because the rapid distribution of albumin into the extravascular volume of patients;
f. The maintenance of the central blood volume and the reduction in the peripheral blood volume in hypovolemic anemic patients;
g. The effect of RBC transfusions to increase the plasma volume in hypovolemic-anemic patients;
h. The degradation of RBC hemoglobin to produce porphyrin and heme without production of carbon monoxide, bilirubin, and urobilinogen;

i. Increase in RBC creatine related to release of muscle creatine from hypoperfused extremities in the hypovolemic patients;

j. Restoration of the peripheral blood volume and peripheral red blood cell volume reduced the edema in the wounds of the extremities which was associated with the increase in the plasma volume following the transfusion of washed allogeneic viable identifiable red blood cells;

k. The debridement of the wounds of the extremities was not associated with increased nonsurgical blood loss. The normal hematocrit values in these wounded servicemen with chronic hypovolemia was associated with the reduced nonsurgical blood loss in the wounds of the extremities;

l. Increased hemolysis of blood treated with adrenochrome in patients with traumatic injuries (Valeri CR Altschule MD, J Trauma 13:678-686, 1973).

The wounded servicemen with normal hematocrit values and normal measurements of arterial blood gases, arterial blood pressure, heart rate, cardiac output, central venous pressure and renal function developed hypotension following general anesthesia for debridement of wounds of the extremities. Investigations in these patients showed that 125I albumin used to measure the plasma volume and the total body hematocrit overestimated the plasma volume, the blood volume, and the red blood cell volume. The 125I albumin with a molecular weight of 68,000 distributed rapidly into the extravascular volume and overestimated the plasma volume, the blood volume, and the red blood cell volume. The plasma volume in these wounded servicemen was accurately measured using ^{131}I or ^{125}I labeled cold agglutinin with a molecular weight of 1,000,000 which was distributed within the plasma volume following infusion (Valeri CR, Altschule MD, Hypovolemic Anemia of Trauma: The Missing Blood Syndrome, CRC Press, Boca Raton, FL 1981).

In the wounded servicemen, the peripheral venous hematocrit did not accurately estimate the red blood cell volume deficit or the blood volume. Studies performed at the NBRL include the following:

a. Radioisotope 51Cr, 99mTc, and 111ln-oxinc labeling were used to measure RBC volume and RBC lifespan in healthy baboons (Valeri CR et al, Transfusion 42:343-348, 2002; Valeri CR et al, Transfusion 43:1366-1373, 2003).

b. Non-radioisotope biotin-X-NHS labeling of RBC was used to measure RBC volume and RBC lifespan in healthy baboons (Valeri CR et al, Transfusion 42:343-348, 2002; Valeri CR et al, Transfusion 43:1366-1373, 2003).

c. In vitro and in vivo studies were done to assess 51Cr elution from fresh and preserved baboon and human RBC. Fresh and preserved autologous baboon RBC were labeled with 51Cr and biotin X-NHS. The 24-hour posttransfusion survival and lifespan of compatible but identifiable human RBC labeled with 51Cr were measured by the automated differential agglutination procedure in patients. These procedures in baboons and in patients permitted the assessment of in vitro and in vivo elution of 51Cr from fresh and preserved RBC (Valeri CR, Transfusion 8:210-219, 1968; Szymanski IO, Valeri CR, Transfusion 10:287-298, 1970; Syzmanski IO, Valeri CR, Brit J Haematol 19:397-409, 1970; Szymanski IO et al, Transfusion 13:13-18, 1973; Valeri CR, Szymanski IO, Vox Sang 24:502-514, 1973; Valeri CR et al, Transfusion 42:343-348, 2002; Valeri CR et al, Transfusion 43:1366-1373, 2003).

d. Radioisotope 51Cr and 111 In-oxine labeling to measure the in vivo recovery 1 to 2 hours after infusion, and lifespan of autologous baboon platelets (Valeri CR et al, Trans Aph Sci 32:275-281, 2005).

e. Non-radioisotope biotin-X-NHS labeling of autologous platelets to measure the in vivo recovery 1 to 2 hours after infusion and lifespan in baboons (Valeri CR et al, Trans Aph Sci 32:275-281, 2005).

f. Functional closed automated method to glycerolize and deglycerolize non-rejuvenated human RBC for storage at 4 C in AS-3 solution for 3 weeks using the Haemonetics Blood Processor ACP215 (Valeri CR et al, Transfusion 41:933-939, 2001; Valeri CR et al, Transfusion 41:928-932, 2001).

g. Storage of autologous RBC in AS-1 or AS-3 at 4 C for 35, 42, 49 and 56 days in healthy volunteers and in patients (Valeri CR et al, Surg Gynec Obstet 166:33-46, 1988; Valeri CR et al, Vox Sang 80:48-50, 2001).

h. Freezing of nonrejuvenated RBC with 40% W/V glycerol and storage at -8OC for 21 years and for up to 37 years (Valeri CR et al, Transfusion 29:429-437, 1989; Valeri CR et al, Vox Sang 79:168-174, 2000).

Survival and function of preserved platelets

1. In vivo recovery 1-2 hours after transfusion, and lifespan of autologous platelets labeled with 51Cr and 111 In-oxine in healthy baboons (Valeri CR et al, Vox Sang 83:347-351, 2002; Valeri CR et al, Trans Aph Sci 32:275-281, 2005).

2. In vivo function of autologous fresh, liquid preserved, and cryopreserved platelets to correct an aspirin-induced thrombocytopathy in normal volunteers and baboons (Handin RI, Valeri CR, N Engl J Med 285:538-543, 1971; Valeri CR, N Engl J Med 290:353-358, 1974; Valeri CR et al, Transfusion 42:1206-1216, 2002).

3. Studies were done to assess the 111 In-oxine circulation and distribution of fresh, liquid preserved, and cryopreserved autologous baboon platelets (Valeri CR et al, Vox Sang 83:347-351, 2002).

4. In patients cryopreserved platelets were more functional than liquid preserved platelets to reduce nonsurgical blood loss and to reduce the need for allogeneic RBC and fresh frozen plasma in patients subjected to cardiopulmonary bypass surgery (Khuri SF et al, J Thorac Cardiovasc Surg 117:172-184, 1999).

5. In baboons autologous platelets stored in the liquid state at 22 C for 48 hours and cryopreserved platelets reduced the bleeding time prolonged by aspirin treatment 18 hours prior to the autotransfusions and increased the thromboxane level at the bleeding time site (Valeri CR et al, Transfusion 42:1206-1216, 2002). Autologous baboon platelets stored at 22 C for 3 days and 5 days did not reduce the bleeding time. The in vivo recovery 1 to 2 hours following transfusion of 111In-labeled autologous cryopreserved platelets was higher than that of autologous baboon platelets stored at 22 C for 5 days (Valeri CR et al, Transfusion 42:1206-1216, 2002).

6. In vitro testing showed that thromboxane A2 production by fresh, liquid preserved, and cryopreserved human platelets following stimulation with a combination of arachidonic acid (AA) and adenosine diphosphate (ADP) correlated to the in vivo function of platelets to produce thromboxane A2 at the bleeding time site (Valeri CR et al, Transfusion 45:596-603, 2005).

7. Platelet thromboxane A2 production by fresh, liquid preserved, and cryopreserved human platelets following treatment with the agonists arachidonic acid (AA) and adenosine diphopshate (ADP) was a better in vitro test of platelets function than platelet aggregation response to the

agonists AA and ADP in vitro (Valeri CR et al, Transfusion 45:596-603, 2005).

8. Platelet morphology did not correlate to the in vivo recovery 1 to 2 hours after infusion and the lifespan of autologous fresh and liquid preserved baboon platelets treated with thrombopoietin or with cytochalasin B and EGTA-AM (Valeri CR et al, Transfusion 44:865-870, 2004).

9. In the baboon platelet p selectin did not correlate to the in vivo survival and function of autologous fresh baboon platelets treated with thrombin. Autologous p selectin positive baboon platelets had normal in vivo recovery and normal lifespan and normal function following infusion (Michelson AD et al, Proc Natl Acad Sci USA 93:11877-11882, 1996).

10. In the baboon autologous platelet GP1b content correlated to the 1 to 2 hour recovery and lifespan following infusion of cryopreserved platelets (Barnard MR et al, Transfusion 39:880-888, 1999).

11. In vitro thromboxane A2 production by human platelets following stimulation with arachidonic acid (AA) and adenosine diphosphate (ADP) and thromboxane A2 level at the bleeding time site correlated to the in vivo function of autologous fresh human RBC and autologous preserved baboon platelets to reduce the bleeding time (Valeri CR et al, Transfusion 41:977-983, 2001; Valeri CR et al, Transfusion 42:1206-1216, 2002; Valeri CR et al, Transfusion 45:596-603, 2005).

12. Hemostatic agents containing poly N-acetylglucosamine and local pressure reduced non-surgical blood loss by local vasoconstriction and activation of RBC, platelets, and clotting proteins to generate thrombin and convert fibrinogen to fibrin at the bleeding time site (Valeri CR et al, J Trauma (suppl) 57:S22-S25, 2004; Fischer TH et al, Biomed Mater 3:1-9, 2008).

13. Human RBC at the bleeding time site activate platelets to produce thromboxane A2 and to reduce the bleeding time (Valeri CR et al, Crit Care Med 23:698-704, 1995).

14. RBC reduce the bleeding time and increase the thromboxane level at the template bleeding time site (Valeri CR et al, Crit Care Med 23:698-704, 1995; Valeri CR et al, Transfusion 41:977-983, 2001).

15. Anemia produces a reversible platelet dysfunction (Valeri CR et al, Transfusion 41:977-983, 2001).

16. Effects of hypothermia, heparin, and anemia to increase the bleeding time and reduce the level of thromboxane A2 at the bleeding time site (Valeri CR et al, Ann Surg 205:175-181, 1987; Valeri CR et al,

Crit Care Med 23:698-704, 1995; Valeri CR et al, Transfusion 41:977-983, 2001).

17. Bleeding time and hematocrit two (2) hours after cardiopulmonary bypass surgery correlated to the nonsurgical blood loss and the need for allogeneic RBC and fresh frozen plasma (Khuri SF et al, J Thorac Cardiovasc Surg 104:94-107, 1992).

18. Hematocrit of 35 V% correlated with reduced nonsurgical blood loss and reduced need for allogeneic RBC and fresh frozen plasma in patients following cardiopulmonary bypass surgery (Khuri SF et al, J Thorac Cardiovasc Surg 104:94-107, 1992).

19. The hematocrit value had a greater effect on the bleeding time and the volume of blood collected at the bleeding time site than did the platelet count (Crowley JP et al, Am J Clin Pathol 108:579-584, 1997; Valeri CR et al, Transfusion 41:977-983, 2001).

20. Transfusion trigger for platelet transfusions should depend upon the hematocrit value, the platelet count, platelet size and platelet hemostatic function and not only on the platelet count (Valeri CR et al, Transfusion 43:1761-1762, 2003).

21. Anemic thrombocytopenic patients should be transfused with RBC to increase the hematocrit to 35 V% to restore platelet hemostatic function before platelet transfusion should be administered to increase the platelet count to an arbitrary platelet count (Valeri CR et al, Transfusion 43:1761-1762, 2003).

Safety and Therapeutic Effectiveness of Resuscitation Solutions

1. The toxicity of Ringer's DL lactate to produce lung injury and acute respiratory distress syndrome (ARDS) in Vietnam was reported by RL Veech in the article on the toxic impact of parenteral solutions on the metabolism of cells: a hypothesis for physiological therapy. Am J Clin Nutri 44:519-551, 1986.

2. Studies in 1999 by Alam HB and associates demonstrated that Ringer's DL lactate produced pulmonary injury in rats subjected to hemorrhagic shock that could explain the acute respiratory distress syndrome (ARDS) observed in Vietnam associated with the infusion of large volume of Ringer's DL lactate crystalloid solution (Koustova E et al, Surgery 134(2):267-274, 2003).

3. The safety and therapeutic effectiveness of Ringer's ketone solution

to resuscitate rodents subjected to hemorrhagic shock, (Valeri CR et al, Trans Aph Sci 39:205-211, 2008).

4. The safety and therapeutic effectiveness of Ringer's ketone solution to treat rodents subjected to hypovolemic anemia and renal ischemia. (Valeri CR et al, Trans Aph Sci 39:205-211, 2008).

5. The therapeutic effects of Ringer's ketone solution to improve both survival and myocardial function in rodents subjected to hemorrhagic shock compared to Ringer's D,L lactate solution (Valeri CR et al, Trans Aph Sci 39:205-211, 2008).

6. The effect of Ringer's ketone solution to improve renal function 48 hours after treatment of hypovolemic anemic rodents subjected to renal ischemia compared to rodents treated with Ringer's DL lactate (Valeri CR et al, Trans Aph Sci 39:205-211, 2008).

7. The stability of filter sterilized large volume Ringer's ketone solution containing 5mM glucose stored at room temperature for 2 years. Ringer's pyruvate solution is unstable prepared using sodium pyruvate or ethyl pyruvate. Heal sterilization of Ringer's ketone solution without glucose produces a stable solution that can be stored at room temperature for 2 years. (Valeri CR et al, Trans Aph Sci 39:205-211, 2008).

<u>Cryopreservation of blood products</u>

1.Publications have reported. on satisfactory storage of RBC frozen with 40% W/V glycerol and storage at -80 C for 37 years, an automated procedure to glycerolize and deglycerolize human RBC for storage at 4 C in AS-3 for 2 weeks using the Haemonetics 215 instrument, human RBC frozen as non-rejuvenated RBC, indated rejuvenated, and outdated rejuvenated RBC, processed in the Haemonetics Blood Processor 215 and stored in AS-3 at 4 C for at least 24 hours with acceptable in vitro and in vivo quality (Valeri CR et al, Transfusion 41:933-939, 2001; Valeri CR et al, Transfusion 41:928-932, 2001; Valeri CR et al, Transfusion 45:1621-1627, 2005; Valeri CR, Ragno G, Trans Aph Sci 34;271-287, 2006; Ragno G, Valeri CR, Trans Aph Sci 35:137-143, 2006).

2. Human platelets can be frozen with 6% DMSO and stored at -80 C for at least 2 years with freeze-thaw-wash recovery values of 70%, in vivo recovery I to 2 hours after transfusion of 35 to 40 V% and lifespan of 7 days (Melaragno AJ et al, Vox Sang 49:245-258, 1985).

3. The frozen, thawed, washed platelets stored at 22 C for as long as 5

hours without agitation have been shown to reduce nonsurgical blood loss and reduced the requirements for allogeneic RBC and fresh frozen plasma compared to liquid preserved platelets stored at 22 C with agitation for a mean of 3.4 days in patients following cardiopulmonary bypass surgery (Khuri SF et al, J Thorac Cardiovasc Surg 117:172-184, 1999).

4. The procedure to freeze, thaw, wash and resuspend the platelets takes 1 1/2 hours. The washing of the platelets frozen with 6% DMSO reduces the DMSO level to 400 mg in the unit. Single donor platelets containing 4 x 10^{11} platelets following the freeze- thaw-wash procedure were resuspended in plasma and stored at 22 C without agitation for 5 hours prior to transfusion (Khuri SF et al, J Thorac Cardiovasc Surg 117:172-184, 1999).

5. Since 1972, the safety and therapeutic effectiveness of platelets frozen with 6% DMSO, stored at -80 C, thawed, washed and resuspended in plasma were studied in:

a. Normal volunteers autotransfused with 51Cr platelets had in vitro freeze-thaw-wash recovery of 70%, 1 to 2 hour in vivo recovery value of 45% and the lifespan of 7 days (Valeri CR et al, Blood 43:131-136, 1974; Valeri CR et al, Transfusion 14:331-337, 1974; Valeri CR, NEJM 290:353-358, 1974).

b. Allogeneic frozen, washed platelets resuspended in plasma were transfused to stable thrombocytopenic patients and the in vivo recovery one to two hours following transfusion was 50% of the value for the fresh platelets. Allogeneic fresh platelets from the same donors who provided the frozen, washed, platelets were transfused into the same stable thrombocytopenic patients. No untoward effects were observed in patients who received the fresh and the frozen and washed platelets resuspended in plasma (Melaragno AJ et al, Vox Sang 49:245-258, 1985).

c. Allogeneic frozen washed platelets resuspended in plasma were transfused into patients following cardiopulmonary bypass surgical procedures and the in vivo recovery of the platelets 1-2 hours following transfusion was reduced compared to the in vivo survival of the liquid preserved platelets stored at room temperature for a mean of 3.4 days with agitation. The frozen washed platelets reduced nonsurgical blood loss and the need for liquid preserved red blood cells and fresh frozen plasma compared to the liquid preserved platelets. No untoward effects were observed in patients transfused with the liquid preserved and frozen washed platelets resuspended in plasma (Khuri SF et al, J Thorac Cardiovasc Surg 117:172-184, 1999).

6. Non leukoreduced platelets and leukoreduced platelets were frozen with 6% DMSO and stored at -80 C and -135 C to assess the effect of white blood cells on the storage of frozen platelets at either -80 C or -135 C for as long as 3 years. The presence of white blood cells did not affect the in vitro recovery of the platelets stored at -80 C and -135 C. The temperature of storage at -135 C significantly improved, the in vitro recovery of platelets compared to storage at -80 C following the freeze-thaw-wash procedure (Valeri CR et al, Transfusion 43:1162-1167, 2003).

7. In March 2000, the procedure to freeze human platelets with 6% DMSO was modified in a manner identical to that used to freeze human RBC with 40% W/V glycerol. In 1981, the NBRL published its study to remove the supernatant glycerol from RBC treated with glycerol to achieve a final glycerol concentration within the RBC of 40±4 gm% prior to freezing. Prior to 1981, RBC treated with 40% W/V glycerol were frozen with the supernatant glycerol (Valeri CR et al, Transfusion 21:138-149, 1981).

8. The removal of the supernatant glycerol from the glycerolized RBC reduces the volume of the unit that is frozen. The reduction in the volume of the frozen glycerol RBC concentrate increases the number of units of frozen RBC that can be stored in a -80 C mechanical freezer (Valeri CR et al, Transfusion 21:138-149, 1981).

9. The removal of the supernatant glycerol from the glycerolized RBC prior to freezing results in:

a. Increase in number of units of frozen glycerolized RBC concentrates that are stored in the -80 C mechanical freezer
b. Reduction in the volume of solutions needed to deglycerolize the RBCs;
c. Reduction in the time required to deglycerolize the RBCs (Valeri CR et al, Vox Sang 45:25-39, 1983; Rosenblatt MS et al, Military Med 159:392-397, 1994).

10. In 1981, the NBRL reported that following the addition of 6.2M glycerol solution to the RBC the supernatant glycerol was removed to produce a glycerolized RBC concentrate with a hematocrit of 60±5 V% (Valeri CR et al, Transfusion 21:138-149, 1981).

11. The removal of the supernatant glycerol was achieved using

the standard blood bank centrifuges which are spun at 1250 X g for 10 minutes with the brake of the centrifuge off to permit a clear separation between the RBC and the supernatant solution to allow for the removal of the supernatant solution to achieve a hematocrit of 60±5 V%. The centrifugation of the glycerolized RBC and the removal of the supernatant glycerol requires time and technical expertise (Valeri CR et al, Transfusion 21:138-149, 1981).

12. In July 2003, the NBRL was successful in the automation of the glycerolization of the RBC with the use of the high separation bowl which is integrally attached to the disposable glycerolizing set used in the Haemonetics ACP215 instrument (Valeri CR et al, Transfusion 45:1621-1627, 2005).

13. The non-automated procedure to glycerolize RBC utilizes two centrifugation procedures and a routine blood bank centrifuge: one centrifugation procedure to prepare a RBC concentrate with a hematocrit of 75±5 V% prior to glycerolization and the second centrifugation procedure to remove the supernatant glycerol to prepare a glycerolized RBC concentrate with a hematocrit of 60±5 V% prior to freezing. Both centrifugation procedures take 2½ hours with technical expertise (Valeri CR et al, Transfusion 21:138-149, 1981).

14. The automated glycerolization procedure using the Haemonetics ACP215 takes only 50 minutes without any technical expertise other than to sterilely attach the tubing of the plastic bag containing the RBC to the disposable glycerolizing set with the attached high separation bowl using the sterile connector device (SCD) (Valeri CR et al, Transfusion 45:1621-1627, 2005).

15. In 1981, the supernatant glycerol was removed from the glycerolized RBC prior to freezing increasing the storage capacity of the -80 C freezer, reduced the volume of the wash solution from 3.2 liters to 1 .6 liters, and reduced the time required to deglycerolize the frozen RBCs (Valeri CR et al, Transfusion 21:138-149, 1981).

16. In March 2000, the supernatant DMSO was removed from the platelets treated with 6% DMSO prior to freezing and storage at -80 C. The thawed, non-washed platelets are diluted with 10 to 20 ml of 0.9% NaCl prior to transfusion (Valeri CR et al, Transfusion 45:1890-1898, 2005).

17. Single donor platelets are treated with 27% DMSO in 0.9% NaCl to achieve a final concentration of 6% DMSO. The DMSO-treated platelets

are transferred into a 300 ml PVC plastic bag and centrifuged at 1250 X g for 10 minutes, which is the same centrifuge speed and time used to concentrate the glycerolized RBC prior to the removal of the supernatant glycerol Valeri CR et al, Transfusion 45:1890-1898, 2005).

18. The supernatant DMSO solution is removed and the platelets rapidly resuspended in 10 ml volume containing 6% DMSO. The 300 ml PVC plastic bag is placed in a polyester plastic bag inserted into a rigid cardboard box and placed at the bottom of a -80 C horizontal mechanical freezer. The platelets are frozen and stored at -80 C (Valeri CR, Transfusion 45:1890-1898, 2005).

19. The 10 ml volume of platelets concentrated to remove the supernatant DMSO prior to freezing in a -80 C mechanical freezer are thawed in a pouch of the Thermogenesis thawing bath for 5 minutes, and then diluted with 10 to 20 ml 0.9% NaCl and stored at 22 C without agitation for 6 hours with a pH of 6.4. The total amount of DMSO in the thawed diluted platelets at the time of infusion is 600 mg (Valeri CR et al, Transfusion 45:1890-1898, 2005).

20. The mean freeze-thaw recovery of platelets treated with DMSO, centrifuged to remove the supernatant DMSO prior to freezing was 90%. The total number of platelets in the frozen, thawed and diluted, product was 3.5 X 10^{11}. The in vivo recovery one to two hours after infusion of 111In-oxine labeled autologous platelets frozen, thawed, non washed diluted with 10 to 20 ml of 0.9% NaCl, and stored at room temperature for 4 hours was 25 to 30% with a lifespan of 7 days for human platelets (Valeri CR et al, Transfusion 45:1890-1898, 2005).

21. The in vitro testing of the frozen, thawed, non-washed, diluted human platelets for platelet surface marker GPIb and. platelet annexin V binding, platelet microparticles and the R time of the thromboelastogram were measured together with the in vivo recovery 1 to 2 hours after transfusion and the lifespan of the platelets (Valeri CR et al, Transfusion 45:1890-1898, 2005).

22. Platelets frozen with 6% DMSO, thawed, washed, and resuspended in plasma had a freeze-thaw-wash recovery of 70% and platelets treated with 6% DMSO, concentrated to remove the supernatant DMSO, frozen, thawed, diluted with 0.9% NaC1 had a freeze thaw recovery of 90% (Valeri CR et al, Transfusion 45:1890-1898, 2005).

23. Platelet microparticles in the freeze-thaw-washed platelets was about 5% and the platelet microparticles in the freeze-thaw non-washed

platelets was about 8% (Valeri CR et al, Transfusion 45:1890-1898, 2005).

24. The platelet GPIb and the platelet annexin V binding were similar for platelets that were frozen, thawed, non-washed and diluted with 0.9% NaCl and platelets that were frozen thawed, washed and resuspended in plasma (Valeri CR et al, Transfusion 45:1890-1898, 2005).

25. Platelet microparticles positive for annexin V were greater in the freeze-thaw-nonwashed saline diluted platelets than in frozen, thawed, washed platelets resuspended in plasma (Valeri CR et al, Transfusion 45:1890-1898, 2005).

26. The R time of the thromboelastogram could not be compared for the frozen, thawed, washed platelets resuspended in plasma and the frozen, thawed, non-washed platelets resuspended in 0.9% NaCl because of the difference in volume of plasma between the two platelet products, (Valeri CR et al, Transfusion 45:1890-1898, 2005; Valeri CR, Ragno G, Trans Aph Sci 35:33-41, 2006).

27. The NBRL has reported that the bleeding time increased significantly following the removal of 2 units of RBC by apheresis in healthy male and female blood donors and the bleeding time decreased significantly after the reinfusion of the 2 units of autologous RBC in the donors (Valeri CR et al, Transfusion 41:977-983, 2001; Valeri CR et al, Transfusion 47:206S-248S, 2007). Significant negative correlations between the bleeding time and the hematocrit ($r=-0.655$, $p<0.001$, $n=40$) and the hemoglobin concentration ($r=-0.655$, $p<0.001$, $n=40$) were observed. Unlike the findings in Apelseth and associates study (Apelseth. TO et al, Transfusion 50:766-775, 2010), in our study using fresh blood obtained from healthy female donors, the bleeding time correlated positively to the maximum amplitude (MA) ($r=0.452$. $p<0.01$, $n=39$) and the angle recorded in the thromboelastogram ($r=0.487$. $p<0.01$,$n=39$) and negatively to the K time in the thromboelastogram ($r=-0485$, $p<0.01$, $n=39$) (Valeri CR, Ragno G, Trans Aph Sci 35:33-41, 2006).

28. Roeloffzen WWH and associates (Transfusion 50:1536-1544, 2010) have reported that blood obtained from patients prior to and after transfusion with fresh and. liquid preserved red blood cells did not detect the effect of red blood cells in clotting of blood measured in the thromboelastogram. The authors reported that an increase in hemoglobin concentration produced by fresh and liquid preserved red blood cells decreased the maximum amplitude (MA); whereas a reduced hemoglobin

concentration increased the maximum amplitude. These data were similar to our data reported in five healthy female donors that were bled 360 ml of red. blood cells and infused with a liter of 0.9% NaCl to produce a reduction in hematocrit and hemoglobin concentrations, an increase in the bleeding time and an increased in the maximum amplitude of the thromboelastogram (Valeri CR, Ragno G, Trans Aph Sci 35:33-41, 2006). Following transfusion of the 2 units of autologous fresh RBC, the bleeding time decreased, the hematocrit and hemoglobin concentrations increased, and the MA decreased (Valeri CR, Ragno G, Trans Aph Sci 35:33-41, 2006). We observed a significant positive correlation between the increased bleeding time and increase in MA of the thromboelastogram of whole blood ($r=0.452$, $p<0.01$, $n=39$). Our study like that of Roeloffzen WWH and associates, showed that red blood cells were not detected in the clotting of blood in the thromboelastogram. Our clinical studies have demonstrated the importance of red blood cells on the bleeding time and nonsurgical blood loss in normal volunteers and in patients (Khuri SF et al, J Thorac Cardiovasc Surg 104:94-107, 1992; Crowley JP et al, Am J Clin Path 108:579-584, 1997; Valeri CR et al, Transfusion 41:977-983, 2001; Valeri CR et al, Transfusion 47:206S-248S, 2007). Roeloffzen WWH and associates report and our data did not detect the effect of red blood cells in clotting of blood measured in the thromboelastogram.

29. Apelseth TO and associates (Transfusion 50:766-775, 2010) reported that the "late maximum amplitude" recorded in the thromboelastogram correlated negativity to the nonsurgical bleeding in anemic thrombocytopenic patients with acute leukemia ($r=-0.494$, $p=0.008$). These authors reported that the function of the transfused platelets can be assessed by the measurement of the "late maximum amplitude" recorded in the thromboelastogram. Apelseth and associates reported that the function of the transfused platelets can be assessed by increase in the late maximum amplitude in the thromboelastogram which was associated with a decrease in nonsurgical blood loss in the anemic thrombocytopenic patients.

Roeloffzen WWH and associates reported the effect of the red blood cells in the clotting of blood in the thromboelastogram in Transfusion (50:1536-1544. 2010). These authors reported that hemoglobin concentrations of less than 10 gm/dl were associated with a significant increase in maximum amplitude of the thromboelastogram whereas a decrease in maximum amplitude of the thromboelastogram was associated with hemoglobin levels of greater than 10 gm/dl (p=0.02). These authors reported no difference in the thromboelastogram tracing whether fresh or preserved red blood cells were transfused to the patients. The findings of Roeloffzen WWH and associates were similar to our findings that an increase in hemoglobin concentration in blood at the time of testing in the thromboelastogram produced a decrease in the maximum amplitude.

Valeri CR, Ragno G in Trans Aph Sci (35:33-41, 2006) reported an increase in bleeding time occurred in normal volunteers subjected to the removal of 360 ml of red blood cells and the infusion of 1.0 liter of 0.9% NaCl; a reduction in the hematocrit values and hemoglobin concentrations; and an increase in maximum amplitude recorded in thromboelastogram. The reinfusion of the fresh autologous red blood cells in the normal volunteers was associated with a decrease in the bleeding time, increase in the hematocrit values and hemoglobin concentrations, and a decrease in the maximum amplitude recorded in the thromboelastogram. The findings of Roeloffzen WWH and associates were similar to our findings that an increase in hemoglobin concentration in blood at the time of testing in the thromboelastogram produced a decrease in the maximum amplitude.

Apelseth and associates have suggested that the "late maximum amplitude" recorded in the thromboelastogram can be used to assess the hemostatic function of platelets transfused to anemic thrombocytopenic patients with acute leukemia. An increase in the "late maximum amplitude" (MA) correlated to a decrease in nonsurgical blood loss (r= -0.494, p=0.008) and documented the hemostatic function of the transfused platelets.

DeLoughery TG at the American Society of Hematology meeting held, Dec. 4 to 7, 2010 at Orlando, FL published in Hematology, pg 470-473, 2010 reported that a decrease in maximum amplitude of the thromboelastogram indicates that platelets should be transfused. In a study reported by Valeri CR and associates in J Trauma 57:S22-S25, 2004 the function of platelets in the clotting of blood was detected by the R time recorded in the thromboelastogram. The suggestion that the thromboelastogram tracing can be used to recommend the blood products that should be administered to correct the hemostatic defect in patients needs data to support this recommendation. The Table in the DeLoughery TG paper reports that specific parameters recorded in the thromboelastogram tracing indicate the need for fresh frozen plasma, cryoprecipitate, platelets and anti-fibrinolytic drugs. The thromboelastogram tracing does not detect the need for red blood cells to restore hemostasis and reduce nonsurgical blood loss in patients. In vitro testing using the thromboelastogram cannot be used to determine the blood products to treat the patient with nonsurgical blood loss. The published data show that the red blood cells have significant effect on the bleeding time and nonsurgical blood loss. Our studies performed at the NBRL reported that thromboelastogram testing could not be used to predict the bleeding time and did not detect the beneficial effect of RBC in the clotting of blood (Valeri CR, Ragno G, Trans Aph Sci 35:33-41, 2006).

30. The recent paper by Stinner DJ and associates in Military Medicine (175:1027-1029, 2010) reported the incidence of late amputation of 15% occurred 12 weeks to 5.5 years following combat related injuries sustained by U.S. military personnel in Afghanistan and Iraq. During the Vietnam War from 1968 to 1974 our laboratory studied and treated over 300 wounded servicemen who returned to the Chelsea Naval Hospital from South Vietnam with severe traumatic injuries to their extremities who were treated by the orthopedic service. These patients with traumatic injuries to their extremities had chronic hypovolemic anemia of trauma with reduction in total blood volume of 30 to 40%, normal hemoglobin and hematocrit values, normal central blood volume but severely reduced blood volume to their extremities. Aggressive transfusion of washed liquid preserved red blood cells and previously frozen deglycerolized red blood cells were administered to restore the peripheral blood volume to their extremities. The repeated red blood cell transfusion treatment to restore to normal the peripheral blood volume to their extremities was associated with repair of the wounded extremities with the rare need for an amputation in the 300 servicemen who were studied for one to two years during their hospitalization at the Chelsea Naval Hospital. The book the Hypovolemic Anemia of Trauma: The Missing Blood Syndrome by Valeri CR and Altschule MD, CRC Press, Boca Raton, FL, 1981, was written to report that chronic hypovolemic patients with traumatic injuries to their extremities need to be transfused red blood cells to restore the peripheral blood volume to repair the injured extremities of these patients. The studies reported in the book demonstrate the limitation of the hematocrit and hemoglobin concentration measurements; the maintenance of the central blood volume and the 30-40% reduction in the peripheral blood volume in patients with chronic hypovolemia in patients with traumatic injuries to their extremities. The repeated transfusions of viable and identifiable compatible red blood cells restored the peripheral blood volume and permitted the healing of their injured extremities without the need for amputations.

CHAPTER 5 REFERENCES

1. Apelseth TO, Bruserud O, Wentzel-Larsen T, Hervig E. Therapeutic efficacy of platelet transfusion in patients with acute leukemia: an evaluation of methods. Transfusion 50:766-775, 2010.
2. Apstein CS, Dennis RC, Briggs L, Vogel WM, Frazer J, Valeri CR. Effect of erythrocyte storage and oxyhemoglobin affinity changes on cardiac function. Am J Physiol 248 (Heart Circ Physiol 17): H508-H515, 1985.
3. Barnard MR, MacGregor H, Ragno G, Pivacek LE, Khuri SF, Michelson AD, Valeri CR. Fresh, liquid-preserved, and cryopreserved platelets: adhesive surface receptors and membrane procoagulant activity. Transfusion 39:880-888, 1999.
4. Biron PE, Howard J, Altschule MD, Valeri CR. Chronic deficits in red-cell mass in patients with orthopaedic injuries (stress anemia). J Bone Joint Surg 54-A:1001-1014,1972.
5. Crowley JP, Metzger JB, Valeri CR. The volume of blood shed during the bleeding time correlates with, the peripheral venous hematocrit. Am J Clin Pathol 108:579-584,1997.
6. Deloughery TG. Logistics of massive transfusion. Transfusion medicine: transfusion support in trauma – military and civilian approaches. Hematology 470-473, 2010.
7. Fischer TH, Valeri CR, Smith CJ, Scull CM, Merricks EP, Nichols TC, Demcheva M, Vournakis JN. Non-classical processes in surface hemostasis: mechanisms for the poly-N-acetyl glucosamine induced alteration of red blood cell morphology and surface prothrombogenicity. Biomed Mater 3(1):1-9, 2008.
8. Fortier NL, Hirsch NM, Valeri CR. Restoration of 2, DPG and ATP in ACD-stored red blood cells. Forsvarsmedicin 5:250-257, 1969.
9. Handin RI, Valeri CR. Hemostatic effectiveness of platelets stored at 22C. NEJM 285:538-543, 1971.
10. Khuri SF, Wolfe JA, Josa M Axford TC, Szymanski I, Assousa S, Ragno G, Patel M, Silverman A, Park M, Valeri CR. Hematologic changes during and after cardiopulmonary bypass and their

relationship to the bleeding tune and nonsurgical blood loss. J Thorac Cardiovasc Surg 104:94-107, 1992.

11. Khuri SF, Healey N, MacGregor H, Barnard MR, Szymanski IO, Birjiniuk V, Michelson AD, Gagnon DR, Valeri CR. Comparison of the effects of transfusions of cryopreserved and liquid-preserved platelets on hemostasis and blood loss after cardiopulmonary bypass. J Thorac Cardiovasc Surg. 117:172-184, 1999.
12. Koustova E, Rhee P, Hancock T, Chen H, Inocencio R, Valeri CR, Alam HB. Ketone and pyruvate Ringer's solutions decrease pulmonary apoptosis in a rat model of severe hemorrhagic shock and. resuscitation. Surgery 134(2):267-274, 2003.
13. Melaragno AJ, Carciero R, Feingold H, Talarico L, Weintraub L, Valeri CR. Cryopreservation of human platelets using 6% dimethylsulfoxide oxide and storage at -80 C. Effects of 2 years of frozen storage at -80 C and transportation in dry ice. Vox Sang 49:245-258, 1985.
14. Michelson AD, Barnard MR, Hechtman HB, MacGregor H, Connolly RJ, Loscalzo J, Valeri CR. In vivo tracking of platelets: circulating degranulated platelets rapidly lose surface P-selectin but continue to circulate and function. Proc Natl Acad Sci USA 93:11877-11882, 1996.
15. Ragno G, Valeri CR: Salvaging of liquid-preserved. 0-positive and 0-negative red blood cells by rejuvenation and freezing. Trans Aph Sci 35:137-143, 2006.
16. Roeloffzen WWH, Kluin-Nelemans HC, Bosman L, de Wolf JTM. Effects of red blood cells on hemostasis. Transfusion 50: 1536-1544, 2010.
17. Rosenblatt MS, Hirsch EF, Valeri CR. Frozen red blood cells in combat casualty care: Clinical and logistica1 considerations. Mil Med 159:392-397, 1994.
18. Stinner DJ, Burns TC, Kirk KL, Scoville CR, Ficke JR, Hsu JR, Late Amputation Study Team. Prevalence of late amputations during the current conflicts in Afghanistan and Iraq. Mil Med 175:1027-1029. 2010.
19. Szymanski IO, Valeri CR. Automated differential agglutination technic to measure red cell survival. II. Survival in vivo of preserved red cells. Transfusion 8:74-83, 1968.

20. Szymanski IO, Valeri CR. Evaluation of double 51Cr technique. Vox Sang 15:287-292,1968.
21. Szymanski IO, Valeri CR. Analysis of erythrocyte survival curves obtained simultaneously by 51Cr and an automated differential agglutination technic. Transfusion 10:287-298, 1970.
22. Szymanski IO, Valeri CR. Factors influencing chromium elution from labeled red cells in vivo and the effect of elution on red-cell survival measurements. Br J Haemat 19:397-409, 1970.
23. Szymanski IO, Lipson CS, Valeri CR. Elution of chromium label firm preserved red blood cells transfused during surgery. Transfusion 13:13-18, 1973.
24. Valeri CR, Fortier NL. Red-cell, 2,3 -diphosphoglycerate and creatine levels in patients with red-cell mass deficits or with cardiopulmonary insufficiency. NEJM 81:1452-1455, 1969.
25. Valeri CR. Observations on the chromium labelling of ACD-stored and previously frozen red cells. Transfusion 8:210-219, 1968.
26. Valeri CR. Viability and function of preserved red cells. NEJM 284:81-88, 1971.
27. Valeri CR, Zaroulis CG. Rejuvenation and freezing of outdated stored human red cells. NEJM 287:1307-1313, 1972.
28. Valeri CR, Cooper AG, Pivacek LE. Limitations of measuring blood, volume with iodinated I 125 serum albumin. Arch Int Med 132:534-538, 1973.
29. Valeri CR, Szymanski IO. Further studies on the rapid and slow components of chromium elution in vivo from preserved erythrocytes. Vox Sang 24:502-514, 1973.
30. Valeri CR, Altschule MD. Hemolysis in vitro of blood obtained from patients with traumatic injuries. J Trauma 13:678-686, 1973.
31. Valeri CR, Feingold, H, Marchionni LD. A simple method for freezing human platelets using 6% dimethylsulfoxide and storage at -80 C. Blood 43:131-136, 1974.
32. Valeri CR, Feingold. H, Marchionni LD. The relation between response to hypotonic stress and the 51Cr recovery in vivo of preserved platelets. Transfusion 14:331-337, 1974.
33. Valeri CR. Hemostatic effectiveness of liquid-preserved and previously frozen human platelets. NEJM 290:353-358, 1974.

34. Valeri CR. Oxygen transport and viability of preserved red blood cells. J Med 5:278-291,1974,
35. Valeri CR and Altschule MD. Hypovolemic Anemia of Trauma: The Missing Blood Syndrome, Chemical Rubber Company Press, Boca Raton, Florida, 1981.
36. Valeri CR, Valeri DA, Anastasi J, Vecchione JJ, Dennis RC, Emerson CP. Freezing in the primary polyvinylchloride plastic collection bag: A new system for preparing and freezing nonrejuvenated and rejuvenated red blood cells. Transfusion 21:138-149. 1981.
37. Valeri CR, Sims KL, Bates JF, Reichman D, Lindberg JR, Wilson AC. An integrated liquid-frozen blood banking system. Vox Sang 45:25-39, 1983.
38. Valeri CR, Pivacek LE, Ouellet R, Gray A. A comparison of methods of determining the 100 percent survival of preserved red cells. Transfusion 24:105-108, 1984.
39. Valeri CR, Landrock RD, Pivacek LE, Gray AD, Fink JG, Szymanski IO. Quantitative differential agglutination method using the Coulter Counter to measure survival of compatible but identifiable red blood cells. Vox Sang 49:195-205, 1985.
40. Valeri CR, Donahue K, Feingold HM, Cassidy GP. Altschule MD. Increase in plasma volume after the transfusion of washed erythrocytes. Surg Gynec Obstet 162:30-36, 1986.
41. Valeri CR, Feingold H, Cassidy G, Ragno G, Khuri S, Altschule MD. Hypothermia-induced reversible platelet dysfunction. Ann Surg 205:175-181, 1987.
42. Valeri CR, Pivacek LE, Palter M, Dennis RC, Yeston N, Emerson CP, Altschule MD. A clinical experience with ADSOL® preserved erythrocytes. Surg Gynec Obstet 166:33- 46, 1988.
43. Valeri CR, Pivacek LE, Gray AD, Cassidy GP, Leavy ME, Dennis RC, Melaragno AJ, Niehoff J, Yeston N, Emerson CP, Altschule MD. The safety and therapeutic effectiveness of human red cells stored at -8O C for as long as 21 years. Transfusion 29:429-437, 1989.
44. Valeri CR, MacGregor H, Cassidy G, Tinney R, Pompei P. Effects of temperature on bleeding time and clotting time in normal male and female volunteers. Crit Care Med 23:698-704, 1995.
45. Valeri CR, Crowley JP, Loscalzo. The red cell transfusion trigger:

has a sin of commission now become a sin of omission? Transfusion 38:602-610, 1998.

46. Valeri CR, Ragno G, Pivacek LE, Cassidy GP, Srey R, Hansson-Wicher M, Leavy ME. An experiment with glycerol-frozen red blood cells stored at -80 C for up to 37 years. Vox Sang 79:168-174, 2000.

47. Valeri CR, Ragno G, Pivacek LE, Srey R, Hess JR, Lippert LE, Mettille F, Fahie R, O'Neill EM, Szymanski IO. A multicenter study of in vitro and in vivo values in human RBCs frozen with 40-percent (wt/vol) glycerol and stored after deglycerolization for 15 days at 4 C in AS-3: assessment of RBC processing in the ACP 215. Transfusion 41:933-939, 2001.

48. Valeri CR, Ragno G, Pivacek LE, O'Neill EM. In vivo survival of apheresis RBCs, frozen with. 40-percent (wt/vol) glycerol, deglycerolized in the ACP215, and stored at 4 C in AS-3 for up to 21 days. Transfusion 41:928-932, 2001.

49. Valeri CR, Cassidy G,. Pivacek LE, Ragno G, Lieberthal W, Crowley JP, Khuri SF, Loscalzo J. Anemia-induced increase in the bleeding time: implications for treatment of nonsurgical blood loss. Transfusion 41 977-983, 2001.

50. Valeri CR, Pivacek LE, Cassidy GP, Ragno G. 24-hour ^{51}Cr post-transfusion survival, ^{51}Cr life span and haemolysis of red blood cells stored at 4 C for 56 days in AS-3. Vox Sang 80:48-50, 2001.

51. Valeri CR, Pivacek LE, Cassidy GP, Ragno G. Volume of RBCs, 24- and 48-hour posttransfusion survivals, and the lifespan of (51)Cr and biotin—X-N-hydroxysuccinimide (NHS)-labeled autologous baboons RBCs: effect of the anticoagulant and blood pH on (51)Cr and biotin-X-NHS elution in vivo. Transfusion 42:343-8, 2002.

52. Valeri CR, MacGregor H, Giorgio A, Ragno G: Circulation and hemostatic function of autologous fresh, liquid preserved, and cryopreserved baboon platelets transfused to correct an aspirin-induced thrombocytopathy. Transfusion 42:1206-16, 2002.

53. Valeri C: Status report on the quality of liquid and frozen red blood cells. Vox Sang 83(1):193-196, 2002.

54. Valeri. CR, Giorgio A, MacGregor H, Ragno G. Circulation and distribution of autotransfused fresh, liquid-preserved and cryopreserved baboon platelets. Vox Sang 83:347-351, 2002.

55. Valeri CR, Srey R, Lane JP, Ragno G. Effect of WBC reduction and storage temperature on PLTs frozen with 6 percent DMSO for as long as 3 years. Transfusion 43(8): 1162-1167, 2003.
56. Valeri CR, MacGregor H, Giorgio A, Srey R, Ragno G. Comparison of radioisotope methods and a non-radioisotope method to measure the RBC volume and RBC survival in the baboon. Transfusion 43(10):1366-1373, 2003
57. Valeri, CR, Khuri S, Ragno G. Role of Hct in the treatment of thrombocytopenic patients. Letter to the Editor, Transfusion 43:1761-1762, 2003.
58. Valeri CR, Ragno G, Marks PE, Kuter DJ, Rosenberg RD, Stossel TP. Effect of thrombopoietin alone and a combination of cytochalasin, B and ethylene and glycol bis (beta-aminoethyl ether) N, N'-tetraacetic acid-AM on the survival and function of autologous baboon platelets stored at 4C for as long as 5 days. Transfusion 44:865-870, 2004.
59. Valeri CR, Srey K, Tilahun D, Ragno G. In vitro effects of polymerized N-acetyl glucosamine (NAG) on the activation of platelets on platelet rich plasma with and without red blood cells. J Trauma (suppl) 57:522-525, 2004.
60. Valeri CR., MacGregor H, Giorgio A, Ragno G: Comparison of radioisotope methods and a non-radioisotope method to measure platelet survival in the baboon. Trans Aph Sci 32(3):275-28l, 2005.
61. Valeri CR, MacGregor H, Ragno G. Correlation between in vitro aggregation and thromboxane A2 production in fresh, liquid-preserved, and cryopreserved human platelets: Effects of agonists, pH, and plasma and saline resuspension. Transfusion 45:596-603, 2005,
62. Valeri CR, Ragno G, Khuri S. Freezing human platelets using 6% DMSO with removal of the supernatant solution prior to freezing and storage at -80C without post-thaw processing. Transfusion 45(12): 1890-1898, 2005.
63. Valeri CR, Ragno G, van Houten P, Rose L, Rose M, Egozy Y, Popovsky M. Automation of the glycerolization of the RBC using the high separation bowl in the Haemonetics ACP215 instrument. Transfusion 45(10): 1621-1627, 2005.
64. Valeri CR, Ragno G. In vitro testing of platelets using the

thromboelastogram, platelet function analyzer, and the clot signature analyzer to predict the bleeding time. Trans Aph Sci 35(1):33-41, 2006.

65. Valeri CR, Ragno G. The survival, and function of baboon red blood cells, platelets, and plasma proteins: A review of the experience from 1972 to 2002 at the Naval Blood Research Laboratory, Boston, Massachusetts. Transfusion, 46(8): 1-42, 2006.
66. Valeri CR, Ragno G. Cryopreservation of human blood products. Trans Aph Sci 34:271-287, with an editorial on pages 267-269, 2006.
67. Valeri CR, Khuri S, Ragno G. Non-surgical, bleeding diathesis in anemic thrombocytopenic patients: Role of temperature, RBC, platelets, and plasma clotting proteins. Transfusion 47:206S-248S, 2007.
68. Valeri CR, Ragno G, Veech RL. Severe adverse events associated with hemoglobin based, oxygen carriers: role of resuscitative fluids and liquid preserved RBC. Trans Aph Sci 39:205-211, 2008.
69. Valeri CR, Ragno G. An approach to prevent the severe adverse events associated with transfusion of FDA-approved blood products. Trans Aph Sci 42:223-233, 2010.
70. Veech, RL. The toxic impact of parenteral solutions on the metabolism of cells: a hypothesis for physiological parenteral therapy. Am J Clin Nutri, 44:519-551, 1986.
71. Vogel WM, Dennis RC, Cassidy GP, Apstein CS, Valeri CR. Coronary constrictor effect of stroma-free hemoglobin solutions Am J Physiol (Heart Circ Physiol 20)251:H413-H420, 1986.
72. Vogel WM, Hsia JC, Briggs LL, Er SS, Cassidy G, Apstein CS, Valeri CR. Reduced coronary vasoconstrictor activity of hemoglobin solutions purified by ATP-agarose affinity chromatography. Life Sciences 41:89-93, 1987.

CHAPTER 6

THE U.S. NAVY'S EXPERIENCE WITH RESUSCITATION OF WOUNDED SERVICEMEN IN VIETNAM USING FROZEN WASHED RED BLOOD CELLS FROM 1966 TO 1974: DEVELOPMENTS FROM THIS EXPERIENCE

As early as the mid-1950's, the U.S. Navy was supporting research in blood preservation research which eventually led to the use of freeze-preserved red blood cells in Vietnam to supplement the liquid blood system. Large numbers of servicemen being rescued from combat areas in Vietnam received previously frozen red blood cells for the emergency treatment of serious wounds. In 1956, the U.S. Navy's Bureau of Medicine and Surgery assigned the Blood Research Laboratory at the Chelsea Naval Hospital the task of evaluating blood cryopreservation (Haynes et al, 1960; Haynes et al, 1962; Tullis et al, 1958). This facility was later named the U.S. Naval Blood Research Laboratory in 1965.

The term "frozen blood" has been used because in early studies using extracellular cryoprotective agents and liquid nitrogen the whole unit of blood was frozen (Valeri, 1966c, 1966d). Today, the components of the whole blood, the red blood cells, platelets and plasma proteins are isolated after blood collection and each stored under conditions most suitable for freeze preservation (Valeri, 1967, 1968b, 1968c, 1969, 1970a, 1970b, 1971, 1972, 1973a, 1973b, 1974c, 1974e, 1974g, 1976a).

Early Attempts to Freeze Blood

In the early 1960's, researchers under the aegis of the Office of Naval Research had attempted to freeze blood with extracellular additives such as polyvinylpyrrolidone; the blood was frozen rapidly in liquid nitrogen and stored in the gas phase of liquid nitrogen at -150 C (Valeri, 1976a). Because blood frozen by this approach did not have to be washed before transfusion it was considered to be the ideal supplement to the supply of liquid-stored whole blood for emergency situations in which large numbers of transfusions were required. However, this advantage was far out-weighed by the disadvantages – the excessive hemolysis and the potential toxicity of the extracellular additive polyvinylpyrrolidone (PVP) with its long-term retention in the body – and so it was deemed an unacceptable approach when the Navy began making plans to use frozen red blood cells in Vietnam.

Freezing Red Blood Cells with Glycerol

Instead cryopreservation method using the intracellular additive glycerol was chosen for the Vietnam study even though washing of the red blood cells was necessary before transfusion. One method utilized a high concentration (40% W/V) glycerol and storage at -80 C; the other method used a low concentration (20% W/V) of glycerol and storage in liquid nitrogen at -150 C (Valeri and Brodine, 1968; Valeri and Runck, 1969a, 1969b; Runck and Valeri, 1969; Valeri et al, 1969a, 1971a). Captain Lewis Haynes, MC, USN, Commanding Officer at the Chelsea Naval Hospital, and CDR Mary T. Sproul, MSC, USN, working in collaboration with the Protein Foundation and Dr. J.L. Tullis and his collaborators, studied red blood cell freezing with 40% W/V glycerol and storage at -80 C. In these studies of human red blood cells, the Cohn Blood Fractionator, a development of Edwin J. Cohn, originally used to isolate albumin was used for the addition and removal of the glycerol cryoprotectant (Tullis et al, 1958; Haynes et al, 1960; Haynes et al, 1962). The red blood cells were glycerolized to a concentration of 40% W/V glycerol, frozen and stored at -80 C, thawed at 37 C, and deglycerolized before transfusion. Glycerolization of the red blood cells in a unit of blood took about 45 minutes, and deglycerolization took about 1 hour. The procedure eventually proved to be impractical for anything but research. Nevertheless, in 1960 these investigators had reported the successful transfusion of 1,000 units of deglycerolized red blood cells to patients at Chelsea Naval Hospital.

Red Cell Washing by Agglomeration (The Huggins Method)

The Cohn Blood Fractionator was used to freeze human red blood cells by a technique using a high concentration of glycerol and slow freezing and thawing, but the process was cumbersome (Valeri, 1965a, 1965b, 1965c, 1966a; Valeri and Henderson, 1964). In May 1964, Dr. C.E. Huggins introduced a new method of red blood cell cryopreservation, which was investigated at the NBRL because the red blood cells could be frozen and stored at -80 C in mechanical freezers, thawed, and washed by a dilutional process in a low ionic solution in a large plastic container without the need for centrifuges. The Naval Blood Research Laboratory worked closely with Dr. Huggins to evaluate his cryopreservation method that utilized reversible agglomeration (Valeri 1966b; Valeri and Bond, 1966; Valeri et al, 1966a, 1966b; Almond and Valeri, 1967b; Valeri et al 1967; Runck et al, 1968; Valeri et al, 1969a, 1969b; Daane and Valeri, 1970).

The Huggins-preserved red cells were evaluated by the 51Cr labeling procedure. The National Academy of Sciences under the leadership of Dr. Max Strumia had established the 51Cr labeling procedure for evaluation of the quality of preserved red blood cells. A 24-hour posttransfusion survival value of at least 70% was established as the criterion for acceptable red blood cell preservation. Autologous preserved red blood cells were labeled with 51Cr before transfusion and measurements were made of the 51Cr-labeled preserved red blood cells and the red blood cell volume of the recipient was estimated from radioiodinated 125I albumin or Evans blue plasma volume and the total body hematocrit to determine the number of viable red blood cells present in the red blood cell product (Valeri 1966b, 1968a; Valeri and Bond, 1966; Valeri et al, 1966a, 1966b; Chaplin et al, 1973; Valeri et al, 1973a). Using this technique, the Naval Blood Research Laboratory studied transfusions of 10 ml aliquots of 51Cr-labeled autologous red blood cells preserved by the Huggins method and washed by reversible agglomeration (Valeri, 1966b; Valeri and Bond, 1966; Valeri et al, 1966a, 1966b).

We discovered that Huggins-preserved red blood cells accumulated B_1C globulin during glycerolization of the red blood cells in the low ionic medium, and it was determined that this could be prevented by adding

Na2EDTA to the glycerol solution (Valeri, 1966b). We also found that the 75 to 100 ml of 0.9% sodium chloride recommended by Huggins for resuspension of the agglomerated red blood cells was not adequate to restore the red blood cell volume to normal, but that a 250 ml volume accomplished satisfactory restoration (Valeri and Bond, 1966; Valeri, 1966a; Valeri et al, 1966b; Valeri et al, 1967). Further, an observed significant hemolysis in the supernatant of the disaggregated red blood cells was corrected by concentrating the red blood cells to a hematocrit of 80 $\pm$ 5 V% by centrifugation prior to transfusion; a colloid or crystalloid solution had to be given when the hematocrit value of the transfused red blood cell concentrate was 80 to 90 V%.

Using an automated differential agglutination (ADA) technique, we found that Cohn-processed red blood cells exhibited 24-hour in vivo recovery values of approximately 90% and an index of therapeutic effectiveness (in vitro recovery (%) multiplied by the 24-hour posttransfusion survival value) of approximately 80% after 7 years of frozen storage at -80 C (Valeri et al, 1970a). Red blood cells frozen using the Cohn process were washed in the Cohn Blood Fractionator by continuous-flow centrifugation using electrolyte solutions. When Huggins freeze-preserved red cells were stored in the frozen state for more than 1½ years, the Huggins dilution/agglomeration wash procedure produced a significant reduction in the in vitro recovery of red blood cells and the red blood cells had decreased 24-hour posttransfusion survival values (Valeri et al, 1970a; Valeri, 1976a). After frozen storage for longer than 2 years, there was intravascular destruction of the compatible nonviable red blood cells. When Huggins-frozen red blood cells were washed with electrolyte solutions in an Arthur D. Little (ADL) reusable polycarbonate bowl, better results were achieved (Valeri et al, 1970a; Runck and Valeri, 1972; Valeri, 1976a).

As part of the Naval Blood Research Laboratory's evaluation of the Huggins method (Huggins, 1965), a small aliquot (10 ml) of autologous chromium-labeled red blood cells were transfused to healthy male volunteers in an attempt to define the pre-freeze, frozen state and post-thaw variables (Valeri 1966b; Valeri and Bond 1966; Valeri et al 1966a; Valeri et al, 1966b; Valeri et al, 1967). Multiple units of homologous compatible identifiable RBC were studied in stable, anemic patients (Almond and Valeri, 1967b). In vivo survivals measured by both the manual Ashby technique and an automated differential agglutination technique using the Technicon

Autoanalyzer showed 24-hour posttransfusion survival values of greater than 70% (Szymanski et al, 1967; Szymanski et al, 1968; Szymanski et al, 1970; Szymanski and Valeri, 1968b; Szymanski and Valeri, 1970; Szymanski and Valeri, 1971). Because of the encouraging results obtained with autologous and homologous transfusions and the relative simplicity of the agglomeration process compared to the Cohn Blood Fractionator procedure, the Huggins technique was selected for field testing in Vietnam. The objectives of the field test were two-fold, to evaluate the Huggins technique and equipment, and to anticipate and deal with the logistics of supporting a frozen blood bank in a combat area.

To supplement the liquid blood system in Vietnam in 1966, the U.S. Naval Blood Research Laboratory was given permission by the Bureau of Medicine and Surgery to provide frozen red blood cells and fresh frozen plasma at the Naval Support Activity, Danang, and aboard the hospital ships USS Repose and USS Sanctuary (Moss 1969; Moss GS et al, 1968). The hardware required for the field testing at these sites included a water-cooled mechanical freezer maintained at -80 C, an air cooled mechanical freezer maintained at -20 C, refrigerated centrifuges, a 6-station Huggins Cytoagglomerator, and a supply of wash solution for deglycerolization (6.8 liters per unit of glycerolized red blood cells). Large PVC plastic bags were used for freezing. Red cell glycerolization, freezing, storage, thawing, and washing were all done in a single large polyvinyl chloride (PVC) plastic bag.

As part of our preliminary investigations, previously frozen red blood cells were sometimes moved from a -80 C refrigerator to storage areas of higher temperature (4 C, -20 C and -30 C) to determine what would happen in the event of an electrical or mechanical failure, and to determine whether more commonly available modes of refrigeration might be suitable for storage of frozen red blood cells (Valeri CR et al, 1967). It was obvious from these studies that a frozen blood bank must have on hand a supply of dry ice or liquid carbon dioxide as a backup for the -80 C refrigerators. Excessive in vitro loss of cellular hemoglobin and unacceptable posttransfusion chromium survival were observed when frozen red blood cells were transferred from a -80 C refrigerator to a 4 C storage area for longer than 24 hours before return to the -80 C refrigerator. These same adverse effects were seen when frozen red blood cells were transferred to -20 C for longer than 3 days or to -30 C for longer than 7 days between periods of -80 C storage. These studies not only showed us the importance of maintaining Huggins freeze-preserved red blood cells at -80 C, but demonstrated that red blood cell agglomeration per se served as an excellent quality control feature. Inadequate agglomeration meant unsatisfactory preservation: the red blood cells leaked intracellular electrolytes, producing an increase in the ionic strength of the environment which prevented agglomeration (Runck et al, 1968; Valeri et al, 1969b). Thus, poor agglomeration of Huggins preserved red blood cells without a documented cause could be assumed to be attributable to an undetected rise in temperature to a critical level.

Most of the red blood cells in the Vietnam study were collected and frozen under a collaborative program with Dr. C.E. Huggins at the Massachusetts General Hospital and Dr. A. Kliman at the Massachusetts Red Cross, working with the U.S. Naval Blood Research Laboratory. The frozen red blood cells were shipped in dry ice at -80 C, first to Oakland, California, and then to Danang, South Vietnam at the Naval Support Activity Hospital and to the hospital ships USS Repose and USS Sanctuary.

Roster of Military Personnel in the Feasibility Study

The U.S. military personnel working on this feasibility study in Vietnam either had previously worked at the U.S. Naval Blood Research Laboratory or had received special training there. Funding for the study was provided by the Naval Medical Research and Development Command of the U.S. Navy's Bureau of Medicine and Surgery. Captain C.E. Brodine, MC, USN, was the coordinator of the feasibility study, and Lieutenant Commander G.S. Moss, MC, USNR was the Officer in Charge of the Surgical Research unit at the Naval Support Activity, Danang, South Vietnam which evaluated the cryopreserved red blood cells.

The first Navy Surgical Research Team was directed by LCDR Gerald Moss, MC, USNR, assisted by Ensign James Bates, MSC, USN, along with enlisted personnel, all of whom were first trained at the Naval Blood Research Laboratory in the proper processing of frozen red cells for transfusion to combat casualties.

The Oakland Naval Hospital, Oakland, CA, and the National Naval Medical Center, Bethesda, MD, also were supplying frozen O positive and O negative red cells to the frozen blood banks in Vietnam.

Adverse Effects of Plasma Protein Fraction

The U.S. Navy Surgical Research Teams serving in Vietnam from 1966 to 1974 were treating combat casualties with a resuscitation fluid consisting of a 25% albumin solution, 5% albumin solution, plasma protein fraction (PPF), isotonic sodium chloride, and Ringer's D,L lactate solution. To reduce the level of supernatant hemoglobin infused into the patients with the previously frozen red cells, it was necessary to concentrate the deglycerolized red cells to hematocrit values of about 80 V% by centrifugation in a refrigerated centrifuge. Volume expanders had to be infused along with the red cell concentrates with hematocrit values of 80 V%. Earlier studies had shown that when plasma protein fraction (PPF) was infused with washed previously frozen red cell concentrates, the incidence of pulmonary dysfunction, morbidity and mortality were greater than when crystalloid solutions consisting of isotonic sodium chloride and Ringer's D,L lactate were used (Carey et al, 1971). The Naval Surgical Research Teams found that they could achieve successful resuscitation in patients in hemorrhagic shock by infusing a combination of crystalloid solution with either liquid-stored whole blood or washed previously frozen red cells (Moss, 1968; 1969).

Frozen Red Cells Shipped to Vietnam

The frozen red cells were shipped to Vietnam in polystyrene containers in dry ice, from the U.S. Naval Blood Research Laboratory at the Chelsea Naval Hospital, Chelsea, MA and from frozen blood banks established at the Oakland Naval Hospital, Oakland, CA and at the National Naval Medical Center, Bethesda, MD. At the Oakland Naval Hospital, CAPT David Rulon, MC, USN was in charge of collecting and freezing red cells and fresh frozen plasma. Additionally, there was in operation at Clark Air Force Base in the Philippines between 1967 and 1972 a frozen blood bank established primarily to provide previously frozen washed red cells for patients undergoing hemodialysis. It was assumed at this time that washed previously frozen red cells would prevent isosensitization to tissue antigens in patients with renal insufficiency. The red cells were shipped in the frozen state to Clark Air Force Base in dry ice and they were thawed and washed there prior to use.

In all the initial studies in Vietnam, only O positive and O negative red cells lacking Kell, Duffy and Kidd antigens were being used, at the recommendation of Dr. Grove-Rasmussen, Director of the Blood Bank at the Massachusetts General Hospital, Boston, MA, who reported that patients usually become isosensitized to Kell, Duffy and Kidd antigens. In subsequent studies, however, O positive and O negative red cells were not tested for Kell, Duffy and Kidd antigens.

The 500 units of selected red cells (O cde/cde, K-, Fya-; O CDe/CDe, K-, Fya-) used in the initial studies were collected with the cooperation of the Massachusetts Chapter of the American Red Cross, the Massachusetts General Hospital, Boston, MA, and the Naval Hospitals at Beaufort, SC and Chelsea, MA. These units were frozen with glycerol within 5 days of collection in ACD (NIH, Formula A); 0.3% Na2EDTA was added to prevent the development of Coombs positive red cells previously observed (Valeri CR, 1966b). The frozen red cells stored at -80 C were shipped in polystyrene foam containers with dry ice to Danang, South Vietnam and to the hospital ships USS Repose and USS Sanctuary via Oakland, CA and Subic Bay, Philippines.

The Huggins freeze-preserved red cells transfused to patients in Vietnam were centrifuged prior to transfusion to remove the supernatant fluid that contained the products of hemolysis. The recipients suffered no adverse reactions from these transfusions.

Disadvantages of Huggins Method

Despite the satisfactory clinical results with Huggins freeze-preserved red cells in Vietnam, we became aware of the limitations of the Huggins method. Cryopreservation research did continue with the goal of providing freeze-preserved O positive and O negative washed red cells as a supplement to the supply of liquid blood.

The Vietnam feasibility study was successful insofar as it provided seriously needed red blood cells for injured servicemen. As regards the Huggins technique, several disadvantages became apparent to us. The volume of wash solutions (6.8 liters) required to prepare one unit of blood was excessive, as was the in vitro loss of red cells during processing (approximately 23-25%), and the time required for processing (approximately 50 minutes). Still, the principal aim of the project had been successful, i.e., to test the practicability of using cryopreserved red blood cells with crystalloid solutions under combat conditions as a supplement to ACD whole blood (Moss, 1968, 1969; Valeri et al, 1968; Moss et al 1968).

In Vitro and In Vivo Testing of Preserved Red Blood Cells

The Naval Blood Research Laboratory, Chelsea, MA, devised the following systematic approach to the clinical evaluation of preserved red blood cells:

1. In vitro testing to establish conditions for optimum recovery of red blood cells after thawing and washing.
2. In vivo survival measurements of 10 ml aliquot autologous transfusions of preserved red blood cells in healthy volunteers. The 10 ml aliquot sample from the preserved unit was labeled with radioactive chromium before autotransfusion, and the red cell volume of the recipient was measured independently using iodinated albumin or Evans blue to measure the plasma volume and the total body hematocrit (peripheral venous hematocrit multiplied by 0.89).
3. Clinical observations in stable, anemic medical patients after multiple units of compatible homologous identifiable preserved red blood cells.
4. Clinical observations in surgical patients after multiple homologous transfusions during and following operations of various types.

Tests were established to determine the posttransfusion survival of cryopreserved red blood cells. In vitro measurements were made of red blood cell adenosine triphosphate (ATP), adenosine diphopshate (ADP), adenosine monophosphate (AMP) levels, total nucleotide level, glutathione level, glutathione stability, hexokinase level, glucose-6-phosphate dehydrogenase level, and glutathione reductase level. In addition, physical and structural measurements were made of: red blood cell indices, density distributions of the red blood cells, osmotic fragility and lipid content, plasma haptoglobin level, red cell affinity for oxygen, red cell compatibility, oxygen content in blood, and the levels of diethylhexylphthalate (DEHP) in liquid preserved and cryopreserved blood products (Lionetti et al, 1964; Lionetti et al, 1966; Valeri et al, 1965a; Valeri et al, 1965b; Valeri and McCallum, 1965a; Valeri and McCallum, 1965b; Almond and Valeri, 1967a; Valeri et al, 1971b; Valeri et al, 1971c; Handin and Valeri, 1971; Valeri et al, 1972e; Valeri et al, 1973b; Contreras et al, 1974; Dennis et al, 1979; Dennis and Valeri, 1980). Extensive studies were performed to determine the number of white blood cells and platelets, and the residual plasma in previously frozen washed red blood cells, the mechanism of removal of these substances and the immunogenicity of previously frozen washed red blood cells (Crowley and Valeri, 1974a; Crowley and Valeri, 1974b; Crowley and Valeri 1974c; Crowley et al, 1974a; Crowley et al, 1975a; Crowley et al, 1977; Valeri, 1976a; Kurtz et al, 1978).

In Vivo Removal After Transfusion of Damaged Red Blood Cells

Under an ONR contract, the U.S. Naval Blood Research Laboratory in collaboration with Boston University studied the manner in which damaged donor red cells are removed from the recipient's circulation immediately after transfusion (Valeri, 1976a). Red cells that are irreversibly damaged during liquid or frozen storage are removed at an accelerated rate from the circulation within the first 24 hours after transfusion (Valeri, 1971). Only by making independent measurements of recipient and donor red blood cells at the time of transfusion is it possible to get an accurate determination of in vivo red blood cell loss. In a collaborative study, Dr. Charles P. Emerson of Boston University Medical School performed the manual differential agglutination studies and Dr. Irma O. Szymanski, a research associate at University Hospital used the Technicon Autoanalyzer to measure the RBC survivals (Szymanski and Valeri, 1968b; Szymanski and Valeri, 1969; Szymanski and Valeri, 1971; Szymanski et al, 1967; Szymanski et al, 1968; Szymanski et al, 1970). With the automated differential agglutination technique, it is possible to make simultaneous measurements in a single recipient of two red blood cell populations preserved by two different methods (Szymanski and Valeri, 1971; Valeri and Altschule, 1981). This approach has been used to evaluate red blood cells preserved by liquid and freezing techniques (i.e. Cohn method, Huggins method, and Pert-Krijnen-Rowe methods) (Valeri et al, 1972a; Valeri, 1976a). Measurements of 218 red blood cell survivals by the automated differential agglutination procedure in patients who had received therapeutic transfusions of washed and non-washed ACD- and CPD-stored whole blood and red blood cell concentrates showed no significant differences related to either the anticoagulant or the washing procedure (Valeri et al, 1972a).

Restoration to Normal States in Surviving Transfused Red Blood Cells

Red blood cells that are reversibly damaged during storage are removed from the circulation at a slower rate, and in these red cells the ATP, 2,3 DPG, potassium ion, and sodium ion levels, which deteriorate during liquid storage, are restored toward normal in vivo (Valeri and Hirsch, 1969; Fortier et al, 1969; Kopriva et al, 1972). The cellular composition and physical characteristics (i.e. osmotic fragility and levels of ATP, 2,3 DPG, potassium ion, and sodium ion) of the donor red blood cells are ultimately affected by the intravascular environment of the recipient (Valeri et al, 1971c). This dynamic interrelation between donor red blood cells and host environment has been demonstrated through the use of an osmotic fragility test using the Danon continuous osmotic fragility test (Valeri et al, 1971c).

Measurements of Red Cell Volume

Between 1962 and 1970 the Naval Blood Research Laboratory made extensive studies in normal healthy volunteers, using 10 ml autologous transfusions of 51Cr labeled red blood cells to measure the red blood cell volume and simultaneously making independent measurements of the red blood cell volume from the iodinated albumin or Evans blue plasma volume and the total body hematocrit (Valeri, 1976a).

The transfusion of compatible but identifiable donor red blood cells makes it possible to obtain measurements of both 24-hour posttransfusion survival values and lifespan values. An automated differential agglutination procedure was used to measure O Rh-positive and O Rh-negative red cell concentrates that were transfused into A, B, and AB recipients; simultaneous and independent measurements were made of the recipients red blood cell volume using 51Cr-labeled autologous red cells (Valeri and Altschule, 1981). The transfused red blood cell concentrates, with hematocrit values ranging from 55 to 60% or from 75 to 90%, had satisfactory survival values, even though there was a coating of the recipient red blood cells caused by isoagglutinins in the residual plasma. In a separate study by Szymanski and Valeri, 1969 in which all or almost all visible plasma was removed from the red cell concentrates before storage at 4 C for as long as 28 days, 24-hour posttransfusion survival values were greater than 70%. The recipient's state of health appeared to be an important factor in the removal of damaged red blood cells from the circulation; the seriously ill patients exhibited a defective removal of damaged red blood cells from the circulation: the seriously ill patients exhibited a defective removal mechanism, whereas otherwise healthy recipients with traumatic injuries removed irreversibly damaged red blood cells promptly (Szymanski and Valeri, 1969).

Blood volume measurements also were shown to be affected by the health of the recipient (Valeri et al, 1973a). In addition, it was shown that red blood cell volume measurements obtained from 51Cr labeled autologous red blood cells gave the most accurate measurement of red cell volume deficit (Valeri et al, 1973a). Moreover, although accurate plasma volume measurements could be obtained with iodinated 125I albumin in healthy patients and in patients with erythrocytosis, this method gave inaccurate measurements in other patients. In patients with traumatic injuries, carcinoma, cardiopulmonary disorders, and other miscellaneous diagnoses, the 125I albumin method overestimated plasma volume. Accurate plasma volumes were measured in 47 patients with traumatic injuries, carcinoma, and various other diagnoses using a cold agglutinin, a macroglobulin with a molecular weight of 1.0M, labeled with radioactive iodine (125I).

Studies of Feasibility of Using Frozen Red Blood Cells to Supplement the Liquid Blood Banking System in a Combat Zone

During the Vietnam conflict, human albumin was being used with whole blood as the primary resuscitation medium, and ACD (NIH, Formula A) was the primary liquid anticoagulant. Blood collected in ACD could be stored for only 3 weeks at 4 C, much of the blood reached its outdating period before it could be used.

Because of the sporadic casualty pattern in Vietnam from 1966 to 1968, the U.S. Navy's Research and Development Command conducted a feasibility study to evaluate the use of frozen red blood cells to supplement the blood banking system in a combat zone.

Frozen red blood cells and fresh frozen plasma served as a supplement to the supply of liquid stored whole blood. In fact, there were such large amounts of blood arriving in Vietnam that about 50 percent of it was being discarded as a result of outdating. In the 1970 to 1972 period the supply of liquid blood was more than sufficient to meet blood requirements, and so the decision was made to terminate the testing of frozen red blood cells. At this point, the U.S. Naval Blood Research Laboratory at Chelsea, Massachusetts, commenced analyses of the vast amount of data it had collected during the Vietnam study.

Data collected during the Vietnam study demonstrated that washed previously frozen red blood cells used in combination with a colloid and crystalloid solutions provided effective treatment for battle casualties (Moss, 1969). The data also showed that red blood cells frozen by the Huggins method used during these studies had in vitro recovery values of only about 70%, possibly 75% under optimum conditions, and a 25% reduction in red blood cell potassium ion related to the low ionic washing procedure. Moreover, the volume of wash solution required was excessive (6.8 liters to wash one unit). The low ionic media used in the washing process to produce agglomeration of the red blood cells in order to separate them from the supernatant resulted in a loss of potassium from the red blood cells, and when frozen red blood cells were stored at -80 C for 2 years or more, damage occurred during the subsequent washing process (Valeri, 1976a). We later substituted electrolyte solutions for the non-electrolyte solutions which were used to wash Huggins-preserved red blood cells. This substitution resulted in better maintenance of red blood cell potassium, improved recovery of red blood cells in vitro, and a reduction in the volume of solutions for red blood cell washing using electrolyte solutions by serial centrifugation or continuous-flow centrifugation (Valeri, 1976a).

Over a 3-year period, between 1966 and 1969, in which the U.S. Navy used more than 2,000 units of frozen processed group O, Rh-positive red blood cells and group O Rh negative red blood cells in Danang, South Vietnam, along with ACD whole blood, close monitoring of the patient showed satisfactory results (Moss, 1969).

Results of the Vietnam study convinced the Navy that the cryopreservation of blood components was an area of investigation worth continuing, not only for military use but for civilian use as well. Thus, studies on cryopreservation of blood products were continued at the Naval Blood Research Laboratory, first while this facility was still a part of the Chelsea Naval Hospital where many wounded servicemen had been sent from Vietnam, and subsequently when the U.S. Naval Blood Research Laboratory entered into a contract with University Hospital at the Boston University School of Medicine, where many important developments were made, not only with red blood cell cryopreservation but with the preservation of other blood components as well.

Studies of Patients with Traumatic Injuries at the Chelsea Naval Hospital

Large numbers of wounded servicemen from Vietnam with musculoskeletal injury complicated by a condition called "stress anemia" were sent to the Chelsea Naval Hospital's Orthopedic Service for treatment and recuperation (Biron et al, 1972). These patients exhibited 30 to 40% reduction in both red blood cell volume and plasma volume associated with normal hemoglobin and hematocrit values in peripheral venous blood. Contrary to what would be expected, these patients did not adapt pathophysiologically to the anemic hypoxia by increasing cardiac output or by red blood cell production, and although they maintained their central red cell volumes, their peripheral red blood cell volumes were markedly reduced (that is, blood volume to the muscle, bone and skin of the extremities and the gastrointestinal tract). A decreased red blood cell oxygen affinity was associated with a significantly increased red blood cell 2,3 diphosphoglycerate level which, in turn, was associated with an increase in systemic arteriovenous difference in oxygen content (Valeri and Fortier, 1969a; Valeri and Fortier, 1969b; Valeri and Collins, 1971a; Valeri and Collins, 1971b). The transfusion of washed liquid preserved red blood cells and previously frozen red blood cells resulted in correction of the red cell volume deficit as well as clinical improvement reflected in an increase perfusion to the extremities (Valeri and Altschule, 1981).

Some patients exhibited recurring red cell volume deficits with or without apparent blood loss, and blood volume measurements were repeated in those who clinically were not doing well. When the red blood cell volumes were calculated from the 125I iodinated plasma volume and the total body hematocrit, overestimations of the red blood cell volume were observed and it was determined that the 51Cr labeling technique should be used for a reliable measurement of the red blood cell volume deficit (Valeri et al, 1973a; Valeri and Altschule, 1981).

Lifespan of Preserved Red Blood Cells

The 24-hour survival value is an accurate indication of the percent of viable red blood cells in the transfusion. In a study of 39 patients with traumatic injuries, 44 long-term red cell survival measurements were made (Szymanski and Valeri, 1971), and estimates of the lifespan and rate of random destruction of the preserved red blood cells in each recipient were made with the use of computer technology.

Accelerated linear removal of red blood cells was seen in severely injured patients, and improved red blood cell survival was associated with improvement in health. The correlation between the lifespan of the transfused red cells and the recipient's general health suggested that the decreased long-term survival noted in these recipients was produced by some extracorpuscular "toxic" factor. Many of these young men, who had serious and poorly healed wounds in one or more extremity, had received immediate medical treatment at an aid station where bleeding was stopped and transfusions were administered. Many of these men required more than 80 units of blood to restore circulation to what appeared to be a normal state, a puzzling dilemma inasmuch as the normal blood volume is about ten units of blood. Equally puzzling was the fact that many of these transfusions were administered when the patient was no longer bleeding (Valeri and Altschule, 1981). The physicians at Chelsea Naval Hospital were faced not only with the problem of what had happened to the missing blood, but what could be done to prevent this phenomenon.

In his care of these patients, Dr. P. Biron of the Orthopedic Service at Chelsea Naval Hospital had observed that the usual treatment was not producing expected results. Debridement of wounds, done under anesthesia, generally a relatively simple procedure designed to remove the dead tissue and stimulate the growth of new tissue, produced life threatening cardiovascular collapse. The initial impression, that the patient's blood volume was reduced, appeared unlikely in view of the large numbers of transfusions administered. Moreover, the usual tests for anemia, the blood hemoglobin and hematocrit levels were normal.

This medical puzzle was presented to the U. S. Naval Blood Research Laboratory. Why were these patients with chronic wounds destroying autologous red blood cells and allogeneic donor red blood cells and why had this problem not been recognized earlier? The Naval Blood Research Laboratory measured the red blood cell volumes, which were found to be low, sometimes by as much as 40%. Frequent transfusions to maintain normal red cell volume greatly improved the health of these patients, i.e. wounds healed, appetite increased, moods became cheerful.

The problem of the "Missing Blood Syndrome" in these wounded servicemen was never completely resolved. However, much data were gathered which will doubtless contribute to the treatment of trauma. The studies performed at the Chelsea Naval Hospital defined the mechanisms responsible for the disorder and indicated its treatment but did not elucidate the cause. Tests were performed in vitro to determine whether red blood cells of patients with traumatic injuries would hemolyze in the presence or absence of adrenochrome, a metabolite of epinephrine. The red blood cells of patients with traumatic injuries were found to have increased spontaneous hemolysis after incubation with adrenochrome at 37 C for up to 48 hours. Increased susceptibility to the hemolytic effect of adrenochrome, an indole metabolite of epinephrine, was observed in one-third of the patients studied (Valeri et al, 1972b; Valeri and Altschule, 1973; Valeri and Altschule, 1981). The data collected during these studies have been published in a book entitled "Hypovolemic Anemia of Trauma: The Missing Blood Syndrome", Chemical Rubber Company Press, Boca Raton, FL, 1981 by CR Valeri and MD Altschule.

Importance of Red Cell 2,3 DPG Levels

While providing blood products for the wounded servicemen at the Chelsea Naval Hospital, the Naval Blood Research Laboratory was alerted to a medical phenomenon that led to what is probably the most important contribution to blood banking made in the past 20 years. The majority of these patients who had significantly reduced red blood cell volumes also had significantly increased RBC 2,3 DPG levels, from 0.8 moles 2,3 DPG/ mole hemoglobin to 2.5 to 3.0 moles 2,3 DPG/mole hemoglobin (Valeri and Fortier, 1969a; Valeri and Fortier, 1969b). Our observation that the compensatory increase in RBC 2,3 DPG from 1-1/2 to 3 times normal, helped improve red cell oxygen delivery in these patients with reductions of 30 to 50 percent the total number of red cells, prompted us to conduct studies to biochemically treat red cells in vitro to achieve this increase in 2,3 DPG with its attendant improved oxygen delivery.

At the Chelsea Naval Hospital physicians treated and studied more than 300 patients sent there with serious wounds incurred in Vietnam. These patients, who exhibited increased levels of red blood cell 2,3 DPG to 1-1/2 to 2-3 times normal, were able to maintain a normal central blood volume (red blood cell volume and plasma volume), but were not able to maintain an adequate peripheral red blood cell volume and plasma volume. Red blood cell production was found to be impaired and red cell survival reduced (Biron et al, 1972). The majority of these patients suffered from a chronic hypovolemic state, with significant reductions in both plasma volume and red blood cell volume, a condition which had gone undiagnosed until the extensive investigation at Chelsea Naval Hospital. The chronic hypovolemic state in these patients caused a reduction in blood flow to the extremities and impaired wound healing of the extremities (Valeri and Altschule, 1981). These patients with chronic hypovolemia but with no evidence of systemic hypotension or decrease in renal function were given transfusions of liquid-stored red blood cell concentrates, liquid-stored washed red blood cells, and previously frozen deglycerolized red blood cells to restore their red cell volume. Red blood cells were frozen by the low glycerol method (20% W/V) as well as by the high glycerol method (40% W/V) were used (Valeri, 1976a).

Red blood cell and plasma volumes were increased after the transfusion of washed liquid stored red blood cells and washed previously frozen red blood cells with hematocrit values of 75 to 80 V%, even though no plasma proteins were infused. All these patients had, while in Vietnam, received liquid stored whole blood and many of them had become alloimmunized. At Chelsea Naval Hospital they agreed to additional transfusion therapy with great reluctance because they feared a recurrence of non-hemolytic transfusion reactions. The washed liquid stored and previously frozen red blood cells produced no such adverse reactions.

Posttransfusion Hepatitis and Cytomegalovirus Infections

Serum samples were collected from 104 patients treated at Chelsea Naval Hospital between 1969 and 1972 were used in a retrospective study for the purpose of determining the incidence of hepatitis B and cytomegalovirus infections following the transfusion of non-washed liquid stored red cells and washed liquid stored and previously frozen red cells (Contreras et al, 1979). The donor blood had not been tested for the hepatitis B antigen prior to transfusion, and the incidence of posttransfusion hepatitis B antigen was about 2.8%. The incidence of antibody to cytomegalovirus was about 22% before transfusion, and 22% of the patients developed complement fixing antibody against cytomegalovirus after transfusion. This retrospective study showed that washed red blood cell products were associated with a lower dose of the hepatitis B antigen, a delayed production of antibody to the hepatitis B antigen, and a lower level of antibody.

Oxygen Transport Function of Preserved Red Blood Cells

It was the observation of Benesch and Benesch and Chanutin and Curnish in 1967 that brought to light the role of red blood cell 2,3 diphosphoglycerate (2,3 DPG) in the delivery of oxygen from the red blood cells to tissue; these observations explained the increased red blood cell oxygen affinity in liquid stored red blood cells with reduced 2,3 DPG levels (Valtis and Kennedy, 1954). The reports of these investigators considered in context with our observations of the "Missing Blood Syndrome" stimulated a study at the Naval Blood Research Laboratory in 1969 in which liquid-stored red blood cells were transfused to anemic patients to evaluate the ability of these recipients to restore red blood cell 2,3 DPG, ATP, potassium, and sodium ion concentrations. The patients were given compatible but identifiable preserved red blood cells that had less than 10% of normal RBC 2,3 DPG and 75% of normal ATP, 80% of normal RBC potassium ion levels, and 100% of normal RBC sodium ion levels. A manual differential agglutination procedure was used to recover the donor red blood cells to make the measurements (Valeri and Hirsch, 1969). RBC ATP levels were found to have increased rapidly within 4 to 6 hours after transfusion at a time when RBC sodium ion levels decreased rapidly. Red blood cell 2,3 DPG levels increased to 50% of normal during the 24-hour posttransfusion period and continued to increase toward a normal level over the 8-10 days following transfusion, and red blood cell potassium ion levels increased at rates similar to RBC 2,3 DPG levels.

These studies also showed correlations between the red blood cell 2,3 DPG level and the in vitro p50 value of the oxyhemoglobin dissociation curve. P50 values were measured by the Bellingham and Huehns method, utilizing a tonometer attached to a cuvette. Samples were obtained from the patient before and after transfusion; the donor red blood cells suspension from the posttransfusion sample was tonometered to assess the oxygen saturation and the P02 tension to define the red cell oxygen affinity at pH 7.2 and a pCO2 of 0 (Valeri and Collins, 1971a; Valeri and Collins, 1971b). The patient's red cell in vivo p50 value was measured before transfusion, and the combined value of the patient and donor red blood cells was measured after transfusion using a Lex-O2-CON galvanic cell to measure the oxygen content of blood (Valeri et al, 1972e).

Measurements of Posttransfusion Survival of Preserved Red Cells

At first we used both a manual and an automated differential agglutination procedure for RBC survival measurements, but once we were sure of the validity of the automated method using the Technicon Auto-Analyzer we used this method (Szymanski and Valeri, 1968b; Szymanski and Valeri, 1971; Szymanski et al, 1967; Szymanski et al, 1968; Szymanksi et al, 1970). Using the automated differential agglutination procedure and ABO, Rh and MN red cell antigens, we were able to make simultaneous survival measurements of ABO and Rh identifiable red cells that had been preserved by two different methods, and to make comparisons of the methods in a single recipient (Szymanski and Valeri, 1971; Valeri and Altschule, 1981).

Using a manual differential agglutination procedure to recover liquid preserved group O red cells after transfusion to group A1 recipients, we studied changes in intracellular levels of 2,3 DPG, ATP, sodium ion, and potassium ion (Valeri and Hirsch, 1969). A significant increase in cellular 2,3 DPG was seen immediately at the completion of a 2 to 3 hour transfusion. Within 3 hours after transfusion, the level of 2,3 DPG was 3.48 umoles per gram of hemoglobin or 0.233 moles per mole of hemoglobin. Within 24 hours after transfusion the level of 2,3 DPG was about 50% the final level, and it then continued to increase gradually for 11 days. The ATP level of the transfused red cells increased rapidly, a change that was associated with a rapid decrease of the donor cell sodium ion content during the 24 hour posttransfusion period. Donor cellular potassium ion concentration increased slowly, and the rate of increase appeared to be related to the increasing level of intracellular 2,3 DPG. There had been previous reports of a relation between the organic phosphates, ATP and 2,3 DPG, and oxyhemoglobin dissociation characteristics of preserved red cells. This study reinforced the impression that the method selected for red cell preservation should be one that maintains the levels of organic phosphates, so that the cells have normal affinity for oxygen at the time of transfusion (Valeri 1969).

Valeri and Collins (1971a, 1971b) later studied the physiologic effects of the transfusion of 3-5 units of washed liquid stored red cells with low 2,3 DPG levels and increased affinity for oxygen in anemic hypoxic patients. There was no change in oxygen consumption immediately after transfusion, but there was a significant decrease in both the arterial blood pH and the systemic arteriovenous difference in oxygen content, and the circulating red cells had an increased affinity for oxygen and a decreased red cell 2,3 DPG level. Within 4 hours after the transfusion, both the arterial pH and the systemic arteriovenous difference in oxygen content had returned toward the pretransfusion levels; the 2,3 DPG level and p50 value of the oxyhemoglobin dissociation curve were restored to normal in vivo within 24 hours. Cardiac index values measured by the indocyanine method were unchanged and in accord with those calculated by the Fick formula prior to and 8 and 24 hours after transfusion. During the 4-hour posttransfusion period, however, the cardiac index calculated by the Fick formula was significantly increased, while the measurement made by the dye method was unchanged (Valeri and Collins, 1971a, 1971b).

Kopriva et al (1972) measured RBC 2,3 DPG, plasma inorganic phosphate, venous blood pH, whole blood lactic acid, plasma creatine, and red cell creatine, in severely injured battle casualties after transfusion of at least 12 units of ACD collected whole blood. The red cell 2,3 DPG level rose rapidly during the first 12 hours posttransfusion, and was within normal range 48 hours after initial sampling. The rapid increase in the red cell 2,3 DPG level during the first 12 hours was associated with an increase in the venous whole blood pH. Between 48 and 120 hours posttransfusion, a higher than normal RBC 2,3 DPG level correlated significantly with an increased RBC creatine level, although the role of creatine was not determined.

Studies at the Naval Blood Research Laboratory, Chelsea, MA conducted between 1968 and 1970 revealed the superiority of the anticoagulant CPD over ACD in maintaining the red cell 2,3 DPG level and ultimately the oxygen transport function during liquid storage at 4 C. When liquid stored red cells with low 2,3 DPG levels are subsequently frozen, the washed previously frozen red cells have impaired oxygen transport function at the time of transfusion. The oxygen transport function of washed previously frozen red cells is also influenced by the composition and pH of the wash solution used to remove the glycerol, the composition and pH of the solution used during post-wash storage of the red cells before transfusion (Valtis and Kennedy 1954; Valeri 1974a, 1974b, 1974c, 1974e, 1974g, 1974h, 1974i, 1975, 1976a; Valeri and Zaroulis, 1972a, 1972b). The cryopreservation process, whether by the high glycerol method at -80 C or by the low glycerol method at -150 C, produces no adverse effect on red cell 2,3 DPG, ATP, p50, potassium ion, or pH (Valeri, 1976a).

The CPD anticoagulant provides good maintenance of red cell 2,3 DPG level for 3 to 5 days of liquid storage at 4 C, and satisfactory maintenance for 7 to 10 days if the unit is stored as a red cell concentrate rather than as a unit of whole blood. During the initial 24 to 48 hours of 4 C storage, CPD-stored red cell concentrates actually were found to exhibit an increase in the RBC 2,3 DPG level (Valeri, 1976a). Red cell concentrates stored in CPD should be frozen within 8 days of collection to ensure normal or only slightly impaired oxygen transport function.

Biochemical Modification to Prepare Red Cells with Normal or Improved Oxygen Transport Function

After extensive study of the RBC 2,3 DPG levels and subsequently the oxygen transport function of liquid stored red blood cells, the Naval Blood Research Laboratory developed a solution to biochemically modify red cells to restore in vitro the 2,3 DPG and ATP levels which deteriorate during 4 C storage. In early studies, two rejuvenation solutions were evaluated: PIGP solution composed of pyruvate, inosine, glucose, and phosphate; and PIGPA solution composed of pyruvate, inosine, glucose, phosphate and adenine (Valeri and Zaroulis, 1972a, 1972b; Valeri, 1974h, 1974i; Valeri, 1976a). PIGPA solution was subsequently modified and called PIPA solution with the removal of glucose. The 50 ml volume of PIPA solution used for rejuvenation of one unit of red cells contains pyruvate 100 mM/l, inosine 100 mM/l, phosphate 100 mM/l, and adenine 5 mM/l, and remains stable at 4 C for at least 1 ½ years.

Initial studies of biochemical modification or "rejuvenation" as it is called were done to ensure that patients in specific clinical situations would receive red cells with normal or improved oxygen transport function important for their clinical care. Now rejuvenation is being used at research centers as well as by certain chapters of the American Red Cross to salvage outdated O positive and O negative red cells that would otherwise be discarded. The cryopreservation of universal donor outdated rejuvenated red cells is seen as a feasible means of supplementing liquid blood banking systems during periods of low donations or high use. A report from the Naval Blood Research Laboratory entitled "Rejuvenation and Freezing of Outdated Stored Human Red Cells" was published in the New England Journal of Medicine in 1972 (Valeri and Zaroulis 1972a). Since that time, the Naval Blood Research Laboratory has published more than 75 papers on the biochemical treatment of indated and outdated red cells.

When indated red cells are biochemically modified, they have 2 to 3 times normal 2,3 DPG levels and improved oxygen transport function. When outdated red cells are biochemically modified, they have 1 ½ to 2 times normal 2,3 DPG and normal or slightly improved oxygen transport function. Although both the high and low glycerol methods can be used to freeze outdated and indated rejuvenated red cells (Valeri 1976a), the high glycerol method is the method of choice at the Naval Blood Research Laboratory.

Red cells with improved oxygen transport function have been administered to stable anemic patients, to patients undergoing cardiopulmonary bypass surgery and treated with hypothermia, and to patients undergoing elective resection of abdominal aneurysm (Dennis et al, 1975, 1978; Valeri, 1976a; Valeri et al, 1978, 1979, 1980a, 1908b, 1980c). These clinical studies have demonstrated the safety and therapeutic effectiveness of outdated and indated rejuvenated red cells.

Stimulated by the results of studies initiated during the Vietnam conflict and continued thereafter, research at the Naval Blood Research Laboratory, Boston, MA has for the past years been directed toward the simplification of the rejuvenation and freezing processes, so that rejuvenated cryopreserved red cells could be provided at a reasonable cost and still maintain their safety and therapeutic effectiveness. A polyvinylchloride (PVC) plastic collection bag has been developed which accomplishes these goals (Valeri et al 1981a, 1981b). This multiple bag collection system is used to collect the blood, separate the components, rejuvenate, glycerolize and freeze and store the red cells. The high glycerol freezing method must be used with this new collection system: the PVC plastic bag will withstand storage at -80 C but will break if stored in liquid nitrogen. Our data show that red cells frozen with 40% W/V glycerol can be stored at -80 C for at least 4 years (Valeri et al 1980a). With this new method of freezing in the primary collection bag, freezer space is better utilized, a smaller amount of wash solution is required, and the potential for contamination is reduced (Valeri et al, 1981a, 1981b).

Platelet Isolation and Preservation

Platelet concentrates used in the treatment of thrombocytopenia are most effective when used within 2 days of collection and storage at 4 C or after 2 days of storage at 22 C. In Vietnam, neither fresh blood nor fresh or liquid stored platelets were readily available for the treatment of wounded servicemen with thrombocytopenia. During that period, the Naval Blood Research Laboratory, Chelsea, MA, was evaluating the cryopreservation of human platelets for use in combat areas. A major need then and now, the isolation of sufficient numbers of platelets for therapeutic effectiveness, is being investigated.

Serial differential centrifugation is one method by which platelets are isolated from a unit of whole blood (Hogman 1974; Valeri 1974j, 1976a, 1976b). Platelet concentrates isolated by serial differential centrifugation must be stored undisturbed at room temperature for a period to resuspend the platelets: the resuspension time is 60 minutes for platelets isolated from CPD blood, and 30 minutes for platelets isolated from ACD blood (Valeri, 1976a). Platelet isolation can be greatly facilitated by adding a reversible platelet inhibitor, such as prostaglandin E1 (PGE1), to the blood (Allen and Valeri, 1974; Valeri, 1974f; Valeri et al, 1972c, 1972d). PGE1 must be added directly to the blood; if added to the blood by way of the anticoagulant, PGE1 will not remain stable during storage. We found that PGE1 increased the recovery of platelets in vivo after storage in the liquid state or in the frozen state with DMSO at -80 C (Valeri et al, 1972c, 1972d).

At the Naval Blood Research Laboratory, platelets isolated from a unit of blood have been frozen in polyvinylchloride and polyolefin plastic bags with good results. The platelets were frozen with 5% DMSO and stored at -150 C in polyolefin plastic bags or with 6% DMSO and stored at -80 C in polyvinylchloride plastic bags, thawed, washed, and stored in autologous plasma (Handin and Valeri 1972; Valeri, 1974d; Valeri and Feingold, 1974; Valeri et al, 1973c, 1974a; Spector et al, 1977a, Zaroulis et al, 1979a). The Naval Blood Research Laboratory has reported on tests to measure the recovery of platelets after freeze-thaw-wash procedures (Handin et al, 1971; Vecchione et al, 1981) as well as on procedures to assess the function of washed previously frozen platelets, including electron microscopy, platelet aggregation patterns, platelet oxygen consumption, platelet ATP and ADP levels, platelet factors 3 and 4 activity, and the platelet release of 14 C serotonin following treatment with increasing concentrations of thrombin (Crowley et al, 1974c; Robblee et al, 1979; Spector et al, 1977c, 1979; Valeri, 1981).

As part of the studies at the Naval Blood Research Laboratory, Chelsea, MA, to evaluate the hemostatic effectiveness of liquid stored and washed previously frozen platelets, normal volunteers were treated with aspirin to increase their bleeding times and then given autologous platelet transfusions (Handin and Valeri 1971; Valeri 1974d, 1974j, 1976b). Platelets stored at 4 C for 24 hours were found to be more hemostatically effective than either fresh platelets or platelets stored at 22 C for 24 hours, but had markedly shortened lifespans. Cryopreserved platelets, on the other hand, i.e. platelets frozen with 6% DMSO and storage at -80 C had both improved hemostatic effectiveness immediately after transfusion and normal lifespan, with in vivo circulation about 50% that of fresh platelets (Valeri, 1974d, 1976a, 1976b). Platelets were isolated from a unit of blood and frozen with 4 or 5% DMSO and stored at -80 C for as long as 8 months, with no evidence of deterioration (Spector et al, 1977a). The defect in aggregation noted in platelets after storage at 22 C for 24 to 48 hours was corrected in vivo during the 24 hour posttransfusion period in a manner similar to that in which the respiratory defect of liquid stored red cells is corrected during the posttransfusion period (Handin and Valeri, 1971; Valeri, 1976a). Platelets stored at 22 C for 24 hours exhibited an in vivo circulation that was about 70% that of fresh platelets; the value was about 50% for platelets stored at 4 C for 24 hours (Valeri, 1974d, 1976a, 1976b).

The Naval Blood Research Laboratory has also used cell separating machines to isolate 6 to 8 units of platelets from healthy volunteers by plateletpheresis (Valeri 1974j, 1976a, 1981). In our early studies with the Haemonetics Blood Processor 10, platelet injury and reduced in vivo circulation were observed (Valeri 1974j, 1976a), but were corrected when modifications were made in the disposable bowl used with this machine (Valeri 1981).

In 1971, the Naval Blood Research Laboratory applied for an IND to study DMSO as a cryoprotectant to freeze platelets, with CAPT C.R. Valeri, MC, USN, as the sponsor, monitor and principal investigator. After more than 10 years of studies involving healthy human volunteers who received washed previously frozen platelets containing 300-400 mg of residual DMSO, we have encountered no serious side effects from the DMSO, nor have periodic eye examinations revealed any evidence of lenticular opacities or other abnormalities.

As helpful as these in vivo studies have been, what we need are dependable in vitro tests that will help predict the viability and function of preserved platelets. Earlier studies showed a correlation between platelet response to hypotonic stress and in vivo circulation (Handin et al, 1970; Valeri et al, 1974b). More recently, the Naval Blood Research

Laboratory has studied quality control tests to determine the safety and therapeutic effectiveness of cryopreserved platelets. Samples obtained from the platelet-DMSO mixture before freezing and from the washed platelet suspension are fixed with 1% glutaraldehyde with phosphate buffered saline. The freeze thaw wash recovery value is then calculated from the platelet count and the volume of platelets frozen and the platelet count and the volume of platelets after washing. Using the H-4 Coulter counter, it is now possible to assess the population and volume distribution histograms and the integrity of the platelets after the freeze thaw wash process. Platelet function can be assessed by the dense body content measured in the 1% glutaraldehyde fixed platelet samples obtained before and after freezing.

Liquid storage of platelets with agitation for 5 days is not a feasible program for military contingency planning (Valeri 1981). Platelet cryopreservation is the best approach. Our recent studies show that multiple units of platelets can be isolated from a single donor by apheresis using a mechanical cell separating system such as the Haemonetics Blood Processor 30, the IBM Blood Processor 2997 with a single or dual separating chamber, or the Fenwal CS-3000; DMSO-cryopreserved platelets (6 to 8 units), either plateletpheresed units or a pool prepared from units of ABO compatible blood, have been shown to be hemostatically effective. Platelet survival values indicate that 2 units of frozen platelets are needed to achieve the same number of platelets in the recipient circulation as one unit of fresh platelets (Zaroulis et al, 1979a).

Platelets have been frozen with 6% DMSO in PVC plastic bags and stored at -80 C for 13 months. After thawing and washing with a crystalloid solution composed of 0.9% NaCl-0.2% glucose-40 mg% inorganic phosphorus, pH 5.0, the residual DMSO was less than 5% of the original concentration (less than 300 mg per unit). The washed platelets resuspended in 50 ml of plasma can be stored at room temperature for 6 to 8 hours before use.

Based on data obtained from ongoing studies at the Naval Blood Research Laboratory, Boston, MA, a Standard Operating Procedure is being prepared for the cryopreservation of human platelets with 6% DMSO and storage at -80 C for 1 year. An application for licensure for this protocol will be submitted by the Surgeon General of the U.S. Navy through the Blood Bank at Bethesda, MD.

The Bureau of Biologics (BOB) recently established DMSO as a processing solution rather than as a drug. As glycerol is used to freeze red cells, DMSO is used to freeze platelets.

Granuloycte Isolation and Preservation

Granulocyte isolation is much more difficult than red cell or platelet isolation; granulocyte preservation also is very difficult (Crowley and Valeri, 1974d; Crowley et al, 1974b; Valeri 1976a, 1981; Lionetti et al 1977, 1978, 1980a, 1980b; Contreras et al, 1978; Roy et al, 1978; Wade et al 1977). Few human studies have been made on granulocyte collection and preservation, much research having been done in guinea pigs, dogs and baboons (Valeri, 1981).

In vitro methods have been established to assess granulocyte function and to test granulocyte oxygen consumption associated with phagocytosis of latex particles (Crowley et al 1975b). Granulocytes have been frozen with a combination of HES and DMSO, together with Normosol-R, glucose and albumin (Lionetti et al, 1980b; Valeri, 1981), but the circulation and function of cryopreserved granulocytes have not been studied in man at the NBRL.

Use of Baboon for In Vivo Studies

In 1972 the Naval Blood Research Laboratory began using the baboon as a model in which to study methods of preserving red cells, platelets and granulocytes (Valeri and Ragno, 2006). The baboon has served as an excellent model for evaluation of preservation of nonrejuvenated and rejuvenated red cells (Herman et al, 1971; Rice et al, 1975; Spector et al, 1977b; Valeri et al, 1975a, 1975b; Zaroulis et al, 1979b). Baboon red cells have been rejuvenated with the same solution and in the same manner as human red cells (Herman et al, 1971; Valeri et al, 1975a, 1975b). The transfusion of baboon red cells with low 2,3 DPG levels to treat acute blood loss has been shown to produce an increased cardiac output, whereas red cells with increased 2,3 DPG produce a decrease in cardiac output (Rice et al, 1975). The baboon has also been used to study the effects of hyperventilation with and without reduction in red blood cell volume on cerebral blood flow and oxygenation (Valeri et al 1975a, 1975b). In vivo studies in the baboon ensure the safety of a procedure before human studies are performed (Valeri and Ragno, 2006; Callow et al, 1980).

Because of the emergency situations in which blood was needed during the Vietnam conflict, consideration was given to the use of blood salvaged during surgery. Baboon studies were done at the Naval Medical Research Institute, Bethesda, MD, in collaboration with the U.S. Naval Blood Research Laboratory, to determine the usefulness of shed blood (Herman et al 1974; Kingsley et al, 1973). Shed blood was collected from baboons during surgery and reinfused with or without preinfusion washing. When the shed blood was not washed before reinfusion there was evidence of disseminated intravascular coagulation, but this problem was not encountered when the red cells were washed before reinfusion into the baboon.

The U.S. Naval Blood Research Laboratory continues to evaluate the safety and therapeutic effectiveness of shed blood and its potential role in the treatment of combat casualties in combat areas. Although the use of shed blood in emergency situations may be worthy of consideration, washing shed blood is a complicated process (Valeri, 1976a). The cryopreservation of autologous red cells and plasma in advance of anticipated surgery is thought to be a more sensible approach (Daane and Valeri, 1970; Daane et al, 1969; Howarth and Valeri, 1973).

Post Vietnam Developments in the Frozen Blood Banking System

The treatment and study of wounded servicemen at the Chelsea Naval Hospital, Chelsea, MA, led to research at the Naval Blood Research Laboratory that resulted in a biochemical modification process which promises to revolutionize blood banking practices. Liquid stored red cells are biochemically modified to increase their 2,3 DPG and ATP levels that deteriorate during storage in the liquid state at 4 C (Valeri 1974h, 1974i, 1976a). Red blood cells so treated are called "rejuvenated red cells".

Liquid preserved red blood cells with reduced 2,3 DPG levels have impaired oxygen transport function immediately after transfusion, and in certain patients this impairment may be critical (Valeri 1976a). Red cells that are biochemically modified within 8 days of liquid storage have 2,3 DPG levels increased to 2 to 3 times normal, ATP levels increased to 1 ½ times normal, and improved oxygen delivery capacity (Valeri et al, 1978, 1979, 1980a, 1980b, 1980c; Valeri, 1976a). In studies made here, excellent in vitro and in vivo characteristics also have been observed in outdated universal O positive and O negative red cells, rejuvenated after liquid storage beyond their acceptable shelf life, frozen with 40% W/V glycerol at -80 C, and stored frozen for as long as 4 years.

More recently, the Naval Blood Research Laboratory has developed a multiple bag collection system which is made commercially and which we use for freezing both nonrejuvenated and rejuvenated red blood cells. The primary bag of this system is used for blood collection, component separation, rejuvenation, cryopreservation, and post wash storage; this process reduces the potential for contamination and is less costly than previously used methods (Valeri et al, 1981a). Red blood cells prepared in this manner have been found to have excellent viability and function.

A label indicating blood type and identification number of the donor is affixed to the primary bag of the collection system at the time of blood collection, and the label remains affixed during frozen storage. The previously frozen red cells are washed before transfusion to remove substances such as the anticoagulant preservative, rejuvenation solution, glycerol solution, products of hemolysis, isoagglutinins, protein and nonprotein plasma components, and at least 95% of the white cells and platelets (Valeri, 1976a).

Outdated O positive and O negative red blood cells have been salvaged by rejuvenation at a time when they would otherwise be discarded (after 28 days of storage in CPD, the most commonly used anticoagulant, or after 35 days of storage in CPDA-1, a newly licensed anticoagulant (Valeri et al, 1979). Outdated rejuvenated red cell concentrates have been frozen with 40% W/V glycerol at -80 C in the PVC plastic primary collection bag, and after 4 years of frozen storage have been washed and stored in a resuspension medium for as long as 3 days at 4 C with only minimal hemolysis (Valeri et al, 1981a).

In our early studies in which red blood cells were rejuvenated with PIGPA solution (pyruvate, inosine, glucose, phosphate and adenine) after 23 days of 4 C storage in ACD, frozen by the high glycerol method and washed before transfusion, oxygen transport function was normal and 24-hour posttransfusion survivals were about 75% (Valeri and Zaroulis. 1972a). The Naval Blood Research Laboratory has since developed a new rejuvenation solution, PIPA solution with excellent results. A 50 ml aliquot of PIPA solution contains pyruvate 100 mM/l, inosine 100 mM/l, phosphate 100 mM/l, and adenine 5 mM/l. This is a clear solution which remains stable at 4 C for at least 18 months.

The low glycerol freezing method was once preferred by many researchers because only 1.5 liters of wash solution were needed to reduce the glycerol concentration to less than 1%, whereas 3.0 liters were needed to wash red cells frozen with the high glycerol method (Valeri 1976a). However, with a new approach developed at the Naval Blood Research Laboratory, only 1.6 liters of wash solution are needed to reduce the glycerol concentration of high glycerol freeze-preserved red blood cells to less than 1%, and hemolysis is less pronounced than with the low glycerol method (Valeri et al, 1981a). Moreover, when the newly developed multiple-bag collection system is used for either nonrejuvenated or rejuvenated red blood cells, the low glycerol method at -150 C cannot be used because the PVC plastic bag will break at the low temperature of liquid nitrogen (Valeri et al, 1981a).

Rejuvenated washed previously frozen red blood cells with improved oxygen transport function have been used successfully in anemic patients with or without cardiopulmonary insufficiency, in patients undergoing cardiopulmonary bypass and hypothermia, and in instances where nonhemolytic transfusion reactions have been a clinical problem (Valeri et al, 1979). Rejuvenated red cells also have been prepared for pediatric transfusion in small aliquots. Four small aliquots of red blood cells are prepared from a single unit of blood because only a small volume of red blood cells is required for each pediatric transfusion (Valeri et al, 1981b).

The goal of a safe and effective method, cost-efficient for wide-scale use, has now been achieved through the intensive research efforts at the Naval Blood Research Laboratory. Another accomplishment has been the satisfactory isolation and preservation of platelets. This involved the collection of a sufficient number of platelets for therapeutic effectiveness in a recipient, and a long-term preservation process that ensures therapeutic effectiveness at the time of transfusion.

In one of the earlier studies at the Naval Blood Research Laboratory, each of the 42 healthy male volunteers was first given 650 ml of aspirin to produce a prolonged bleeding time, and then treated with liquid-stored or previously frozen autologous platelet. A platelet concentrate stored in the liquid state at 4 C for 24 hours before transfusion reduced the bleeding time, but platelet survival was decreased (Valeri, 1974d). Platelets stored in the liquid state at 22 C for 24 hours and platelets frozen and stored with 5% DMSO at -150 C for 24 hours before transfusion had normal lifespans but did not reduce the bleeding time. However, platelets that were frozen and stored with 6% DMSO at -80 C for 24 hour exhibited both hemostatic effectiveness and normal lifespans (Valeri 1974d, 1976b).

The Naval Blood Research Laboratory recently directed a collaborative study to determine the feasibility of maintaining cryopreserved red cells and platelets at various military sites. The blood products used in these studies were processed in the newly developed multiple bag collection system consisting of an 800 ml primary collection bag and 3 empty transfer packs integrally attached by a line of tubing connected by an adaptor port. The primary bag is used for red cell preparation, one transfer pack is used for preparation of a platelet concentrate, and one for platelet-poor fresh frozen plasma. The third transfer pack is used for collection of the supernatant glycerol solution from the red cell concentrate before freezing. The hematocrit level of the red cell concentrate during storage in the liquid state at 4 C should be 75 to 80 V%.

The feasibility study was carried out as follows: a 450 ml volume of blood was drawn into the primary collection bag containing either CPD or CPDA-1 anticoagulant, both of which are licensed by the Bureau of Biologics. Two other anticoagulants, CPDA-2 and CPDA-3, unlicensed Investigational New Drugs, also were studied. The blood components, red blood cells, platelets, and plasma were prepared within 8 hours of collection.

Units of red blood cells that were frozen with 40% W/V glycerol within 3 to 5 days of 4 C storage are referred to as nonrejuvenated frozen red blood cells. Units that were rejuvenated with PIPA solution after they had reached their outdating period are referred to as outdated rejuvenated frozen red blood cells. Rejuvenation restores to normal or increases the red cell 2,3 DPG and ATP levels which fall during liquid storage; red blood cells rejuvenated after only a few days of storage have higher 2,3 DPG and ATP levels than units rejuvenated after longer periods of storage. It is practical to freeze only O positive and O negative red blood cells since they can be transfused to any patient. A, B and AB red blood cells are not practical for freezing; their limited transfusion usefulness would not warrant it. Nonrejuvenated frozen red blood cells have been stored at -80 C for as long as 10 years. Outdated rejuvenated frozen red cells have been stored at -80 C for as long as 4 years.

The hematocrit level of previously frozen red cells, both nonrejuvenated and rejuvenated during post-wash storage at 4 C should be about 40 V%. Red cells have been stored at this hematocrit level in a sodium chloride glucose phosphate solution at 4 C for 3 days, with acceptable survival values of at least 70% and normal or only slightly impaired oxygen transport function after transfusion (Valeri et al, 1981a). Before transfusion, the red cells are centrifuged to remove the supernatant and to adjust the hematocrit to a concentration of 80 V%. Frozen red cells in the primary PVC plastic bag have been shipped in polystyrene foam containers with dry ice with no adverse effects.

In June 1979 the Naval Blood Research Laboratory directed a collaboration study of a Frozen Blood Banking System at the PACOM Blood Program Office, Okinawa, Japan. Equipment used at a Frozen Blood Bank for rejuvenation and freezing of red cells includes: rejuvenation solution, water bath, glycerol solution, modified Eberbach shakers for addition of the glycerol, refrigerated centrifuges, -80 C mechanical freezers, and washing solutions. Each air-cooled mechanical freezer holds 250 frozen units. In studies at the Naval Blood Research Laboratory, we have used both the IBM Blood Processor which utilized automated serial centrifugation, and the Haemonetics Blood Processor 115, a continuous-flow centrifugation wash process. With 8 Haemonetics Blood Processor 115's, two technicians can wash 16 units of red cells in 1 hour. The wash solution consists of a 50 ml volume of 12% sodium chloride and 2.0 liters of 0.9% sodium chloride, 0.2% glucose, and 40 mg/dl inorganic phosphorus solution.

The Naval Blood Research Laboratory has also made significant progress in platelet cryopreservation: platelets frozen with DMSO in PVC plastic bags have been stored at -80 C for at least 1 year. After thawing, the platelets are washed with a sodium chloride glucose phosphate solution to remove about 95% of the DMSO, and the washed platelets can be stored resuspended in plasma at 22 C for 6 hours before transfusion.

Platelets have been obtained by isolation from individual units of blood and pooled before cryopreservation. Multiple units of platelets also have been collected from a single donor by apheresis using either the Haemonetics Blood Processor 30 or the IBM Blood Processor 2997. Platelets frozen with 6% DMSO at 2-3 C per minute in a -80 C mechanical freezer have been shown to have freeze-thaw-wash recovery values of 75% and in vivo survival of 50% of fresh platelets. Two units of cryopreserved platelets are needed to achieve the same number of circulating platelets as one unit of fresh platelets. Platelets frozen as described above and washed before transfusion have improved hemostatic function immediately after transfusion.

In more recent studies at the Naval Blood Research Laboratory, ABO compatible platelets isolated from individual units of blood have been pooled and frozen with DMSO in the same type PVC plastic bag as is used for red cell freezing. The DMSO is diluted with sodium chloride solutions to prepare a 27% solution of DMSO for cryopreservation of the pooled ABO compatible platelets. The post-wash resuspension medium is prepared from a 50 ml sample of plasma obtained from one of the units before pooling and frozen in a 150 ml PVC plastic bag alongside the pooled platelets in an aluminum container stored in a -80 C mechanical freezer. The platelets are thawed in a water bath maintained at 42 C in 2 to 2.5 minutes, and then washed with 250 ml of a solution containing 0.9% sodium chloride, 0.2% glucose, 40 mg/dl inorganic phosphorus, with a pH of 5.0. After concentration by centrifugation to remove the supernatant solution, the platelets are resuspended in the 50 ml of thawed plasma prior to transfusion.

Studies at the Naval Blood Research Laboratory have shown that cryopreserved nonrejuvenated and rejuvenated O positive and O negative red cells and cryopreserved group O pooled or single donor apheresed platelets stored at -80 C in mechanical freezers would be an important supplement to the current blood banking system.

The account presented here from 1966 to 1982 describes the manner in which red blood cells, platelets, and plasma were preserved for use in resuscitation procedures. Much remains to be done as regards the use of blood products in trauma resuscitation. Alert readers of publications from this and other laboratories are aware of the unsolved problems:

1. The adverse effects of administering human albumin solution in patients subjected to trauma in man (Lucas et al, 1978, 1979; Johnson et al, 1979; Dahn et al, 1979).
2. A demonstrably worsening clinical condition produced by plasma protein fraction given to wounded men in Vietnam (Carey 1971).
3. Evidence suggestive of a toxic factor in the genesis of the post trauma "missing blood syndrome", with no decision as to whether the cause of this toxic factor is the presence of massive wounds or something generated in the liquid stored blood that is given in large amounts to the wounded (Cannon et al, 1918; Valeri and Altschule 1981). The massive transfusion of liquid stored blood

did not cure the "missing blood syndrome" but the transfusion of washed red blood cells appeared to alleviate it.

4. The evidence that "aged" human plasma damages the lungs (Mayer et al, 1981).

CHAPTER 6 REFERENCES

1. Allen JE, Valeri CR. Prostaglandins in hematology. Arch Int Med 133:86-96, 1974.
2. Almond DV, Valeri CR. Relationship between lipid fractions of erythrocytes and their in vivo survival following preservation with glycerol and the slow freeze technic. Transfusion 7:10-16, 1967a.
3. Almond DV, Valeri CR. The in vivo effects of deglycerolized agglomerated erythrocytes transfused in multiple units to stable anemic patients. Transfusion 7:95-104, 1967b.
4. Benesch R, Benesch RE. The effect of organic phosphate from the human erythrocyte on the allosteric properties of hemoglobin. Biochem Biophys Res Comm 26:162-167, 1967.
5. Biron PE, Howard J, Altschule MD, Valeri CR. Chronic deficits in red-cell mass in patients with orthopaedic injuries (stress anemia). J Bone Joint Surg 54-A:1001-1014, 1972.
6. Callow AD, Ledig CB, O'Donnell TF, Kelly JJ, Rosenthal D, Korwin S, Hotte C, Kahn PC, Vecchione JJ, Valeri CR. A primate model for the study of the interaction of ^{111}In-labeled baboon platelets with dacron® arterial prostheses. Ann Surg 191:362-366, 1980.
7. Cannon WB. Acidosis in cases of shock, hemorrhage and gas infection. JAMA 70:531-535, 1918.
8. Cannon WB, Fraser J, Hooper AN. Some altercations in distribution and character of blood in shock and hemorrhage. JAMA 70:526-531, 1918.
9. Carey LC, Lowery BD, Cloutier CT. Hemorrhagic Shock in Current Problems in Surgery, January 1971, Yearbook Medical Publishers, Inc., Chicago.
10. Chanutin A, Curnish RR. Effects of organic and inorganic phosphates on the oxygen equilibrium of human erythrocytes. Arch Biochem 121:96-102, 1967.
11. Chaplin H, Jr., Jaffe ER, Lenfant C, Valeri CR. Preservation of Red Blood Cells. National Academy of Sciences, Washington, DC, 1973.
12. Contreras TJ, Sheibley RH, Valeri CR. Accumulation of di-2-

ethylhexyl phthalate (DEHP) in whole blood, platelet concentrates, and platelet poor plasma. Transfusion 14:34-46, 1974.

13. Contreras TJ, Hunt SM, Lionetti FJ, Valeri CR. Preservation of human granulocytes. III. Liquid preservation studied by electronic sizing. Transfusion 18:46-53, 1978.
14. Contreras TJ, Lang DJ, Pivacek LE, Valeri CR. Occurrence of HB_sAg, anti-HB_s, and anti-CMV following the transfusion of blood products. Transfusion 19:129-136, 1979.
15. Crowley JP, Valeri CR. The purification of red cells for transfusion by freeze preservation and washing. I. The mechanism of leukocyte removal from washed, freeze-preserved red cells. Transfusion 14:188-195, 1974a.
16. Crowley JP, Valeri CR. The purification of red cells for transfusion by freeze preservation and washing. II. The residual leukocytes, platelets, and plasma in washed, freeze-preserved red cells. Transfusion 14:196-202, 1974b.
17. Crowley JP, Valeri CR. The purification of red cells for transfusion by freeze-preservation and washing. III. Leukocyte removal and red cell recovery after red cell freeze-preservation by the high or low glycerol concentration method. Transfusion 14:590-594, 1974c.
18. Crowley JP, Valeri CR. Recovery and function of granulocytes stored in plasma at 4 C for one week. Transfusion 14:574-580, 1974d.
19. Crowley JP, Skrabut EM, Valeri CR. Immunocompetent lymphocytes in previously frozen washed red cells. Vox Sang 26:513-517, 1974a.
20. Crowley JP, Rene A, Valeri CR. The recovery, structure, and function of human blood leukocytes after freeze-preservation. Cryobiology 11:395-409, 1974b.
21. Crowley JP, Rene A, Valeri CR. Changes in platelet shape and structure after freeze preservation. Blood 44:599-603, 1974c.
22. Crowley JP, O'Donnell M, Sell KW, Valeri CR. The purification of red cells for transfusion by freeze-preservation and washing. IV. The use of micropore filtration to reduce the residual HL-A antigenicity of previously frozen, washed red cells. Transfusion 15:34-38, 1975a.
23. Crowley JP, Skrabut EM, Valeri CR. The determination of

leukocyte phagocytic oxidase activity by measurement of the initial rate of stimulated oxygen consumption. J Lab Clin Med 86:586-594, 1975b.

24. Crowley JP, Wade PH, Wish C, Valeri CR. The purification of red cells for transfusion by freeze-preservation and washing. V. Red cell recovery and residual leukocytes after freeze-preservation with high concentrations of glycerol and washing in various systems. Transfusion 17:1-7, 1977.
25. Daane TA, Valeri CR, Barton RK. Autotransfusions of previously frozen blood in elective gynecologic surgery. Am J Obstet Gynecol 105:394-399, 1969.
26. Daane TA, Valeri CR. Autologous transfusions of previously frozen blood in elective surgery. Lab Med 1:34-35, 1970.
27. Dahn MS, Lucas CE, Ledgerwood AM, Higgins RF, Negative ionotropic effect of albumin resuscitation for shock. Surgery 235-241, 1979.
28. Dennis RC, Vito L, Weisel RD, Valeri CR, Berger RL, Hechtman HB. Improved myocardial performance following high 2,3 diphosphoglycerate red cell transfusions. Surgery 77:741-747, 1975.
29. Dennis RC, Hechtman HB, Berger RL, Vito L, Weisel RD, Valeri CR. Transfusion of 2,3 DPG-enriched red blood cells to improve cardiac function. Ann Thoracic Surg 26(1):17-26, 1978.
30. Dennis RC, Bechthold D, Valeri CR. In vitro measurement of p50--the pH correction, and use of frozen red blood cells as controls. Crit Care Med 7:385-390, 1979.
31. Dennis RC, Valeri CR. Measuring percent oxygen saturation of hemoglobin, percent carboxyhemoglobin and methemoglobin, and concentrations of total hemoglobin and oxygen in blood of man, dog, and baboon. Clin Chem 26:1304-1308, 1980.
32. Fortier NL, Hirsch NM, Valeri Cr. Restoration of 2,3 DPG and ATP in ACD-stored red blood cells. Forsvarsmedicin 5:250-257, 1969.
33. Handin RI, Valeri CR. Hemostatic effectiveness of platelets stored at 22 C. NEJM 295:538-543, 1971.
34. Handin RI, Valeri CR. Improved viability of previously frozen platelets. Blood 40:509-513, 1972.
35. Handin RI, Fortier NL, Valeri CR. Platelet response to hypotonic

stress after storage at 4 C or 22 C. Transfusion 10:305-309, 1970.

36. Handin RI, Lawler KC, Valeri CR. Automated platelet counting. Am J Clin Path 56:661-664, 1971.
37. Haynes LL, Tullis JL, Pyle HM, Sproul MT, Wallach S, Turville WC. Clinical use of glycerolized frozen blood. JAMA 173:1657-1663, 1960.
38. Haynes LL, Turville WC, Sproul MT, Zemp JW, Tullis JL. Long-term blood preservation – a reality. J Trauma 2:2-9, 1962.
39. Herman CM, Rodkey FL, Valeri CR, Fortier NL. Changes in the oxyhemoglobin dissociation curve and peripheral blood after acute red cell mass depletion and subsequent red cell mass restoration in baboons. Ann Surg 74:734-743, 1971.
40. Herman CM, Kingsley JR, Valeri CR, Peters H, Cole BC, Fouty WJ. Autotransfusion for treatment of experimental hemorrhagic shock. In Acute Fluid Replacement in the Therapy of Shock, Ed Malinin et al, Stratton Intercontinental Medical Book Corp, New York, 1974, pp 111-117.
41. Hogman C, Krijnen HW, Valeri CR. Platelet preservation and transfusion. Proc Int Soc Blood Transf Meeting, Amsterdam 1974.
42. Howarth HC, Valeri CR. Autotransfusion of frozen red blood cells in elective oral surgery. J Oral Surg 31:40-43, 1973.
43. Huggins CE. Frozen blood: theory and practice. JAMA 193:941-945, 1965.
44. Johnson SD, Lucas CE, Gerrick SJ, Ledgerwood AM, Higgins RF. Altered coagulation after albumin supplements for treatment of oligemia. Arch Surg 114:379-383, 1979.
45. Kingsley JR, Valeri CR, Peters H, Cole BC, Fouty WJ, Herman CM. Citrate anticoagulation and on-line cell washing in intraoperative autotransfusion in the baboon. Surg Forum 24:258-260, 1973.
46. Kopriva CJ, Ratliff JL, Fletcher JR, Fortier NL, Valeri CR. Biochemical and hematological changes associated with massive transfusion of ACD-stored blood in severely injured combat casualties. Ann Surg 176:585-589, 1972.
47. Kurtz SR, Van Deinse WH, Valeri CR. The immunocompetence of residual lymphocytes at various stages of red cell cryopreservation

with 40% W/V glycerol in an ionic medium at -80 C. Transfusion 18:441-447, 1978.

48. Lionetti FJ, Valeri CR, Bond JC, Fortier NL. Measurement of hemoglobin binding capacity of human serum or plasma by means of dextran gels. J Lab Clin Med 64:519-528, 1964.
49. Lionetti FJ, Valeri CR, Bond JC, Kivowitz C, Weinman E. Nucleotides in frozen glycerolized erythrocytes. Transfusion 6:116-123, 1966.
50. Lionetti FJ, Hunt SM, Lin PS, Kurtz SR, Valeri CR. Preservation of human granulocytes. II. Characteristics of granulocytes obtained by counterflow centrifugation. Transfusion 17:465-472, 1977.
51. Lionetti FJ, Hunt SM, Mattaliano RJ, Valeri CR. In vitro studies of cryopreserved baboon granulocytes. Transfusion 18(6):685-692, 1978.
52. Lionetti FJ, Lin PS, Mattaliano RJ, Hunt SM, Valeri CR. Temperature effects on shape and function of human granulocytes. Exp Hemat 8:304-317, 1980a.
53. Lionetti FJ, Hunt SM, Valeri CR. Isolation of human blood phagocytes by counterflow centrifugation elutriation: In: Methods of Cell Separation, Vol. 3, edited by N Catsimpoolas, Plenum Press, New York, 3:141-152, 1980b.
54. Lucas CE, Weaver D, Higgins RF, Ledgerwood AM, Johnson SD, Bouwman DL. Effects of albumin versus non-albumin resuscitation on plasma volume and renal excretion. J Trauma 18:564-570, 1978.
55. Lucas CE, Ledgerwood AM, Higgins RF. Impaired salt and water excretion after albumin resuscitation for hypovolemic shock. Surgery 86:544-549, 1979.
56. Mayer JE, Kersten TE, Humphrey EW. Effects of transfusion of emboli and aged plasma on pulmonary capillary permeability. J Thorac Cardiovasc Surg 82:358-364, 1981.
57. Moss GS. The use of balance salt solution in the resuscitation of battle casualties in Vietnam. In: Intermedes Proceedings-Combined Injuries and Shock, pp 291-302, 1968.
58. Moss GS, Valeri CR, Brodine CE. Clinical experience with the use of frozen blood in combat casualties. New Engl J Med 278:748-752, 1968.
59. Moss GS. Massive transfusion of frozen preserved red cells in

combat casualties: report of three cases. Surgery 66:1008-1013, 1969.

60. Rice CL, Herman CM, Kiesow LA, Homer LD, John DA, Valeri CR. Benefits from improved oxygen delivery of blood in shock therapy. J Surg Res 19:193-198, 1975.
61. Robblee LS, Shepro D, Vecchione JJ, Valeri CR. Increased thrombin sensitivity of human platelets after storage at 4C. Transfusion 19:45-52, 1979.
62. Roy AJ, Ramirez M, Valeri CR. Stability of neutrophil-releasing activity of plasma obtained from leukapheresed rats and stored at 4 C. Transfusion 18:734-737, 1978.
63. Runck AH, Valeri CR. Recovery of glycerolized red blood cells frozen in liquid nitrogen. Transfusion 9:297-305, 1969.
64. Runck AH, Valeri CR. Measurement of the rate of red cell oxygenation and deoxygenation in the Technicon AutoAnalyzer. In Proc Technicon International Congress, New York, NY, 1970, pp 477-479.
65. Runck AH, Valeri CR. Continuous-flow centrifugation washing of red blood cells. Transfusion 12:237-244, 1972.
66. Runck AH, Valeri CR, Sampson WT. Comparison of the effects of ionic and non-ionic solutions on the volume and intracellular potassium of frozen and non-frozen human red cells. Transfusion 8(1):9-18, 1968.
67. Spector JI, Yarmala JA, Marchionni LD, Emerson CP, Valeri CR. Viability and function of platelets frozen at 2 to 3 C per minute with 4 or 5 percent DMSO and stored at -80C for 8 months. Transfusion 17:8-15, 1977a.
68. Spector JI, Zaroulis CG, Pivacek LE, Emerson CP, Valeri CR. Physiologic effects of normal-or-low-oxygen affinity red cells in hypoxic baboons. Am J Physiol 232(1):H79-84, 1977b.
69. Spector JI, Skrabut EM, Valeri CR. Oxygen consumption, platelet aggregation and release reactions in platelets freeze-preserved with dimethylsulfoxide. Transfusion 17:99-109, 1977c.
70. Spector JI, Flor WJ, Valeri CR. Ultrastructural alterations and phagocytic function of cryopreserved platelets. Transfusion 19:307-312, 1979.
71. Szymanski IO, Valeri CR. Evaluation of double 51Cr technique. Vox Sang 15:287-292, 1968a.

72. Szymanski IO, Valeri CR. Automated differential agglutination technic to measure red cell survival. II. Survival in vivo of preserved red cells. Transfusion 8:74-83, 1968b.
73. Szymanski IO, Valeri CR, McCallum LE, Emerson CP, Rosenfield RE. Automated differential agglutination technic to measure red cell survival. I. Methodology. Transfusion 8:65-73, 1968.
74. Szymanksi IO, Valeri CR. Clinical evaluation of concentrated red cells. NEJM 280:281-287, 1969.
75. Szymanski IO, Valeri CR. Factors influencing chromium elution from labeled red cells in vivo and the effect of elution on red cell survival measurements. Br J Haemat 19:397-409, 1970a.
76. Szymanski IO, Valeri CR. Analysis of erythrocyte survival curves obtained simultaneously by 51Cr and an automated differential agglutination technic. Transfusion 10:287-298, 1970b.
77. Szymanski IO, Valeri CR. Lifespan of preserved red cells. Vox Sang 21:97-108, 1971.
78. Szymanski IO, Dean HM, Valeri CR, Bougas JA, Desforges JF. Measurement of erythrocyte survival during open-heart surgery. Transfusion 10:163-170, 1970.
79. Szymanski IO, Valeri CR, Emerson CP. Effect of chromium elution on red cell survival measurements. In Proc XII Congr Int Soc Blood Transf, Moscow, USSR, August 17-23, 1969, Bibl Hemat 38:268-269, Karger, Basel, 1971.
80. Szymanski IO, Lipson CS, Valeri CR. Elution of chromium label from preserved red blood cells transfused during surgery. Transfusion 13:13-18, 1973.
81. Tullis JL, Ketchel MM, Pyle HM, Pennel RB, Gibson JG, Tinch RJ, Driscoll SG. Studies on the in vivo survival of glycerolized and frozen human red blood cells. JAMA 168:399-404, 1958.
82. Valeri CR. Effect of resuspension medium on in vivo survival and supernatant hemoglobin of erythrocytes preserved with glycerol. Transfusion 5:25-35, 1965a.
83. Valeri CR. Observations on recipient plasma hemoglobin concentration after transfusion with glycerolized frozen blood. Transfusion 5:36-53, 1965b.
84. Valeri CR. The in vivo survival, mode of removal of the non-viable cells, and the total amount of supernatant hemoglobin in deglycerolized, resuspended erythrocytes. I. The effect of the

period of storage in ACD at 4C prior to glycerolization. II. The effect of washing deglycerolized, resuspended erythrocytes after a period of storage at 4C. Transfusion 5:273-285, 1965c.

85. Valeri CR. In vivo survival and supernatant hemoglobin of autologous, deglycerolized, resuspended erythrocytes processed using centrifugation. III. The effect of the length of storage at -80 C. Transfusion 6:112-115, 1966a.
86. Valeri CR. The in vivo survival of Coombs positive autologous erythrocytes produced by agglomeration. Transfusion 6:247-253, 1966b.
87. Valeri CR. Clinical evaluation of erythrocytes preserved for prolonged periods. Military Med 131:705-710, 1966c.
88. Valeri CR. Frozen blood. NEJM 275:425-431, 1966d.
89. Valeri CR. Comparative evaluation of freezing techniques. Int. Working Conf. on the Freeze-Preservation of Blood, November 28-December 1, 1967, ONR Report DR-143, pp 125-140, 1967.
90. Valeri CR. Observations on the chromium labelling of ACD-stored and previously frozen red cells. Transfusion 8:210-219. 1968a.
91. Valeri CR. Clinical effectiveness of concentrated liquid stored red cells and previously frozen red cells. Maryland State Med J 17:59-64, 1968b.
92. Valeri CR. Preservation of human red blood cells. Bull NY Acad Med 44:3-17, 1968c.
93. Valeri CR. Can current blood-bank practices be improved? Medical Counterpoint 1:25-28, 1969.
94. Valeri CR. Recent advances in the preservation of human red cells by glycerol freezing technics. In Proc Int Symp on Modern Problems of Blood Preservation, March 1969, Frankfurt, Germany G Fischer, Verlag, Stuttgart, pp 125-137, 1970a.
95. Valeri CR. International Forum – In routine blood transfusion, should ACD anticoagulants be replaced by CPD or ACD adenine solutions? If so, by which one? Vox Sang 19:560-561, 1970b.
96. Valeri CR. Viability and function of preserved red cells. NEJM 284:81-88. 1971.
97. Valeri CR. Overview of red cell preservation and survival. In Proc of the Conference on Red Cell Preservation. National Research Council – National Academy of Sciences Division of

Medical Sciences, June 5-6, 1972, Washington, DC, pp 265-297, 1972.

98. Valeri CR. Preservation of red cells: research and technology. In Proc Int Coll Surg Meting, 18th World Congress, May 1972, Rome, Italy, Excerpta Medica, 1972, pp 540-542 1973a.

99. Valeri CR. Status report: blood preservation. Contemporary Surgery 2:21-31, 1973b.

100. Valeri CR. Factors influencing the 24-hour posttransfusion survival and the oxygen transport function of previously frozen red cells preserved with 40 percent W/V glycerol and frozen at -80 C. Transfusion 14:1-15, 1974a.

101. Valeri CR. Oxygen transport and viability of preserved red blood cells. J Med 5:278-291, 1974b.

102. Valeri CR. Use of blood components in the treatment of hemorrhagic shock. In Acute Fluid Replacement in the Therapy of Shock, Ed. Malinin et al, Stratton Intercontinental Medical Book Corp, New York, pp 119-1137, 1974c.

103. Valeri CR. Hemostatic effectiveness of liquid-preserved and previously frozen human platelets. NEJM 290:353-358, 1974d.

104. Valeri CR. Cryopreservation of red cells – freeze preservation of non-rejuvenated red cells, indated rejuvenated red cells, and outdated rejuvenated red cells. In Proc of the Australian Society of Blood Transf Symposium, Canberrra, September 30, 1974, pp 58-90, 1974e.

105. Valeri CR. Prostaglandins and their role in clinical medicine. Chicago Clin Chem 7:32-34, 1974f.

106. Valeri CR. Quality of red cell transfusion. Giornale di Medicina Militare 124:7-18, 1974g.

107. Valeri CR. Metabolic regeneration of depleted erythrocytes and their frozen storage. In The Human Red Cell In Vitro, Ed. TJ Greenwalt and GA Graham, Grune & Stratton, New York, London, pp 281-321, 1974h.

108. Valeri CR: Oxygen transport function of preserved red cells. In Clinics in Haematology, Anaemia and Hypoxia, ED L Garby, Vol 3, No 3, October 1974, pp 649-688, 1974i.

109. Valeri CR: Therapeutic effectiveness of human platelets freeze-preserved with dimethylsulfoxide at -80 C. Presented at the International Society of Blood Transfusion Meeting. In Platelet

Preservation and Transfusion, Ed CF Hogman, HW Krijnen, and CR Valeri, 1974, pp 41-50, 1974j.

110. Valeri CR. Simplification of the methods for adding and removing glycerol during freeze-preservation of human red blood cells with the high or low glycerol methods: Biochemical modification prior to freezing. Transfusion 5:195-218, 1975.

111. Valeri CR. Blood Banking and the Use of Frozen Blood Products. Chemical Rubber Company, Boca Raton, FL, 1976a.

112. Valeri CR. Circulation and hemostatic effectiveness of platelets stored at 4 C or 22 C: Studies in aspirin-treated normal volunteers. Transfusion 16:20-23, 1976b.

113. Valeri CR. The current state of platelet and granulocyte cryopreservation. Crit Rev Clin Lab Sci 14(1):21-74, 1981.

114. Valeri CR, Henderson ME. Recent difficulties with frozen glycerolized blood. JAMA 188:1125-1131, 1964.

115. Valeri CR, McCallum LE. Relationship between glutathione stability and in vivo survival of autologous, deglycerolized resuspended red blood cells. Nature 205:561-563, 1965a.

116. Valeri CR, McCallum LE. The age of human erythrocytes lost during freezing and thawing with glycerol using the Cohn Fractionator. Transfusion 5:421-426, 1965b.

117. Valeri CR, Bond JC. Observations on the preservation of autologous human erythrocytes using glycerol, slow-freeze technic and agglomeration. Transfusion 6:254-262, 1966.

118. Valeri CR, Brodine CE. Current methods for processing frozen red cells. Cryobiology 5:129-135, 1968.

119. Valeri CR, Runck AH. Long term frozen storage of human red blood cells: Studies in vivo and in vitro of autologous red blood cells preserved up to six years with high concentrations of glycerol. Transfusion 9:5-14, 1969a.

120. Valeri CR, Runck AH. Viability of glycerolized red blood cells frozen in liquid nitrogen. Transfusion 9:306-313, 1969b.

121. Valeri CR, Fortier NL. Red-cell 2,3-diphosphoglycerate and creatine levels in patients with red-cell mass deficits or with cardiopulmonary insufficiency. NEJM 81:1452-1455, 1969a.

122. Valeri CR, Fortier NL. Red-cell mass deficits and erythrocyte 2,3 DPG levels. Forsvarsmedicin 5:212-218, 1969b.

123. Valeri CR, Hirsch NM. Restoration in vivo of erythrocyte

adenosine triphosphate, 2,3-diphosphoglycerate, potassium ion, and sodium ion concentrations following the transfusion of acid-citrate-dextrose-stored human red blood cells. J Lab Clin Med 73:722-733, 1969.

124. Valeri CR, Collins FB. The physiologic effect of transfusing preserved red cells with low 2,3-diphosphoglycerate and high affinity for oxygen. Vox Sang 20:397-403, 1971a.

125. Valeri CR, Collins FB. Physiologic effects of 2,3-DPG-depleted red cells with high affinity for oxygen. J Appl Physiol 31:823-827, 1971b.

126. Valeri CR, Kopriva CJ. Oxyhemoglobin dissociation curve in hemorrhagic and septic shock. In The Fundamental Mechanisms of Shock Ed LB Hinshaw and BG Cox, Plenum Publishing Corp, New York, pp 177-193, 1972.

127. Valeri CR, Zaroulis CG. Rejuvenation and freezing of outdated stored human red cells. NEJM 287:1307-1313, 1972a.

128. Valeri CR, Zaroulis CG. Cryopreservation and red cell function. In Progress in Transfusion and Transplantation, Ed. PJ Schmidt, 1972, pp 343-365, 1972b.

129. Valeri CR, Altschule MD. Hemolysis in vitro of blood obtained from patients with traumatic injuries. J Trauma 13:678-686, 1973.

130. Valeri CR, Szymanski IO. Further studies on the rapid and slow components of chromium elution in vivo from preserved erythrocytes. Vox Sang 24:502-514, 1973.

131. Valeri CR, Feingold H. Hemostatic effectiveness of liquid-preserved platelets stored at 4 C or 22 C and freeze-preserved platelets stored with 5 percent DMSO at -150 C or stored with 6 percent DMSO at -80 C. In Platelets : Production, Function, Transfusion and Storage, Ed MG Baldini and S. Ebbe, Grune & Stratton, New York, 1974, pp 377-393.

132. Valeri CR, Altschule MD. Hypovolemic Anemia of Trauma, The Missing Blood Syndrome, Chemical Rubber Company, Boca Raton, FL, 1981.

133. Valeri CR, Mercado-Lugo R, Danon D. Relationship between osmotic fragility and in vivo survival of autologous deglycerolized resuspended red blood cells. Transfusion 5:267-272, 1965a.

134. Valeri CR, Bond JC, Fowler K, Sobucki J. Quantitation of serum

hemoglobin binding capacity using cellulose acetate membrane electrophoresis. Clin Chem 11:581-588, 1965b.

135. Valeri CR, Bond JC, McCallum LE. Relationships between metabolic state and (1) in vivo survival and (2) density distribution of previously frozen human erythrocytes. Transfusion 6:543-553, 1966a.
136. Valeri CR, McCallum LE, Danon D. Relationships between in vivo survival and (1) density distribution, (2) osmotic fragility of previously frozen, autologous, agglomerated, deglycerolized erythrocytes. Transfusion 6:554-564, 1966b.
137. Valeri CR, Runck AH, McCallum LE. Observations on autologous, previously frozen, deglycerolized, agglomerated, resuspended red cells. I. Effect of storage temperatures. II. Effect of adenine supplementation of glycerolized red cells prior to freezing. Transfusion 7(2) : 105-16, 1967.
138. Valeri CR, Brodine CE, Moss GS. Use of frozen blood in Vietnam. Proc XIth Congr Int Soc Blood Transf, Sydney, 1966, Bibl Haemat 29 :735-738, 1968.
139. Valeri CR, Runck AH, Brodine CE. Recent advances in freeze-preservation of red blood cells. JAMA 208:489-492, 1969a.
140. Valeri CR, Runck AH, Sampson WT. Effects of agglomeration on human red blood cells. Transfusion 9:120-134, 1969b.
141. Valeri CR, Szymanski IO, Runck AH. Therapeutic effectiveness of homologous erythrocyte transfusions following frozen storage at -80C for up to seven years. Transfusion 10:102-112, 1970a.
142. Valeri CR, Bougas JA, Talarico L, Emerson CP, Didimizio T, Pivacek, L. Behavior of previously frozen erythrocytes used during open-heart surgery. Transfusion 10:238-246, 1970b.
143. Valeri CR, Runck AH, Brodine CE : Current status of freeze-preservation of human red blood cells with glycerol. In Proc XII Congr Int Soc Blood Transf, Moscow USSR, August 17-23, 1969, Bibl Haemat 38 :249-261, Karger Basel, 1971a.
144. Valeri CR, Hirsch NM, Runck AH, Szymanski IO, Johnson LH. An automated low ionic antiglobulin test. Transfusion 11:110-116, 1971b.
145. Valeri CR, Szymanski IO, Pivacek LE. Effects of the host on transfused preserved red blood cells. J Med 2:228-247, 1971c.
146. Valeri CR, Zaroulis CG, Fortier NL . Peripheral red cells as a

functional biopsy to determine tissue oxygen tension. In Proc IV Annual Alfred Benzon Symposium, Munksgaard, Copenhagen, pp 650-675, 1971d.

147. Valeri CR, Szymanski IO, Zaroulis CG. 24-hour survival of ACD- and CPD-stored red cells. I. Evaluation of nonwashed and washed stored red cells. Vox Sang 2:289-308, 1972a.
148. Valeri CR, Altschule MD, Pivacek LE. The hemolytic action of adrenochrome, an epinephrine metabolite. J Med 3:20-40, 1972b.
149. Valeri CR, Zaroulis CG, Rogers JC, Handin RI, Marchionni LD. Use of prostaglandins in the preparation of blood components. In Prostaglandins in Cellular Biology, Ed PW Ramwell and BB Pharriss, Plenum Press, New York, pp 5-25, 1972c.
150. Valeri CR, Zaroulis CG, Handin RI, Marchionni LD. Prostaglandins in the preparation of blood components. Science 175:539-542, 1972d.
151. Valeri CR, Zaroulis CG, Marchionni LD, Patti KJ. A simple method for measuring oxygen content in blood. J Lab Clin Med 79:1035-1040, 1972e.
152. Valeri CR, Cooper AG, Pivacek LE. Limitations of measuring blood volume with iodinated I 125 serum albumin. Arch Int Med 132:534-538, 1973a.
153. Valeri CR, Contreras TJ, Feingold H, Sheibley RH, Jaeger RJ. Accumulation of di-2-ethylhexyl phthalate (DEHP) in whole blood, platelet concentrates, and platelet-poor-plasma. I. Effect of DEHP on platelet survival and function. Environmental Health Perspectives, Jan 1973, pp 103-118, 1973b.
154. Valeri CR, Feingold H, Marchionni LD, Rogers JC. Hemostatic effectiveness of preserved human platelets. In Erythrocytes, Thrombocytes and Leukocytes, Ed E Gerlach, K Moser, E Deutch, W Wilmans, G Thieme, Stuttgard, 1973, pp 312-316, 1973c.
155. Valeri CR, Feingold, H, Marchionni LD. A simple method for freezing human platelets using 6% dimethylsulfoxide and storage at -80 C. Blood 43:131-136, 1974a.
156. Valeri CR, Feingold H, Marchionni LD. The relation between response to hypotonic stress and the ^{51}Cr recovery in vivo of preserved platelets. Transfusion 14:331-337, 1974b.
157. Valeri CR, Rorth M, Zaroulis CG, Jakubowski MS, Vescera S.

Physiologic effects of hyperventilation and phlebotomy in baboons: Systemic and cerebral oxygen extraction. Ann Surg 181:99-105, 1975a.

158. Valeri CR, Rorth M, Zaroulis CG, Jakubowski MS, Vescera S. Physiologic effects of transfusing red blood cells with high or low affinity for oxygen to passively hyperventilated, anemic baboons: Systemic and cerebral oxygen extraction. Ann Surg 181:106-113, 1975b.

159. Valeri CR, Weisel RD, Dennis RC, Mannick JA, Berger RL, Hechtman HB. Oxygen transport function of preserved red blood cells and myocardial performance. Fourth Int Congr on Red Blood Cell Metabolism and Function. In The Red Cell, Ed GJ Brewer, Alan R Liss, Inc, New York, Vol 21, 1978, pp 597-614.

160. Valeri CR, Valeri DA, Dennis RC, Vecchione JJ, Emerson CP. Biochemical modification and freeze-preservation of red blood cells: A new method. Crit Care Med 7:439-447, 1979.

161. Valeri CR, Zaroulis CG, Vecchione JJ, Valeri DA, Anastasi J, Pivacek LE, Emerson CP. Therapeutic effectiveness and safety of outdated human red blood cells rejuvenated to restore oxygen transport function to normal, frozen for 3 to 4 years at -80 C, washed, and stored at 4 C for 24 hours prior to rapid infusion. Transfusion 20:159-170, 1980a.

162. Valeri CR, Zaroulis CG, Vecchione JJ, Valeri DA, Anastasi J, Pivacek LE, Emerson CP. Therapeutic effectiveness and safety of outdated human red blood cells rejuvenated to improve oxygen transport function, frozen for about 1.5 years at -80 C, washed, and stored at 4 C for 24 hours prior to rapid infusion. Transfusion 20:263-276, 1980b.

163. Valeri CR, Yarnoz M, Vecchione JJ, Dennis RC, Anastasi J, Valeri DA, Pivacek LE, Hechtman HB, Emerson CP, Berger RL. Improved oxygen delivery to the myocardium during hypothermia by perfusion with 2,3 DPG-enriched red blood cells. Ann Thoracic Surg 30:527-535, 1980c.

164. Valeri CR, Valeri DA, Anastasi J, Vecchione JJ, Dennis RC, Emerson CP. Freezing in the primary polyvinylchloride plastic collection bag: A new system for preparing and freezing nonrejuvenated and rejuvenated red blood cells. Transfusion 21:138-149, 1981a.

165. Valeri CR, Valeri DA, Gray A, Melaragno AJ, Vecchione JJ,

Dennis RC, Emerson CP. Cryopreserved red blood cells for pediatric transfusion: Frozen storage of small aliquots in polyvinyl chloride (PVC) plastic bags. Transfusion 21:527-536, 1981b.

166. Valeri CR, Ragno G. The survival and function of baboon red blood cells, platelets, and plasma proteins: A review of the experience from 1972 to 2002 at the Naval Blood Research Laboratory, Boston, Massachusetts. Transfusion 46(8): 1-42, 2006.

167. Valtis DJ, Kennedy AC. Defective gas-transport function of stored red blood cells. Lancet 1:119-125, 1954.

168. Vecchione JJ, Chomicz SM, Emerson CP, Valeri CR. Cryopreservation of human platelets isolated by discontinuous-flow centrifugation using the Haemonetics Model 30 Blood Processor. Transfusion 20:393-400, 1980.

169. Vecchione JJ, Chomicz SM, Emerson CP, Valeri CR. Enumeration of previously frozen platelets using the Coulter counter, phase microscopy, and the Technicon optical system. Transfusion 21:511-516, 1981.

170. Wade PH. Skrabut EM, Vinciguerra L, Valeri CR. In vitro function of granulocytes isolated from blood of normal volunteers using continuous-flow centrifugation in the IBM-Aminco Centrifuge and adhesion-filtration leukapheresis using nylon fiber. Transfusion 17:136-140, 1977.

171. Zaroulis CG, Spector JI, Emerson CP, Valeri CR. Therapeutic transfusions of previously frozen washed human platelets. Transfusion 19:371-378, 1979a.

172. Zaroulis CG, Pivacek LE, Lowrie GB, Valeri CR. Lactic acidemia in baboons after transfusion of red blood cells with improved oxygen transport function and exposure to severe arterial hypoxemia. Transfusion 19:420-425, 1979b.

CHAPTER 7

COLLABORATIONS BETWEEN DR. SHUKRI KHURI AT THE WEST ROXBURY VA HOSPITAL AND THE NBRL FROM 1982 TO 2008

From 1982 to 2008 numerous collaborative studies were performed between the NBRL and the West Roxbury Veterans Hospital, West Roxbury, MA to investigate the following factors in the treatment of patients: hypothermia; hemodilution; heparin; bleeding time; hematocrit; platelet count, platelet size and platelet function; plasma oncotic protein – albumin; plasma clotting proteins – factors V, VIII, fibrinogen, von Willebrand's factor; plasma opsonic protein – fibronectin; and hemostatic agents and local pressure.

Prior to the collaborative studies with the West Roxbury Veterans Hospital, studies performed at the NBRL reported the following:

a. Limitation of the hematocrit to assess red blood cell volume in hypovolemic anemic patients
b. Need to measure RBC volume using the 51Cr radioisotope
c. Plasma volume cannot be measured in patients using iodinated albumin with a molecular weight of 68,000. Plasma volume in patients needs to be measured using iodinated cold agglutinin with a molecular weight of 1,000,000
d. Normal central blood volume and decreased peripheral blood volume in chronic hypovolemic anemic patients
e. Increased RBC 2,3 DPG and RBC creatine levels in patients with cardiopulmonary insufficiency and in patients with red blood cell volume deficits

f. Aspirin treated normal volunteers and baboons to assess platelet function of autologous non-aspirin treated preserved platelets
g. Survival and function of preserved RBC
h. Survival and function of preserved platelets
i. Oxygen transport of preserved RBC to maintain normal pO2 tensions in heart, brain and kidney
j. Hemostatic function of RBC
k. Washed RBC increased RBC volume, plasma volume, and the blood volume in chronic hypovolemic patients
l. Peripheral blood volume includes blood in the gastrointestinal tract and blood in the muscle, bone and skin of the extremities in hypovolemic patients
m. Resuscitation fluids: safety and therapeutic effectiveness of Ringer's D,L lactate, Ringer's L lactate and Ringer's ketone solutions
n. Hemostatic agents and local pressure

The following clinical events were investigated at the NBRL prior to our collaborative studies with West Roxbury Veterans Administration Hospital:

a. Acute renal failure in four (4) patients in July 1962 at the Chelsea Naval Hospital transfused with frozen deglycerolized RBC
b. Hemoglobinuria associated with frozen deglycerolized RBC at the Chelsea Naval Hospital
c. Aggregates of outdated albumin used to resuspend the deglycerolized RBC at the Chelsea Naval Hospital produced acute renal failure in 4 patients
d. Limitations of 125I-labeled albumin to measure the plasma volume in patients and the need to use 131I-labeled cold agglutinin to measure the plasma volume in patients
e. Hypotension occurred following anesthesia in wounded servicemen during debridement of wounds of the extremities
f. Chronic hypovolemic anemia in patients with traumatic injuries with normal hematocrits and hemoglobin concentrations and "stress anemia"
g. Transfusion of patients with normal hematocrits with 3 to 5 units of liquid washed RBC and previously frozen deglycerolized RBC increased the peripheral blood volume

h. Normal central blood volume in patients with 30 to 40% reduction in total blood volume in chronic hypovolemic anemic patients
i. Transfusion of chronic hypovolemic anemic patients with normal hematocrit values increased the peripheral blood volume to increase blood volume to gastrointestinal tract, bone, muscle, skin of the extremities without an increase in the central blood volume
j. RBC DPG and RBC creatine levels were increased in patients with red blood cell deficits and in patients with cardiopulmonary insufficiency
k. Hypoxia produced release of creatine from ischemic muscles which was taken up by RBC and RBC 2,3 DPG level increased to compensate for the reduction in the red blood cell volume

The publications of Koch CG et al in NEJM (358:1229-1239, 2008) and Edgren G et al in Transfusion (50:1185-1195, 2010) supported the NBRL recommendation that liquid preserved RBC should be stored at 4 C for no more than 2 weeks to ensure their survival and function in patients to provide high pO2 tensions in brain, heart, and kidneys and to exert a hemostatic effect to reduce nonsurgical blood loss (Valeri CR et al, Surg Gynec Obstet 166:33-46, 1988; Valeri CR, Vox Sang 83;193-196, 2002; Valeri CR et al, Transfusion 47:206S-248S, 2007; Valeri CR, Ragno G, Trans Aph Sci 42:223-233, 2010).

Use of resuscitation solutions containing Ringer's ketone instead of Ringer's D,L lactate and Ringer's L-lactate to treat patients subjected to traumatic injuries (Koustova E et al, Surgery 134:267-274, 2003; Jaskille A et al, J Am Coll Surg 202:25-35, 2006; Ayuste ES et al, J Trauma 60:52-63, 2006; Valeri CR et al, Art Cells, Blood Subst, Biotech 34:601-606, 2006; Valeri CR et al, Trans Aph Sci 39:205-211, 2008).

Use of oral ketone esters to improve physical performance, improve cognitive function, decrease appetite, treat neurological diseases like Parkinson's disease, amyotrophic lateral sclerosis (ALS) and Alzheimer's disease (Kashiwaya Y et al, Am J Cardiol 80:50A-64A, 1997; Kashiwaya Y et al, J Biol Chem 285:25950-25956, 2010; Kashiwaya Y et al, Proc Natal Acad Sci USA 97:5440-5444, 2000; Kashiwaya Y et al, Neurobiology of Aging 1-10, 2012).

Our experience at NBRL from 1965 to 1974 was related to the study of 300 wounded Vietnam servicemen at Chelsea Naval Hospital, the deployment of a frozen blood bank at Danang, South Vietnam between

1966 to 1974, and the study of patients, normal volunteers, baboons and dogs at the Chelsea Naval Hospital treated with autologous and allogeneic preserved RBC, platelets and plasma. These studies were published in two books: Blood Banking and the Use of Frozen Blood Products in 1976 by CRC Press, Boca Raton, FL, and the Hypovolemic Anemia of Trauma: The Missing Blood Syndrome in 1981 by CRC Press, Boca Raton, FL. The US NBRL was relocated to a government owned building at 615 Albany Street between 1974 to 1979. In 1979, the US NBRL was disestablished and became a government owned, contractor operated laboratory. The contract was negotiated with BUMC from January 1979 to September 30, 2003. In October 2003, the contract was negotiated with New England Medical Research Institute (NEMRI) at the West Roxbury VA Hospital. From 2004 to 2010, the NBRL, Inc., a non-profit research laboratory was provided government and commercial funds to study frozen platelets, resuscitation solutions, and hemostatic agents to treat casualties.

In 1982, our laboratory became associated with Dr. S. Khuri at the WRVA Hospital which was stimulated by Mr. Fred Levan who wanted to assess membrane oxygenators produced by Terumo Company and bubble oxygenators in patients subjected to extracorporeal bypass surgery.

I met Dr. Khuri at a symposium on myocardial function. The NBRL had collaborated with Dr. Carl Apstein and Dr. Mark Vogel to study the function of human RBC in isolated perfused rabbit hearts. The NBRL was funded by the Naval Medical Research and Development Command (NMRDC) of the U.S. Navy's Bureau of Medicine and Surgery and the Office of Naval Research to assess methods using liquid and freezing procedures to preserve red blood cell survival and function, preserve platelet survival and function, the survival and function of preserved plasma proteins, and the survival and function of autologous adult stem cells isolated from peripheral blood to treat naval personnel subjected to radiation exposure assigned to nuclear submarines.

From 1962 to 1974 all studies performed at the Chelsea Naval Hospital on patients and normal volunteers did not require IRB approval and signed informed consent forms. With the disestablishment of the Chelsea Naval Hospital renamed Boston Naval Hospital our research laboratory was permitted to relocate in the Boston area to continue our research. The 615 Albany Street site was deeded by University Hospital at Boston

University Medical Center to the U.S. Navy for one dollar in April 1975. This building was donated to University Hospital by the New England Nuclear Company, Boston, MA.

In 1981, Fred Levan who was a salesman for Terumo Company requested a clinical site to evaluate the difference between the membrane oxygenator and the bubble oxygenator. Dr. Khuri refused to consider the study which would be supported by Terumo Company unless the NBRL was involved with the study.

Several meetings were held to discuss the clinical study. The study of the bleeding diathesis in patients subjected to extracorporeal bypass surgery provided a unique opportunity to study patients at the West Roxbury Veterans Hospital using IRB approved protocols and signed informed consent. In the design of the study, Dr. Khuri wanted to study all the factors that could be measured to assess hemostasis during cardiopulmonary bypass surgery. Khuri stated he wanted to study everything related to nonsurgical blood loss.

Shukri did not believe anything. Shukri needed data which he could analyze himself. The NBRL had published that preserved RBC must survive and function, preserved platelets must survive and function, and that the hematocrit was falsely elevated in hypovolemic anemic patients and the platelet count provided no information with regards platelet hemostatic function. Dr. Khuri did not agree with any of these reports.

Our laboratory had published extensively on the methods to measure the survival and function of RBC preserved in the liquid state and frozen state and the assessment of the survival and function of fresh and preserved autologous platelets in normal volunteers and baboons prior to and following treatment with aspirin to produce a platelet dysfunction manifested by an increase in the bleeding time and decreased thromboxane level at the bleeding time site.

Our studies demonstrate that the platelets were dysfunctional in hypothermic baboons and hypothermic normal volunteers. Our data demonstrated that the bleeding time was an excellent in vivo test to assess the hemostatic mechanism in normal volunteers, baboons, and patients. Published papers using meta-analysis by Rodgers RP and J Levin and by Lind SE in 1991 had reported that the bleeding time was not useful to assess surgical and nonsurgical bleeding diathesis in patients (Lind SE, Blood 77:2547-2552, 1991; Rodgers RP and J Levin, Semin Thromb Hemost 16:1-20, 1991).

Dr. S. F. Khuri conducted the first study to assess non-surgical blood loss in patients subjected to extracorporeal bypass surgery after the heparin was neutralized by protamine. The blood loss from nonsurgical sites was collected intraoperatively and postoperatively. In these studies, the surgeons needed to report whether the blood loss was due to a surgical cause. This was a major contribution to provide for the first time quantitative assessments of non-surgical blood loss which occurred in patients. This approach permitted the study of patients with nonsurgical blood loss and the role of temperature, hematocrit, platelet count, platelet size, and platelet function, and plasma clotting proteins, oncotic proteins, and opsonic proteins. These studies were conducted between 1982 to 2008 and resulted in several peer-reviewed publications.

These papers demonstrate that nonsurgical blood loss was related to temperature, hematocrit, the number, size, and hemostatic function of the platelets and the dose of heparin administered to increase the activated clotting time (ACT) to 400 seconds.

Khuri SF and associates reported that in studies conducted over the past 15 years at the West Roxbury VA Hospital in patients undergoing CPB surgical procedures nonsurgical blood loss and the need for allogeneic red blood cells and allogeneic FFP were significantly reduced when normothermic CPB was used, heparin dose reduced, heparin-coated surfaces used, and a hematocrit of 35 V% was maintained after cardiopulmonary bypass surgery (Khuri SF et al, In Management of Bleeding in Cardiovascular Surgery, Ed. R. Pifarre, Hanley and Belfus, Philadelphia, 145-160, 2000).

A recent monograph by Valeri CR, S Khuri, G Ragno published. in the Transfusion 47;2065-248S, 2007, reported 45 peer-reviewed publications on nonsurgical bleeding diathesis in anemic and thrombocytopenic patients; role of temperature, RBC, platelets and plasma clotting proteins.

This monograph summarized 45 peer-reviewed publications by the NBRL and WRVA Hospital.

1. Effect of temperature on the bleeding time in patients subjected to cardiopulmonary bypass surgery.
2. The hemostatic effect of RBC to reduce the bleeding time, reduce nonsurgical blood loss, and increase the thromboxane A2 level at the bleeding time site.
3. Heparin effect to produce a platelet dysfunction demonstrated by an increase in bleeding time and reduction in the shed blood level of thromboxane A2 at the bleeding time site.

4. Heparin induced fibrinolytic activity to support the rationale for use of anti-fibrinolytic agents during cardiopulmonary bypass surgery like epsilon aminocaproic acid (EACA) and tranexamic acid.
5. Platelet dysfunction during cardiopulmonary bypass was due to extrinsic factors. Platelet dysfunction during cardiopulmonary bypass was associated with hypothermia and the inhibition of thromboxane production.
6. Dr S.F. Khuri and associates have reported in the J Thorac Cardiovasc Surg 117:172-184, 1999 a clinical study to assess the function of preserved platelets in anemic thrombocytopenic patients by the measurement of non surgical blood loss in patients associated with cardiopulmonary bypass procedures. In this clinical study, the function of frozen washed platelets was compared to liquid preserved platelets stored at room temperature for a mean of 3.4 days with agitation. The in vivo survival of the frozen platelets was reduced compared to the liquid preserved platelets, but the frozen platelets function to reduce nonsurgical blood loss was significantly better than that of the liquid preserved platelets. This prospective randomized clinical study assessed the function of liquid preserved and previously frozen platelets on nonsurgical blood loss in patients.
7. Khuri SF and his associates studied the use of polymerized N-acetyl glucosamine bandage with external pressure to restore hemostasis at the bleeding time site when the arterial catheter was removed following catheterization for diagnostic and therapeutic interventions. A prospective randomized, blinded study to control the external pressure applied at the bleeding time site and the time required to stop the bleeding at the catheter site were assessed (Najjar SF et al, J Trauma 57:S38-S41, 2004). The polymerized N- acetyl glucosamine bandage significantly reduced the bleeding at the catheter site compared to the control bandage.

Does the bleeding time correlate to nonsurgical blood loss? (Duke WW, JAMA 60:1185-1192, 1910; reprinted JAMA 250:1201-1209, 1983; Rodgers RP and J Levin, Semin Thromb Hemost 16:1-20, 1991; Lind SE, Blood 77:2547-2552, 1991). Duke reported that the bleeding time correlated to nonsurgical blood loss, whereas Rodgers and Levin and Lind reported that surgical, and nonsurgical blood loss did not correlate to the bleeding time. Our data over the past 45 years have shown that the bleeding time correlated to nonsurgical bleeding and all the in vitro tests that were performed did not correlate to the in vivo bleeding time measurement. In our experience the hemostatic defect in patients cannot be assessed by in vitro tests but can be assessed by the in vivo bleeding time measurement (Valeri CR and G Ragno, Trans Aph Sci 35:33-41, 2006).

Hypothermia produced adverse effects on the function of RBC, platelets and plasma clotting proteins (Valeri CR et al, Ann Surg 205:175-181, 1987; Michelson AD et al, J Thromb Haemost 5:633-640, 1994; Michelson AD et al, Br J Hematol 104:64-68, 1999; Valeri CR et al, J Thorac Cardiovasc Surg 104:108-116, 1992; Valeri CR et al, Crit Care Med 23:698-704,1995; Valeri CR et al, Transfusion 41:977-983, 2001).

Heparin affected platelet hemostatic function and fibrinolytic activity (Khuri SF et al, Ann Thorac Surg 60:1008-1014, 1995; Upchurch GR et al, Am J Physiol 271:H528-534, 1996). Hematocrit of 35 V% was needed to reduce nonsurgical blood loss (Valeri CR et al, Transfusion 47:206S-248S, 2007). The bleeding time was affected by the platelet production of thromboxane at the bleeding time site (Valeri CR et al, J Thorac Cardiovasc Surg 104:108-116, 1992; Valeri CR et al, Crit Care Med 23:698-704, 1995; Valeri CR et al, Transfusion 41:977-983, 2001). Local pressure released endothelin from the endothelial cells to vasoconstrict the nonsurgical bleeding time site (Ikeda Y et al, J Surg Res 102:215-220, 2002; Favuzza J and HB Hechtman, J Trauma Suppl 57:S42-S44, 2004).

Platelet dysfunction produced by hypothermia was reversible and corrected by normothermia. Platelet dysfunction produced by heparin was reversible and corrected by neutralization with protamine sulphate and metabolism. Platelet dysfunction produced by anemia was reversible and corrected by increase in hematocrit to 35 V%. Platelet dysfunction produced by aspirin was irreversible and corrected, by non-aspirinated functional platelets. Hypothermia, heparin and anemia produce a platelet dysfunction associated with a prolongation of the bleeding time and a reduction in shed blood level of thromboxane at the bleeding time site.

Platelets and red blood cells interact at the bleeding tine site to produce thromboxane a potent vasoconstrictor and platelet aggregating substance. Adenosine diphopshate (ADP) and arachidonic acid (AA) are provided by the viable and functional platelets and viable and functional red blood cells at the bleeding time site. Viable and functional RBC at the bleeding time site are activated to undergo morphologic changes into stomatocytes to expose RBC phosphatidylserine and generate thrombin. In a similar manner viable and functional platelets and functional plasma clotting proteins are activated to generate thrombin to convert fibrinogen to fibrin to reduce nonsurgical bleeding. Local pressure at the bleeding time site release endothelin from endothelial cells which is a potent vasoconstrictor substance to reduce nonsurgical blood loss.

Studies were done to document the therapeutic effect of RBC:

1. Restoration of RBC volume, plasma volume, and blood volume in hypovolemic anemic patients by transfusion of compatible and identifiable viable washed and nonwashed RBC.
2. Maintenance of the central blood volume and restoration of peripheral blood volume by transfusion of washed and non-washed red blood cells to hypovolemic anemic patients.
3. Hemostatic effect of RBC: RBC and platelet interaction at the nonsurgical bleeding site
    a. Aggregation response in vitro to agonists assessed in platelet rich plasma.(Born GVR, Nature (Lond) 194:927-929, 1962; Born GVR, J Physiol (Lond) 162:67-68, 1962).
    b. Aggregation response in vitro to agonists assessed in whole blood. (Marcus AJ, N Engl J Med 280:1213-1220, 1278-1294, 1330-1335, 1969; Marcus AJ, Ann NY Acad Sci 485:369-373, 1986).
4. Oxygen delivery to maintain critical p02 tension of the brain, heart and kidney
5. RBC role to maintain blood pH.
6. Heparin produces a platelet dysfunction to prolong the bleeding time, reduces the level of thromboxane at the bleeding time site, and produces an increase in fibrinolytic activity. The dose of heparin utilized to anticoagulate the patient is monitored by the activated clotting time (ACT).

Nonsurgical blood loss associated with cardiopulmonary bypass surgery is reduced by dose of heparin administered; the use of heparin coated surfaces; use of antifibrinolytic agents like epsilon aminocaproic acid (EACA) and tranexamic acid, and RBC transfusion to increase the hematocrit to 35 V%.

Khuri SF and associates have published a study to quantitate the nonsurgical blood loss associated with survival and function of liquid preserved and cryopreserved allogeneic platelets in hemodiluted anemic thrombocytopenic patients. The measurement of nonsurgical blood loss in patients during cardiopulmonary bypass demonstrated that cryopreserved platelets significantly reduced nonsurgical blood loss compared to liquid preserved platelets stored at room temperature for a mean of 3.4 days. Nonsurgical blood loss and need for RBC and fresh frozen. plasma transfusions were significantly reduced in patients transfused with frozen washed platelets (Khuri SF et at, J Thorac Cardiovasc Surg 117:172-184, 1999).

Viability and function of red blood cells, platelets, and plasma proteins affect the nonsurgical blood loss in patients.

1. Viability and function of RBC
2. Viability and hemostatic function of platelets
3. Function of plasma proteins. Complement activation occurs during cardiopulmonary bypass. Hypothermia, heparin, and hemodilution attenuate complement activation during cardiopulmonary bypass (Moore FD Jr et al, Ann Surg 208:95-103, 1988).
4. Hypothermia adversely affects the function of RBC, platelets, and plasma clotting proteins.

Treatment of anemic thrombocytopenic patients require viable and functional RBC to increase the hematocrit to restore platelet function. Anemic thrombocytopenic patients need to be transfused with viable and functional RBC to achieve a hematocrit of 35 V% prior to prophylactic and therapeutic platelet transfusions. Transfusion trigger for platelets need to consider the patient's hematocrit, platelet court, platelet size and platelet hemostatic function.

The following publications report the relationship between bleeding time and surgical and nonsurgical blood loss:

- Duke WW. The relation of blood platelets to hemorrhagic disease. JAMA 60:1185-1192, 1910; JAMA 250:1201-1209, 1983
- Quick AJ. Salicylates and bleeding time: the aspirin tolerance test. Am Med Sci 252:265-269, 1966
- Harker LA, Slichter SJ. The bleeding time as a screening test for the evaluation of platelet function. N Engl I Med 287:155-159. 1972.
- Lind SE. The bleeding time does not predict surgical bleeding. Blood 77:2547-2552, 1991.
- Rodgers RP, Levin J. A critical reappraisal of the bleeding time. Semin Thromb Hemostat 16:1-20, 1991.

Hemostatic effects of viable RBC with normal oxygen transport function were studied to assess:

a) Scavenging of endothelial cell nitric oxide (NO) by oxidation to perioxynitrite, nitrite and nitrate and binding of nitric oxide to hemog1obin (NO Hb and SNOHb)

b) Restoration of platelet function by reduction in nitric oxide and decrease in platelet cyclic guanosine monophosphate (cGMP) level

c) RBC, like platelets, provide both adenosine diphopshate (ADP) and arachidonic acid (AA) for platelet production of thromboxane at the bleeding site.

d) RBC increases blood viscosity and shear stress which may release ADP to aggregate platelets and. may release endothelin, a potent vasoconstrictor substance from endothelial cells

e) RBC may be stimulated in vivo by agonists like derivatives of subendothelial glycocalyx to produce thrombin at the bleeding site.

f) In the administration of erythropoietin (EPO) to patients with renal disease, the dose of erythropoietin to increase the hematocrit to 33 to 35 V% produces optimum hemostasis, improvement in platelet function, and minimize adverse events.

g) Transfusion trigger for prophylactic and therapeutic platelet transfusion in anemic thrombocytopenic patients need to assess hematocrit value, platelet count, platelet size, and platelet hemostatic function.

In 1998, the Nobel Prize in Medicine and Physiology was awarded to RF Furchgott, LJ Ignarro and F Murad. These recipients described the role of endothelial nitric oxide to vasodilate blood vessels and to inhibit platelet function. Endothelial derived relaxing factor (EDRF). was reported by Furchgott RF et al in Vasodilator Mechanisms, edited by PM Vanhoutte and S Vatner Basel, Karger 1-15, 1983, LJ Ignarro reported that nitric oxide was the endothelial relaxing factor In Proc Natl Acad Sci USA 84:9265-9269, 1987. F Murad reported that nitric oxide increased platelet cGMP which inhibited platelet function in Adv in Pharmacol Vol 26, F Murad ed, Academic Press Chapter 3, pp 19-33, 1994. Yanagisawa M and associates reported the role of endothelin and proendothelin as potent vasoconstrictor peptides produced by endothelial cells. Yanagisawa M and associates in Nature (London) 332:411-415, 1988 (Furchgott RF, D Jothiananandan and PD Cherry, Endothelial depended responses: The last three years. In Vasodilator Mechanisms edited by PM Vanhoutte and S Vatner, Basel Karger, 1983, pp 1-15; Ignarro LJ, BM Buza, KS Wood et al, Endothelium derived relaxing factor produced and released from artery and vein is nitric oxide. Proc Natl Acad Sci 84:9265-9269, 1987; Murad F,

Adv In Pharmacol, Vol 26, F Murad ed, Academic Press Chapt 3, pp 19-33, 1944; Yanagisawa M, et al Nature (London), 332:411-415, 1988.

RBC like platelets provide adenosine diphosphate (ADP) and arachidonic acid (AA) to stimulate the production of thromboxane by the platelets, a potent vasoconstriction substance and platelet aggregating agent. Silliman CC and associates in Transfusion 51:2549-2554, 2011 have reported the release of arachidonic acid from red blood cells during storage at 4 C. These data indicated that red blood cells can provide both arachidonic acid (AA) and adenosine diphosphate (ADP) for platelets to produce thromboxane,

At the BT site tissue substances like calcium and derivatives of subendothelial glycocalyx like N-acetyl glucosamine may stimulate the platelets and red blood cells to expose phosphatidylserine which accumulates factor Va and factor Xa to produce prothrombinase activity to convert prothrombin to thrombin and fibrinogen to fibrin.

Activation of RBC by the agonists at the bleeding time site produce thrombin to convert fibrinogen to fibrin demonstrate the prothrombotic effect of RBC. The hematocrit to minimize the prothrombotic effect of RBC is about 35 V% which minimizes shear stress on the endothelial cells. The release at the nonsurgical bleeding site of thromboxane from the platelets and endothelin from the endothelial cells vasoconstrict the microcirculation.

The generation of thrombin by activation of the platelet and RBC convert the fibrinogen to fibrin to restore hemostasis. Nonsurgical bleeding diathesis in anemic thrombocytopenic patients is affected by temperature, hematocrit, platelet (Plt) count, Plt size, Plt function, and plasma clotting proteins factors V and VIII, von Willebrand's factor and fibrinogen. Bleeding time (BT) correlates to nonsurgical blood loss but does not correlate to surgical blood loss which requires surgical intervention and hemostatic agents with local external pressure to stop the bleeding. BT correlates with the volume of shed blood collected at the bleeding time site. The BT and the volume of shed blood collected at the BT site correlates better with the hematocrit than the Plt count in normal volunteer and in patients with bleeding disorders.

The RBC transfusion trigger should minimize the relative shear stress on endothelial cells to minimize nitric oxide production which inhibits platelet function. Anemic thrombocytopenic patients with a bleeding diathesis are usually transfused prophylactic platelet transfusions which usually alloimmunize the patients after at least 3 allogeneic transfusions and may produce transfusion related acute lung injury (TRALI). Indications for platelet transfusions require that the platelets function to reduce the bleeding time and reduce and prevent nonsurgical blood loss.

The current FDA standard for acceptable therapeutic effectiveness of preserved platelets is that they have in vivo recovery of 66% of fresh platelets and a lifespan of at least 50% of fresh platelets. There is no FDA requirement that the preserved platelets function. The basic assumption is that the in vivo recovery and lifespan of the preserved platelets correlate with their in vivo function.

Studies in aspirin treated normal volunteers, in aspirin treated baboons, and in hemodiluted anemic thrombocytopenic patients following CPB surgery demonstrated that the in vivo recovery and survival of preserved platelets did not correlate with their function to reduce the bleeding time, reduce nonsurgical blood loss, and increase the level of thromboxane at the bleeding time site. Khuri SF and associates reported that cyropreserved platelets had reduced in vivo recovery but improved hemostatic function compared to liquid preserved platelets in patients following cardiopulmonary bypass surgery.

Autologous baboon platelets stored at 22 C for 48 hours had in vivo recovery values similar to those stored for I8 hours, and they significantly reduced the bleeding lime (BT) and increased the shed blood thromboxane B2 (TxB2) level after transfusion in aspirin treated baboons.

Autologous baboon platelets stored at 22 C for 72 hours had in vivo recovery values similar to those stored for 18 hours, but the BT was not corrected after transfusion, although there was a significant increase in the shed blood TxB2 level in aspirin treated baboons.

Autologous cryopreserved baboon platelet significantly reduced the BT and significantly increased the shed blood TxB2 level after transfusion in aspirin treated baboons. Cryopreserved autologous baboon platelets had better in vivo survival and function than the 5-day liquid stored autologous baboon platelets.

The survival of autologous fresh, liquid preserved and cryopreserved baboon platelets did not correlate with their function to reduce an increased bleeding time in baboons treated with aspirin (Valeri CR et al, Transfusion 42:1206-1216, 2002). The hematocrit, hemoglobin concentration, platelet count, and mean platelet volume (MPV) did not change significantly during the 48 hour period for each of the eight groups of baboons that were studied. Acceptable platelet preservation should ensure that both survival and hemostatic function are maintained. Current guidelines only require satisfactory platelet survival. Studies have demonstrated that platelet survival does not correlate with platelet hemostatic function,

The hematocrit of 35 V% produces minimal shear stress on endothelial cells to minimize nitric oxide production which inhibits platelet hemostatic function. The optimum treatment of anemic thrombocytopenic patients should be to transfuse the patients with red blood cells to increase the hematocrit value to 35 V% to restore platelet hemostatic function prior to prophylactic and therapeutic transfusion of platelets which should circulate and function to restore hemostasis.

Aspirin treated normal volunteers have been studied by other investigators to assess the survival and hemostatic function of non-aspirin treated autologous washed and filtered platelets by measuring the in vivo recovery and lifespan and the reduction in the bleeding time in aspirin treated normal volunteers (Valeri CR et al, Transfusion 47:206S-248S, 2007; Pineda AA et al, Transfusion 29:524-527, 1989; Brecher ME et al, Transfusion 30:718-721, 1990).

RBC must circulate and function to release oxygen at high tissue p02 tension to scavenge the endothelial cell produced nitric oxide—a potent vasodilator substance that inhibits platelet function by increasing platelet cyclic GMP 1evel. The reduction in endothelial nitric oxide restores platelet hemostatic function by reduction in platelet GMP level and stimulates the release of endothelial cell endothelin, a potent vasoconstrictor substance to reduce blood flow to the bleeding time site (Valeri CR et al, Transfusion 47:206S-248S, 2007).

Anemic thrombocytopenic patients have both reduced number of platelets and dysfunctional platelets. Treatment of anemic thrombocytopenic patients require viable and functional RBC to increase the hematocrit to restore platelet hemostatic function. Anemic thrombocytopenic patients need to be transfused with viable and functional RBC to achieve a hematocrit of 35 V% prior to prophylactic and therapeutic platelet transfusions. Transfusion trigger for platelets need to consider the patient's hematocrit, platelet count, platelet size and the platelet hemostatic function (Valeri CR and G Ragno, Trans Aph Sci 42:223-233, 2010).

CHAPTER 7 REFERENCES

1. Apstein CS, Dennis RC, Briggs L, Vogel WM, Frazer J, Valeri CR. Effect of erythrocyte storage and oxyhemoglobin affinity changes on cardiac function. Am J Physiol 248 (Heart Circ Physiol 17): H508-H5 15, 1985.
2. Ayuste EC, Chen B, Koustova E, Rhee P, Ahuja N, Chen Z, Valeri CR, Spaniolas K, Mehrani T, Alam HB. Hepatic and pulmonary apoptosis after hemorrhagic shock in swine can be reduced through modifications of conventional Ringer's solution. J. Trauma 60(l):52-63, 2006.
3. Barnard MR, MacGregor H, Ragno G, Pivacek LE, Khuri SF, Michelson AD, Valeri CR. Fresh, liquid-preserved,, and cryopreserved platelets: adhesive surface receptors and membrane procoagulant activity. Transfusion 39:880-888. 1999.
4. Born GVR. Aggregation of blood platelets by adenosine phosphate and its reversal. Nature (Lon) 194:927-929, 1962.
5. Born GVR. Quantitative investigations into the aggregation of blood platelets. J Physiol (Lon), 162:67-68, 1962.
6. Brecher ME, Pineda AA, Zylistra-Halliug VW, Chowdhury S, Fontrorn LA. In vivo viability and functional integrity of filtered platelets. Transfusion 30:718-721, 1990.
7. Duke WW. The relation of blood platelets to hemorrhagic disease. JAMA 60:1185-1192, 1910; reprinted JAMA 250:1201-1209, 1983.
8. Edgren G, Kamper-Jorgensen M, Eloranta S, Rostgaard K, Custer B, Ullum H, Murphy EL, Busch MP, Reilly M, Me1bve M, Hialgrim H, Nyren O. Duration of red blood cell storage and survival of transfused patients. Transfusion 50:1185-1195, 2010.
9. Favuzza J, Hechtman HB. Hemostasis in the absence of clotting factors. J Trauma: 57:S42-S44, 2004.
10. Furchgott RF, Jothiananandan D, Cherry PD. Endothelium-dependent responses: The last three years. In: Vasodilator Mechanisms, edited by PM Vanhoutte and S Vatner, Basel:Karger, 1983, pp 1-15.
11. Harker LA, Slichter SJ. The bleeding time as a screening test for the evaluation of platelet function. N Engl J Med 287:155-159, 1972.

12. Ignarro LJ, Buga GM, Wood KS, et al.. Endothelium-derived relaxing factor produced and released from artery and vein in nitric oxide. Proc Natl Acad Sci USA 84:9265- 9269, 1987.
13. Ikeda Y, Young LH, Vournakis JN, Leifer AM. Vascular effects of poly-N-acetylglucosamine in isolated rat aortic rings, J Surg Res 102:215-220, 2002,
14. Jaskille A, Koustova E, Rhee P, Britten-Web I, Chen H, Valeri CR, Kirkpatrick JR, Alam HB. Hepatic apoptosis following hemorrhagic shock in rats can be reduced through modifications of conventional Ringer's solution. J Am Coll Surg 202(1) :25-35, 2006.
15. Kashiwaya Y, King MT, Veech RL. Substrate signaling by insulin: a ketone bodies ratio mimics insulin action in heart. Am J Cardiol 80:50A-64A, 1997.
16. Kashiwaya Y, Pawlosky R, Markis W, King MT, Bergman C, Srivastava S, Murray A, Clarke K, Veech RL. A ketone ester diet increases brain malonyl-CoA and uncoupling proteins 4 and 5 while decreasing food intake to the normal Wistar rat. J Biol Chem 285:25950-25956, 2010.
17. Kashiwaya Y, Takeshima T, Mora T, Mori N, Nakashima K, Clarke K, Veech RL. D-beta-hydroxybutyrate protects neurons in models of Alzheimer's and Parkinson's disease. Proc Natl Acad Sci USA 97:5440-5444, 2000.
18. Kashiwaya Y, Bergman C, Lee J-H, Wan R, King MT, Mughal MR, Okun E, Clarke K, Mattson MP, Veech RL. A ketone ester diet exhibits anxiolytic and cognition-sparing properties, and lessens amyloid and tau pathologies in a mouse model of Alzheimer's disease. Neurobiology of Aging xxx:1-10, 2012.
19. Kestin AS, Valeri CR, Khuri SF, Loscalzo J, Ellis PA, MacGregor H, Birjiniuk V, Ouimet H, Pasche B, Nelson MJ, Benoit SE, Rodino L, Barnard MR, Michelson AD. The platelet function defect of cardiopulmonary bypass. Blood 82:107-117, 1993.
20. Khuri SF, Wolfe JA, Jose M, Axford TC, Szymanski I, Assousa S, Ragno G. Patel M, Silverman A, Park M, Valeri CR. Hematologic changes during and after cardiopulmonary bypass and their relationship to the bleeding time and nonsurgical blood loss. J Thorac Cardiovasc Surg 104:94-107, 1992.
21. Khuri SF, Valeri CR, Loscalzo J, Weinstein MJ, Birjiniuk V,

Healey NA, MacGregor H, Doursounian. M, Zolkewitz MA. Heparin causes platelet dysfunction and induces fibrinolysis before cardiopulmonary bypass. Ann Thorac Surg 60:1008-1014, 1995.
22. Khuri SF, Healey N, MacGregor H, Barnard MR, Szymanski 10, Birjiniuk V, Michelson AD, Gagnon DR, Valeri CR. Comparison of the effects of transfusions of cryopreserved and liquid-preserved platelets on hemostasis and blood loss after cardiopulmonary bypass. J Thorac Cardiovasc Surg 117:172-184, 1999.
23. Khuri SF, Valeri CR, Treanor P. Bleeding during and after cardiopulmonary bypass: a 15-year institutional experience. In Management of Bleeding in Cardiovascular Surgery, Ed. R. Pifarre. Hanley & Belfus. Philadelphia, 2000:145-160.
24. Koch CG, Li L, Sessler DI, et al. Duration of red cell storage and complications after cardiac surgery. N EnglJ Med 358:1229-1239, 2008.
25. Koustova E, Rhee P, Hancock T, Chen H, Inocencio R, Valeri CR, Alam HB. Ketone and pyruvate Ringer's solutions decrease pulmonary apoptosis in a rat model of severe hemorrhagic shook and resuscitation. Surgery 134(2):267-274, 2003.
26. Lind SE. The bleeding time does not predict surgical bleeding. Blood 77:2547-2552, 1991.
27. Marcus AJ. Platelet function. N Engl J Med 280:1213-1220, 1278-1284, 1330-1335, 1969.
28. Marcus AJ. The role of thrombin in transcellular metabolism of eicosanoids. Ann NY Acad Sci 485:369-373, 1986.
29. Melaragno AJ, Carciero R, Feingold H, Talarico L, Weintraub L, Valeri CR. Cryopreservation of human platelets using 6% dimethylsulfoxide and storage at -8OC. Effects of 2 years of frozen storage at -8OC and transportation in dry ice. Vox Sang 49:245-258, 1985.
30. Michelson AD, MacGregor H, Barnard MR., Kestin. AS, Rohrer MJ, Valeri CR. Reversible inhibition of human platelet activation by hypothermia in vivo and in vitro. J Thrombosis and Haemostasis 5:633-640, 1994.
31. Michelson AD, Barnard MR Khuri SF. Rohrer MJ, MacGregor H, Valeri CR. The effects of aspirin and hypothermia on platelet function in vivo. Br J Haematol 104:64-68, 1999.
32. Moore FD Jr, Warner KG, Assousa S, Valeri CR, Khuri SF. The

effects of complement activation during cardiopulmonary bypass. Attenuation by hypothermia, heparin and hemodilution. Ann Surg 208:95-103, 1988.

33. Murad F. Regulation of cytosolic guanylyl cyclase by nitric oxide. The nitric oxide-cyclic GMP signal transduction system. In: Cyclic GMP synthesis. Metabolism and Function Adv in Pharmacol, Vol 26, F Murad, ed., Academic Press, Chapter 3, pp 19-33, 1994.
34. Najjar SF, Healey NA, Healey CM, McGarry T, Khan B, Thatte HS, Khuri SF. Evaluation of poly-N-acetyl glucosamine as a hemostatic agent in patients undergoing cardiac catheterization: a double-blind, randomized study. J Trauma 57 Suppl:38-41, 2004.
35. Pineda Z, Zylstra VW, Clare DE, Dewanjee MK, Forstrom LA. Viability and functional integrity of washed platelets. Transfusion 29:524-527, 1989.
36. Prins ML. Cerebral metabolic adaptation. a ketone metabolism after brain injury. J Cereb Blood Flow Metab 28:1-16, 2008.
37. Quick AJ. Salicylates and bleeding time: the aspirin tolerance test Am J Med 252:265-269, 1966.
38. Rodgers RP, Levin J. A critical reappraisal of the bleeding time. Semin Thromb Hemostat 16:1-20, 1991.
39. Silliman CC, Moore EE, Keiher MR, Khan SY, Gellar L, Eizi DJ. Identification of lipids that accumulate during the routine storage of prestorage leukoreduced red blood cells and cause acute lung injury. Transfusion 51:2549-2554, 2011.
40. Valeri CR. Blood Banking and the Use of Frozen Blood Products. Chemical Rubber Company Press, Boca Raton, Florida, 1976.
41. Valeri CR and Altschule MD. Hypovolemic Anemia of Trauma: The Missing Blood Syndrome. Chemical Rubber Company Press, Boca Raton, Florida, 1981.
42. Valeri C. Status report on the quality of liquid and frozen red blood cells. Vox Sang 83(1):193-196, 2002.
43. Valeri CR, Ragno G. The survival and function ofbaboon red blood cells, platelets, and plasma proteins: A review of the experience from 1972 to 2007 at the Naval Blood Research Laboratory, Boston, Massachusetts, Transfusion 46(8): 1-42, 2006.
44. Valeri CR, Ragno G. Cryopreservation of human blood products. Trans Aph Sci 34:271-287, with an editorial, on pages 267-269, 2006.

45. Valeri CR, Ragno G. Role of RBC and platelet transfusion in the treatment of anemic and thrombocytopenic patients. Letter to the Editor. Transfusion 46:1210-1211, 2006.
46. Valeri CR. Ragno G. The effects of preserved red blood cells on the severe adverse events observed in patients infused with hemoglobin based oxygen carriers. Art Cells, Blood Subst and Biotech 36 (1):3-1 8, 2008.
47. Valeri CR, Ragno G. An approach to prevent the severe adverse events associated with transfusion of FDA-approved blood products. Trans Aph Sci 42:223-233, 2010.
48. Valeri CR, Feingold H, Cassidy G, Ragno G, Khuri S, Altschule MD. Hypothermia induced reversible platelet dysfunction. Ann Surg 205:175-181, 1987.
49. Valeri CR, Pivacek LE, Palter M, Dennis RC, Yeston N, Emerson CP, Altschule MD. A clinical experience with ADSOL® preserved erythrocytes. Surg Gynec Obstet 166:33-46, 1988.
50. Valeri CR, MacGregor H, Pompei F, Khuri SF. Acquired abnormalities of platelet function. Letter to the Editor, NEJM 324:1670. 1991.
51. Valeri CR, Khabbaz K, Khuri SF, Marquardt C, Ragno G, Feingold H, Gray AD, Axford T. Effect of skin temperature on platelet function in patients undergoing extracorporeal bypass. J Thorac Cardiovasc Surgery 104:108-116, 1992.
52. Valeri CR, MacGregor H, Cassidy G, Tinney R. Pompei F. Effects of temperature on bleeding time in normal male and female volunteers. Crit Care Med 23:698-704, 1995.
53. Valeri CR, Crowley JP, Loscalzo J. The red cell transfusion, trigger: has a sin of commission now become a sin of omission? Transfusion 38:602-610, 1998.
54. Valeri CR, Cassidy G, Pivacek LE, Ragno G, Lieberthal W, Crowley JP, Khuri SF, Loscalzo J. Anemia-induced increase in the bleeding time: implications for treatment of nonsurgical blood loss. Transfusion 41:977-983. 2001.
55. Valeri CR, Ragno G, Pivacek LE, Dennis RC, Hechtman HB, Khuri SF. Survival and function of baboon RBCs released from clotted blood and washed before autologous transfusion. Transfusion 41:1384-1389, 2001.
56. Valeri CR, Dennis RC, Ragno G, Pivacek LE, Hechtman HB,

Khuri SF. Survival, function, and hemolysis of shed red blood cells processed as nonwashed blood and washed red blood cells. Ann Thoracic Surg 72:1598-1602, 2001.

57. Valeri CR, Giorgio A, MacGregor H, Ragno G. Circulation and distribution of autotransfused fresh, liquid preserved, and cryopreserved baboon platelets. Vox Sang 83:347-351, 2002.
58. Valeri CR, MacGregor H, Giorgio A, Ragno G: Circulation and hemostatic function of autologous fresh, liquid preserved, and cryopreserved baboon platelets transfused to correct an aspirin-induced thrombocytopathy. Transfusion 42:1206-1216, 2002.
59. Valeri CR, Khuri S, Ragno G. Role of Hct in the treatment of thrombocytopenic patients. Letter to the Editor. Transfusion 43:1761-1762, 2003,
60. Valeri CR, Srey R, Lane JP, Ragno G. Effect of WBC reduction and storage temperature on PLTs frozen with 6 percent DMSO for as long as 3 years. Transfusion 43(8): 1162-1167, 2003.
61. Valeri CR, Ragno G, Khuri S. Freezing human platelets using 6% DMSO with removal of the supernatant solution prior to freezing and storage at -8OC without post-thaw processing. Transfusion 45(12): 1890-1898, 2005.
62. Valeri CR, Dennis RC, Ragno G, MacGregor H, Menzoian JO, Khuri SF. Limitations of the hematocrit to assess the need for RBC transfusion in hypovolemic anemic patients. Transfusion 46:365-371 2006.
63. Valeri CR, MacGregor H, Ragno G, Healey N, Fonger J, Khuri SF. Effects of centrifugal and roller pumps on the survival of autologous red blood cells in cardiopulmonary bypass surgery Perfusion 21:291-296. 2006.
64. Valeri CR, Ragno G, Veech RL. Effects of the resuscitation fluid and the HBOC excipient on the toxicity of the HBOC: Ringer's D, L-Lactate, Ringer's L-Lactate, and Ringer's ketone solutions. Art Cells, Blood Subst and Biotech 34(6): 601-606. 2006.
65. Valeri CR, Ragno G. In vitro testing of platelets using the thromboelastogram, platelet function analyzer, and the clot signature analyzer to predict the bleeding time. Trans Aph Sci 35(1):33-41, 2006.
66. Valeri CR. Khuri S, Ragno G. Non-surgical bleeding diathesis in anemic thrombocytopenic patients: Role of temperature, RBC,

platelets, and plasma clotting proteins. Transfusion 47:206S-248S, 2007.

67. Valeri CR, Ragno G, Veech RL. Severe adverse events associated with hemoglobin based oxygen carriers: role of resuscitative fluids and liquid preserved RBC. Trans Apher Sci 39:205-211, 2008.
68. Upchurch GR., Valeri CR, Khuri SF, Rohrer MJ, Welch GN, MacGregor H, Ragno G, Francis S. Rodino LJ, Michelson AD, Loscalzo J. Effect of heparin on fibrinolytic activity and platelet function in vivo. Am J Physiol. 271 (Heart Circ Physiol 40): H528- H534, 1996.
69. Yanagisawa M, Kurihara H, Simura S, et al. A novel vasoconstrictor peptide produced by vascular endothelial cells. Nature 332:411-415, 1988.
70. Veech RL, Rogeness GA, Weil-Malherbe H. Formation of protoporphyrin from hemoglobin in vitro. Biochem J 105:1209-1217, 1967.
71. Vogel WM, Hsia JC, Briggs LL, Er SS. Cassidy G, Apstein CS, Valeri CR. Reduced coronary vasoconstrictor activity of hemoglobin solutions purified by ATP-agarose affinity chromatography. Life Sciences 41:89-93, 1987.

CHAPTER 8

EFFECTS OF RESUSCITATION FLUIDS, THE HBOC EXCIPIENTS, AND LENGTH OF STORAGE OF RBC AT 4 C ON THE TOXICITY OF HBOCS

Hemoglobin based oxygen carriers (HBOC) are resuspended in media referred to as excipients. The primary excipient for HBOC has been Ringer's D,L lactate solution supplemented with antioxidants to prevent the formation of methemoglobin during the storage period. Investigators have reported cardiac arrhythmias in animals and patients infused with Ringer's D,L lactate solution. Studies have also shown that D-lactate stimulates human granulocytes to generate oxygen free radicals (Chan L et al, Integr Physiol Behav Sci 29:383-394, 1992; Veech RL, Prostaglandins, Leukotrienes, and Essential Fatty Acids 70:309-319, 2004). These studies led the manufacturers to modify the Ringer's D,L-lactate solution so that it now contains only the L isomer of lactate. However, a publication by Cross HR and associates has shown, that L-lactate inhibits glycolysis impairing resuscitation of the isolated rat hearts (Cross HR et al, J Mol Cell Cardiol 27:1369-1381, 1995).

Based on several studies, Veech RL has recommended replacing the 28 mM Ringer's D,L lactate with 28 mM Na-D-betahydroxybutyrate (Na BHB). Na BHB has been shown to reduce the generation of oxygen free radicals by mitochondria and human granulocytes. The original Ringer's ketone solution suggested by RL Veech contained a physiological concentration of glucose (5 mM).

A commercial company prepared one liter filter sterilized Ringer's ketone-glucose solution in ethox plastic bags under GMP conditions. Ringer's ketone-glucose solution cannot be sterilized by conventional heat sterilization or gamma irradiation. Sterile and endotoxin free Ringer's ketone-glucose solution has been prepared using filter sterilization. Chemical composition and high pressure liquid chromatography testing demonstrated that filter-sterilized Ringer's ketone-glucose solution tested in tissue culture of mouse proximal tubular cells sensitive to endotoxin demonstrated that Ringer's ketone-glucose solution did not adversely affect the survival of the cells in either aerobic and anaerobic conditions. Alam HB and associates, Koustova E and associates, Jaskille A and associates, and Ayuste EC and associates demonstrated that Ringer's ketone-glucose solution administered to rats and swine subjected to hemorrhagic shock produced significantly less apoptosis in the lung than Ringer's D,L lactate solutions (Alam HB et al, J Am Coll Surg, 193:255-263, 2001; Koustova E et al, Surgery 134:267-274, 2003; Jaskille A et al, J Am Coll Surg 202:25-35, 2006; Ayuste EC et al, J Trauma 60:52-63, 2006).

Ringer's D,L lactate has been the principle resuscitation fluid used in clinical medicine and the excipient for HBOC for more than 30 years (Pope A et al, 1999). Ringer's D,L lactate solution causes neutrophil activation which produce reactive oxygen intermediates which are inhibited by ketones (Rhee P et al, J Trauma 44:313-319, 1998; Rhee P et al, Crit Care Med 28:74-78, 2000; Sato N et al, Life Sci 51:113-119, 1992).

Both Ringer's D,L lactate and HBOC produce oxygen free radicals and the presence of both may be responsible for the severe adverse events reported with the use of HBOC in patients subjected to elective orthopedic and cardiopulmonary bypass surgical procedures. Orthopedic and cardiopulmonary bypass patients who received HBOC in Ringer's D,L lactate excipient have also been resuscitated with Ringer's D,L lactate solution. Oxygen free radicals generated by Ringers D,L lactate and HBOC may oxidize nitric oxide in endothelial cells causing the vasoconstrictor effects reported following the infusion of HBOC. In addition, generation of oxygen free radicals activate nuclear fragment-kappa beta (NF-kb) and the apoptotic cascade. The combination of Ringer's D,L lactate resuscitation fluid and the HBOC in the Ringer's D,L lactate excipient may be responsible for the severe adverse events observed in the clinical studies of HBOC (Valeri CR et al, Art Cells, Blood Subst and Biotech, 34:601-606, 2006).

Scientific rational for use of Ringer's ketone solution include the maintenance of ATP production which is essential for cell survival and function. NaDBHB (NaD betahydroxy-butyrate) produces ATP by mitochondria and prevents oxygen free radical damage. By acting as a substrate during acute and chronic stress NaDBHB maintain cell survival and function and modifies disease progression (Veech RL, Prostaglandins, Leukotrienes, and Essential Fatty Acids 70:309-319, 2004).

Studies done by our laboratory demonstrated that final heat sterilization or irradiation could not be used to sterilize the Ringer's ketone solution containing 5 mM glucose without adversely affecting betahydroxybutyrate and glucose levels. The large volume (1L) Ringer's ketone-glucose solution was prepared by a commercial company using final filter sterilization (Valeri CR et al, Art Cells, Blood Subst and Biotech, 34:601-606, 2006).

Recent studies have demonstrated that removal of the 5 mM glucose allowed for heat sterilization of the solution with no adverse effects on betahydroxybutyrate levels. Ringer's ketone solution is now prepared without glucose to allow for heat sterilization.

Insulin Resistance and Therapeutic Effect of NaBeta Hydroxybutyrate Solution

Severe injury, infection and hemorrhage all cause insulin resistance. Insulin resistance blocks glucose entry into cells and conversion of pyruvate to acetyl CoA impairing cellular energy production. Lactate fluids decrease energy production by inhibiting the production of pyruvate and pyruvate entry into mitochondria.

Na-DB-hydroxybutyrate containing fluids increase metabolic efficiency and bypass blocks caused by insulin resistance and lactate. D-lactate containing fluids can cause neurological dysfunction, cardiac arrhythmias, and death (DeCosta JM, Am J Med Sci 61:18-52, 1971; Carr DB et al, Am Neurol 11:195-197, 1982; Veech RL, Am J Clin Nutr 44:519-551, 1986; Chan L et al, Integr Physiol Behav Sci 29:383-394, 1992). Ringer's D,L lactate solution was a major cause of DaNang lung in Vietnam wounded casualties. Ringer's D,L lactate was used to prevent hyperchloremic acidosis (Hartmann AF, JAMA, 103:1349-1354, 1934) and Ringer's D,L lactate became the standard for treatment of hemorrhagic shock (Shires T, Arch Surg 88:688-693, 1964). After publication of several papers in 1986 on the toxicity of Ringer's D,L lactate solution, Baxter Laboratories provided Ringer's L lactate solution in 1987.

Acute respiratory distress syndrome observed in wounded casualties by administration of Ringer's D,L lactate can be prevented by the use of Ringer's ketone solution. The ketone body Na D-B-hydroxybutyrate improves cardiac efficiency, maintains brain function and can be incorporated into improved resuscitation fluids (Veech RL et al, IUBMB Life 51:241-247, 2001).

Sato K and associates reported that Ringer's ketone solution increased myocardial function of isolated perfused rat heart by 28% while decreasing oxygen consumption compared to Ringer's D,L 1actate solution (FASEB J 9:651-658, 1995). Clarke K and associates have reported that feeding ketone esters to rats increased physiological and cognitive performance. Masuda R and associates have reported Na-D-B-hydroxybutyrate was neuroprotective against hypoxia in serum free hippocampal cultures (Neurosci Res 8:501-509, 2005). Hu ZG and associates have reported that ketogenic diet reduced apoptosis in traumatic brain injury (Am Clin Lab Sci 39:76-83, 2005). Effective therapy of traumatic brain injury can be achieved by closure of the mitochondrial permeability transitional pore by administration of cyclosporine A or by oral administration of ketone body esters (Veech RL et al, IUBMB Life 64:203-207, 2012). Prins ML has reported on cerebral metabolic adaptation and ketone metabolism after brain injury (J Cerebral Blood Flow and Metabolism 28:1-16, 2008).

Longnecker DE et al recommended investigation of NaD-B-hydroxybutyrate resuscitation fluids at the Institute of Medicine Committee on Fluid Resuscitation for Combat Casualties and Civilian Injuries in 1999. Methods for the production of Na-D-B-hydroxybutyrate parenteral fluids have been developed at NIH by Veech RL and associates. In collaboration with Dr. Richard Veech Ringer's ketone solutions were prepared by NBRL, Boston, MA which are stable following storage at room temperature for 2 years.

Severe Adverse Events Associated with Hemoglobin Based Oxygen Carriers: Role of Resuscitation Fluids and Liquid Preserved RBC.

Severe adverse events have been observed following the infusion of hemoglobin based oxygen carriers in patients subjected to elective orthopedic procedures, cardiopulmonary bypass surgery, and vascular surgical procedures. Along with all three of the hemoglobin based oxygen carriers, the patients received Ringers' D,L-lactate as the resuscitation fluid, Ringer's D,L lactate in the excipient medium for the stroma free hemoglobin, and liquid preserved red blood cells that had been stored at 4 C for longer than 2 weeks. The Ringer's D,L lactate solution has been shown to be toxic in both animals and patients. The current formulation of Ringer's lactate contains only the L-isomer which has been shown in animals to be less toxic than the D-isomer of lactate. In a recent publication morbidity and mortality have been reported and associated with the length of storage of red blood cells at 4 C for longer than 2 weeks in patients subjected to reoperative cardiac surgery (Koch CG et al, N Engl J Med 358:1229-1239, 2008). Current clinical adverse events may be associated with the composition of the resuscitation solution (Ringer's lactate), the composition of the excipient medium (Ringer's lactate) for the hemoglobin based oxygen carrier, and the length of storage of the liquid preserved red blood cells infused with the hemoglobin based oxygen carrier.

Two billion dollars have been spent on attempts to develop FDA-approved safe and therapeutically effective hemoglobin based oxygen carriers. Despite the significant investment, the clinical studies performed in elective orthopedic surgical patients by the Biopure Company; in patients subjected to cardiopulmonary bypass surgery by Hemosol Company; and in patients subjected to major vascular surgical procedures by Northfield Company have all been associated with severe adverse events. The hemoglobin based oxygen carriers evaluated by these three companies were produced by three different procedures. The combination of the Ringer's D,L lactate resuscitation fluids administered to these patients, the Ringer's D,L lactate excipient medium for the HBOC, and the length of storage at 4 C of the liquid preserved RBC infused into these patients who received HBOC may be responsible for the severe adverse events that have been observed.

At an Institute of Medicine meeting held in September 1998 in Washington, DC on resuscitation fluids, the toxicity of Ringer's D,L lactate was discussed. Veech RL has reported the cardiac arrhythmias observed in experimental animals and patients infused with Ringer's D,L lactate solution (Veech RL, Am J Clin Nutri 44:519-551, 1986). Veech also reported that D- lactate generated oxygen free radicals and that the metabolism of L-lactate inhibited glycolysis and recommended the use of Ringer's ketone solution which contains 28mM D betahydroxybutyrate instead of the 28 mM of D,L-lactate. Subsequent studies by Alam, Koustova, Jaskille, Ayuste and associates have reported that Ringer's D,L-lactate resuscitation solution produced pulmonary apoptosis and. intracellular adhesion molecule-1 expression in rats and swine and that those adverse effects were attenuated by the use of Ringer's ketone solution and Ringer's pyruvate solution (Alam HB et al, J Am Coll Surg 193:255-263, 2001; Koustova E et al, Surgery 134:267-274, 2003; Jaskille A et al, J Am Coll Surg 202:25-35, 2006; Ayuste EC et al, J Trauma 60:52-63, 2006). Ringer's ketone solution is stable at room temperature for at least 2 years, but Ringer's sodium pyruvate and Ringer's ethyl pyruvate solutions are not stable at room temperature.

In studies conducted during the Vietnam war, young wounded servicemen who were resuscitated with large volumes of crystalloid solutions consisting of Ringer's D,L lactate developed acute respiratory distress syndrome referred to as "Danang Lung" In recent studies in which rats subjected to hemorrhagic shock were resuscitated with Ringer's D,L lactate, the toxic effects of this solution were observed in the production of pulmonary, liver and intestinal apoptosis. Rhee and associates have reported that Ringer's D,L lactate stimulates human granulocytes to produce oxygen free radicals but that Ringer's L-lactate is less effective (Rhee P et al, J Trauma 44:313-319, 1998; Rhee P et al, Crit Care Med 28:74-78, 2000).

A study was performed at the NBRL with the collaboration of Dr. Carl S. Apstein and his associates to assess the therapeutic effect of Ringer's ketone solution or Ringer's D,L lactate solution on myocardial function in rats The cardiac output was reduced from 60 ml/min to 5-7 ml/min and then the rats were resuscitated with Ringer's ketone solution or Ringer's D,L lactate solution for 2 hours followed by stimulation with dobutamine. Both mortality and cardiac output were measured.

The data show that treatment with Ringer's ketone solution produced significant improvement in survival in rodents at one hour and two hours following hemorrhage and resuscitation. Cardiac output was significantly improved at one hour and two hours in rats resuscitated with Ringer's ketone solution compared to Ringer's D,L lactate solution (Valeri CR et al, Trans Aph Sci 39:205-211, 2008).

In another study at the NBRL, the therapeutic effect of resuscitative solutions to treat hypovolemic anemic rats subjected to renal ischemia was studied in collaboration with. Dr. W. Lieberthal and his associates. A protocol was developed in rats subjected to hypovolemic anemia and renal ischemia to assess the therapeutic effects of blood, Ringer's D,L-lactate, and Ringer's ketone solution.

The left kidney of the rat was removed and. the right renal artery was clamped for 30 minutes. During the period of renal ischemia, 8 ml of blood were removed in 10 minutes from the three groups except for the sham-treated rats. The 8 ml of heparinized blood was reinfused over a 10-minute period or 15 ml of solution of Ringer's ketone or Ringer's D,L lactate. The renal artery clamp was removed after 30 minutes of renal ischemia. Twenty-four (24) hours and 2 days following the resuscitation of the hypovolemic anemic rats with renal ischemia, the glomerular filtration rate (GFR), the hematocrit, the weight and ratio of inulin in the urine and plasma were measured.

The glomerular filtration rate and the inulin urine to plasma ratio 48 hours following resuscitation were the highest for rats treated with Ringer's ketone solution. The weight loss in the rats treated with Ringer's ketone was similar to the sham treated rats. Greater weight loss was observed in the rats treated with blood and Ringer's D,L-lactate. The Ringer's ketone solution improved renal function 48 hours after resuscitation when compared to the Ringer's D,L lactate solution. The observation on the improvement of renal function in hypovolemic anemic rats subjected to renal ischemia by the use of Ringer's ketone solution supports the preparation of ketone containing resuscitation solutions (Valeri CR et al, Trans Aph Sci 39:205-211, 2008).

Following the documentation of Ringer's D,L-lactate's toxicity and the demonstration that the D-isomer of lactate was primarily responsible for its toxicity, Ringer's DL-lactate solution was modified and the solution now contains only the L-isomer of lactate, i.e. Ringer's L-lactate. However, there has been no clinical testing of the modified solution. Cross HR and associates have reported that L-lactate inhibited glycolysis and impaired resuscitation of isolated rat hearts. Recent studies in rodents and swine subjected to hemorrhagic shock showed that resuscitation with Ringer's L-lactate solution reduced the hepatic and pulmonary apoptosis observed with Ringer's D,L-lactate solution (Cross HR et al, J Mol Cell Cardiol 27:1369-1381, 1995).

Any resuscitation fluid and excipient medium used in the HBOC should minimize the oxygen free radicals that may be formed by the HBOC. Betahydroxybutyrate, when substituted for D,L lactate in the

Ringer's ketone solution, has been shown to attenuate the pulmonary apoptosis in rat lungs following hemorrhagic shock and resuscitation. Moreover, betahydroxybutyrate reduces the generation of oxygen free radicals by mitochondria and by human granulocytes (Veech RL, Am J Clin Nutri 44:519-551, 1986; Sato N et al, Life Sci 51:113-118, 1992).

Data have shown that Ringer's D,L-lactate resuscitation fluid combined with at HBOC that contains Ringer's D,L-lactate in the excipient may be responsible for reported severe adverse events observed in clinical studies of HBOC. Diabetic patients have a higher incidence of severe adverse events than non-diabetic patients following cardiopulmonary bypass surgery when they received HBOC in the Ringer's D,L-lactate excipient. In these patients, Ringer's D,L-lactate was used as the hemodi1ution solution and it was infused as the resuscitation solution. Ringer's D,L-lactate is toxic and it should not he used as the excipient for HBOC. Ringer's L-lactate should be more extensively evaluated, both as a resuscitation fluid and as an excipient medium for the HBOC. The therapeutic effectiveness of Ringer's ketone solution in resuscitation has been demonstrated in animals and it appears to be an ideal resuscitative fluid and excipient medium for HBOC (Veech RL et al, IUBMB Life 51:241-247, 2001).

The Effects of Preserved Red Blood Cells on the Severe Adverse Events Observed in Patients Infused with Hemoglobin Based Oxygen Carriers

The severe adverse events observed in patients who received hemoglobin based oxygen carriers (HBOCs) were associated with the Ringer's D,L-lactate resuscitation solution administered and in the excipient used in the HBOCs containing Ringer's D,L-1actate and the length of storage of the preserved RBC administered to the patient at the time that the HBOCs were infused. The quality of the red blood cells preserved in the liquid state at 4 C and that of previously frozen RBCs stored at 4 C with regard their survival, function, therapeutic effectiveness and safety need to be assessed. Severe adverse events have been observed related to the length of storage of the liquid preserved RBC stored at 4 C for longer than 2 weeks prior to transfusion (Koch CG, N Engl J Med 358:1229-1239, 2008). The current methods to preserve RBC in the liquid state in additive solutions at 4C maintain their survival and function for only 2 weeks. The freezing of red blood cells with 40% W/V glycerol and storage at -80 C allows for storage at -80 C for 10 years and following thawing, deglycerolization and storage at 4 C in the additive solution (AS-3 Nutricel) for 2 weeks with acceptable 24 hour posttransfusion survival, less than 1% hemolysis, and moderately impaired oxygen transport function with no associated adverse events. Frozen deglycerolized RBCs are leukoreduced and contain less than 5% of residual plasma and non-plasma substances. Frozen deglycerolized RBCs are the ideal RBC product to transfuse patients receiving HBOCs (Valeri CR and G Ragno, Art Cells, Blood Subst, Biotech 36:3-18, 2008).

Role of nitric oxide in the prevention of severe adverse events associated with blood products

The reduction in vitro of nitric oxide binding to the globin portion of hemoglobin (SNOB) in fresh and liquid preserved red blood cells has been reported to be responsible for the severe adverse events (SAEs) associated with red blood cell transfusion. No in vivo data were reported that the reduction in SNOHb in red blood cells produced severe adverse events (SAEs) in recipients.

Several articles have reported severe adverse events associated with the transfusion of FDA-approved RBC blood products. The recent article published in the Proceedings of the National Academy of Sciences Journal by Bennett-Guerrero E and associates (104:17063-l7068, 2007) reported on the possible relationship between the decrease in nitric oxide occurring in. fresh and liquid preserved RBC in AS-3 additive solution (Nutricel) during storage at 4 C for 42 days and the mortality and morbidity associated with the transfusion of FDA approved RBC products. Since this article received extensive coverage in the lay press, we felt that it was important to comment on our experiences at the NBRL in human and baboon studies which suggest that the reduction in S-nitrosohemoglobin is a reversible defect that is corrected following transfusion. The preservation of red blood cel1, platelets, and plasma using freeze preservative procedures has not been associated with the severe adverse events that are currently observed with the transfusion of FDA-approved blood products.

The recent publication by Bennett-Guerrero E and associates suggests that nitric oxide bound to the cysteine of the globin portion of hemoglobin (S-NOHb) within red blood cells is rapidly lost in fresh CPD whole blood and in CP2D RBC stored in the additive solution AS-3 (Nutricel) at 4 C for 42 days. The authors speculate that this depletion of S-nitrosohemoglobin (S-NOHb) produces severe adverse events in recipients related to vasoconstriction of the microcirculation. They report no data to document that this in vitro decrease in S-NOHb is irreversible and is not corrected following the transfusion of the red blood cells (Valeri CR and Ragno G, Trans Aph Sci 39:241-245, 2008).

Valeri CR and RL Veech have reported on the unrecognized effects of the volume and composition of the resuscitation fluid used during the administration of blood products in Transfusion and Apheresis Science 46:121-123, 2012. Recent publications have reported the severe adverse events associated with blood products but have not considered the effect of the volume and composition of the resuscitative fluids infused with the blood products.

Injury leads to cellular reaction characterized by insulin resistance during which glucose cannot enter muscle and fat cells. In all cells, mitochondrial pyruvate dehydrogenase (PDH) activity is decreased during insulin deficiency leaving cells deficient in substrates needed to power the Krebs cycle and make ATP.

D-B-hydroxybutyrate, a normal ketone body metabolite, enters cells on the monocarboxylate transport channel mimicking the action of insulin and bypassing the enzymatic block at PDH. Metabolism of ketone bodies increases efficiency of mitochondrial energy production and cellular ATP level.

Infusion of 250 ml of 600 mM Na-D-B-hydroxybutyrate solution, with the same osmotic strength as the hypertonic NaCl solution currently being used, would correct insulin resistance, provide energy substrates for cells to produce ATP, correct the tendency of injured tissue to swell due to decreased energy of ionic gradients, and correct acidosis observed in hemorrhage (Table 1).

Substrate resuscitation

Carl W. Walter in 1937 established the first blood bank in Boston. He later developed the plastic blood bag for collection of blood thus facilitating the separation of blood components. In 1986, he wrote a succinct summary of a paper which outlined the toxicity of current parenteral fluid published by RL Veech in the Am J Clin Nutri 44:519-551, 1986. "The prescribing of parenteral fluids has become so routine that most physicians have become oblivious to the toxic impact of current practices on the cellular metabolism of their patients. Few physicians recognize the iatrogenic threat of replacement of body fluids based solely on volumetric and caloric needs. Understanding the metabolic and ionic organization of cells can provide the physician means to use parenteral fluids to control the inherent metabolic energy of cells. Application of new insight into physical chemistry and. metabolic properties of the cell can enhance the physician's therapy in critically ill patients."

Injury of any sort, leads to a cellular reaction which is characterized by insulin resistance (Li L and ML Messina, Trends Endocrin Metab 20:429-435, 2009). During insulin resistance, glucose cannot enter muscle and fat cells nor can the cell metabolize the lactate given in lactated Ringer's solutions. More importantly in all cells the mitochondrial pyruvate dehydrogenase (PDH) activity is decreased during insulin deficiency leaving the cell deficient in substrates needed to power the Krebs cycle to make ATP (Sharma P et al, J Emerg Trauma Shock 2:67-72, 2009). The normal ketone body metabolite D-B-hydroxybutyrate enters cells via the monocarboxylate transport channel mimicking the action of insulin (Kashiwaya Y et al, Am J Cardiol 80:50A-64A, 1997) and bypassing the enzymatic block at PDH. Even more importantly, the metabolism of ketone bodies, increases the efficiency of mitochondrial energy production and increases the energy contained within the ATP molecule that is the delta G ATP (Sato K et al, FASEB J 9:651-658, 1995).

In order to prevent dilutional coagulopathy, one needs to use low volume substrate based resuscitation fluids which are capable of correcting the metabolic and physical chemical abnormalities and the energy deficit of the injured cell. The most important metabolic defect needing correction in injured patients receiving blood products and parenteral fluids is insulin resistance (Li L and JL Messina, Trends Endocrinal Metab 20:429-435, 2009). Insulin resistance prevents the injured cell from metabolizing the lactate produced by glycolysis as well as the lactate administered in Ringer's lactate fluids. This metabolic defect of insulin resistance in the cells of injured patients can be overcome by the administration of Na-D-B-hydroxybutyrate containing solutions which bypass the metabolic block at PDH and increase the energy of the ATP molecule (Kashiwaya Y et al, Am J Cardiol 80:50A-64A, 1997; Sato K et al, FASEB J 9:651-658, 1995). The six most important metabolic effects of ketone bodies can be summarized as follows (Sato K et al, FASEB J 9:651-658, 1995):

1. Increases the concentration of Krebs cycle substrates depleted by insulin resistance
2. Production of mitochondrial NADH required as a substrate for electron transport
3. Oxidation of coenzyme Q thus lowering the production of free radicals
4. Increases the energy of ATP hydrolysis, delta G ATP
5. Mimics the action of insulin
6. Overcomes the blockade of insulin resistance

Such a solution would be one compromised of 600 mM Na-D-B-hydroxybutyrate which would have the same osmotic strength as the current hypertonic NaCl solutions being used. It could be provided in 250 ml volumes. Infusion of D-B-hydroxybutyrate containing solutions would correct the insulin resistance of injury (Li L and JL Messina, Trends Endocrinal Metab 20:429-435, 2009) provide energy substrates for the cell to produce ATP (Sato K et al, FASEB J 9:651-658, 1995) correct the tendency of injured tissue to swell due to the decrease in the energy of ionic gradients (Veech RL et al, IUBMB Life 54:241-252, 2002) as well as correcting the acidosis often accompanying hemorrhage.

Table 1*

High volume resuscitation and low volume resuscitation fluids.

Component (mM)	Ringer's lactate	Ringer's ketone	Normal saline	Hypertonic saline	Hypertonic ketone
Na+	130	130	155	603	600
K+	4	4			
Ca++	3	3			
Cl-	109	109	155	603	
Lactate	28				
D-B-hydroxy butyrate		28			600
pH	6.5	6.5	6.5	6.5	6.5
Osmolarity (mOsm/l)	275	275	310	1200	1200
Sterilization	Heat	Heat	Heat	Heat	Heat

*with permission from Elsevier, Inc., 360 Park Avenue South, New York, NY 10010-1710, USA

CHAPTER 8 REFERENCES

1. Alam HB, Austin B, Koustova E, Rhee P. Resuscitation induced pulmonary apoptosis and intracellular adhesion molecule-1 expression in rats are attenuated by the use of ketone Ringer's solution. J Am Coll Surg 193:255-263, 2001.
2. Ayuste EC, Chen H, Koustova E, Rhee P, Ahuja N, Chen Z, Valeri CR, Spaniolas K, Mehrani T, Alam HB. Hepatic and pulmonary apoptosis after hemorrhagic shock in swine can be reduced through modifications of conventional Ringer's solution, J Trauma 60(1):52-63, 2006.
3. Bennett-Guerrrero E, Veldman TH, Doctor A, Telep MS, Otel TL, Beld TS. et al. Evolution of adverse changes in stored RBCs. Proc Natl Acad Sci 104:17063-17068, 2007.
4. Carr DB et al. D-lactate acidosis stimulating hypothalamic syndrome. Am Neurol 11:195-197, 1982.
5. Chan L, Slater J, Hasbargen J, Herndon ON, Veech, RL, Wolf S. Neurocardiac toxicity of racemic D,L-lactate fluids. Integr Physiol Behav Sci 29:383-394, 1992.
6. Cross HR, Clarke K, Opie LH, Radda OK. Is lactate induced myocardial ischemic injury mediated by decreased pH or increased intracellular lactate? J Mol Cell Cardiol 27:1369-1381, 1995.
7. DeCosta JM, On irritable heart. Am J Med Sci.61:18-52, 1971.
8. Hartmann AF. D-L lactate was used to prevent hyperchloremic acidosis. JAMA 103:1349-1354, 1934.
9. Hu ZG et al. Ketogenic diet reduces apoptosis in traumatic brain injury, Am Clin Lab Sci 39:76-83, 2005.
10. Jaskille A, Koustova B, Rhee P. Britten-Web 3, Chan H, Valeri CR, Kirkpatrick JR, Alam HB. Hepatic apoptosis following hemorrhagic shock in rats can be reduced through modification of conventional Ringer's solution. J Am Coll Surg 202(l):25-35, 2006.
11. Kashiwaya Y, King MT and Veech RL. Substrate signaling by insulin: a ketone bodies ratio mimic insulin action in heart. Am J Cardiol 80:50A-64A, 1997.
12. Koch CG, Li L, Sessler DL, et al. Duration of red cell storage and complications after cardiac surgery. N Engl J Med 358:1229-1239, 2008.

13. Koustova E, Rhee P, Hancock T, Chen, H Inocencio R, Jaskille A, Hanes W, Valeri CR, Alam RB. Ketone and pyruvic Ringer's solution decreases pulmonary apoptosis in a rat model of severe hemorrhagic shock and resuscitation. Surgery 134:267-274, 2001.
14. Li L and Messina JL. Acute insulin resistance following injury. Trends Endocrinol Metab 20:429-435, 2009.
15. Longnecker DE et al. Recommend investigation of Na-D-B-hydroxybutyrate resuscitation fluids. Institute of Medicine Committee on Fluid Resuscitation for Combat Casualties and Civilian Injuries. National Academy Press, 1999.
16. Masuda R, et al. D-B-hydroxybutyrate is neuroprotective against hypoxia in serum free hippocampal primary cultures. Neurosci Res 8:501-509, 2005.
17. Prins ML. Cerebral metabolic adaptation and ketone metabolism after brain injury. J Cerebral Blood Fluids and Metabolism 28:1-16, 2008.
18. Pope A, French G, Longnecker DF. Fluid resuscitation: state of the science for treating combat casualties and civilian injuries. National Academy Press: Washington, DC 1999.
19. Rhee P, Burris D, Kaufmann C, Pikoulis M, Austin B, Ling G, Harviel D, Waxman K. Lactated Ringer's solution resuscitation causes neutrophils activation after hemorrhagic shock. J Trauma 44:313-319, 1998.
20. Rhee P, Wang D, Ruff P, Austin B, DeBreux S, Wolcott K, Burris D, Ling G, Sun L. Human neutrophils activation and increased adhesion by various resuscitation fluids. Crit Care Med 28:74-78, 2000.
21. Sato N, Shimizu H, Shimomura Y, Suwa K, Mort M, Kobayashi L. Mechanism of inhibitory action of ketone bodies on the production of reactive oxygen intermediates (ROIS) by polymorphonuclear leukocytes. Life Sci 51:113-119, 1992.
22. Sato K, Kashiwaya Y, Keon CA, Tsuchiya N, King MT, Radda GK, Chance B, Clarke K, Veech RL. Insulin, ketone bodies, and mitochondrial energy transduction. FASEB J 9:651-658, 1995.
23. Sharma P, Benford B, Li ZZ, Ling GS. Role of pyruvate dehydrogenase complex in traumatic brain injury and measurement of pyruvate dehydrogenase enzyme by dipstick test. J Emerg Trauma Shock 2:67-72, 2009.

24. Shires T. Ringer's D-L lactate became standard treatment of hemorrhagic shock. Arch Surg 88:688-693, 1964.
25. Valeri CR, Ragno G. The effects of preserved red blood cells on the severe adverse events observed in patients infused with hemoglobin based oxygen carriers. Art Cells, Blood Subst and Biotech 36 (1):3-1 8, 2008.
26. Valeri CR, Ragno G. Role of nitric oxide in the prevention of severe adverse events associated with blood products. Trans Apher Sci 39:241-245, 2008.
27. Valeri CR, Ragno G, Veech RL. Effects of the resuscitation fluid and the HBOC excipient on the toxicity of the HBOC: Ringer's D, L-Lactate, Ringer's L-Lactate, and Ringer's ketone solutions. Art Cells, Blood Subst and Biotech 34(6):601-606, 2006
28. Valeri CR, Ragno G, Veech RL. Severe adverse events associated with hemoglobin based oxygen carriers: role of resuscitative fluids and liquid preserved RBC. Trans Apher Sci 39:205-211, 2008.
29. Valeri CR, Veech RL. The unrecognized effects of the volume and composition of the resuscitation fluid used during the administration of blood products. Trans Aph Sci 46:121-123, 2012.
30. Veech RL. The toxic impact of parenteral solutions on the metabolism of cell: a hypothesis for physiological parenteral therapy. Am J Clin Nutri 44:519-551, 1986.
31. Veech RL, Chance B, Kashiwaya Y, Lardy HA, Cahill GFJ. Ketone bodies, potential therapeutic uses. IUBMB Life 51:241-247, 2001.
32. Veech RL, Kashiwaya Y, Gates DN, King MT, Clarke K. The energetic of ion distribution: the origin of the resting electric potential of cells. IUBMB Life 54:241- 252, 2002.
33. Veech RL. The therapeutic implications of ketone bodies: the effects of ketone bodies in pathological conditions, ketosis, ketogenic diet, redox states, insulin resistance, and mitochondrial metabolism. Prostaglandins, Leukotrienes, and Essential Fatty Acids 70:309-319, 2004.
34. Veech RL, Valeri CR, Van Itallie TB. The mitochondrial permeability transition pore provides a key to the diagnosis and treatment of traumatic brain injury. IUBMB Life 64:203-207, 2012.

CHAPTER 9

CRYOPRESERVATION OF HUMAN BLOOD PRODUCTS

Frozen blood is a misnomer, blood is not frozen only its components, i.e. red blood cells, platelets, pluripotential hematopoietic mononuclear adult stem cells and plasma.

RBC are frozen with glycerol, platelets are frozen with DMSO, pluripotential mononuclear adult stem cells are frozen with DMSO with or without hydroxyethyl starch (HES) and plasma is frozen without a cryoprotectant.

Rate of freezing and temperature of storage in the frozen state have been investigated using instruments to control the rate of freezing with liquid nitrogen and freezing by storage in a -80 C mechanical freezer; frozen storage in the gas phase of liquid nitrogen at -150 C or in liquid nitrogen at -197 C; and frozen storage in a -80 C mechanical freezer with air-cooled dual cascade compressors (Valeri CR, CRC Press, Boca Raton, FL, 1976; Valeri CR and Ragno G, Trans Aph Sci 34:271-287, with an editorial on pages 267-269, 2006; McCullough J et al, Transfusion 50:808-819, 2010).

For the past 45 years, NBRL has evaluated cryopreservation of RBC, platelets, pluripotential hematopoietic mononuclear adult stem cells, and plasma to treat military personnel subjected to traumatic injuries and those exposed to radiation injury. The goal of this research funded by the US-Navy's Bureau of Medicine and Surgery, the Office of Naval Research and the Congress of the United States was to provide frozen blood products to supplement the liquid preserved blood products.

NBRL has demonstrated that RBC preserved with 40% W/V glycerol, platelets preserved with 6% DMSO, pluripotential hematopoietic mononuclear adult stem cells preserved with 10% DMSO and plasma can all be frozen and stored at -80 C. RBC frozen with 40% W/V glycerol can be stored at -80 C for at least 21 years and as long as 37 years (Valeri CR et al, Transfusion 29:429-437, 1989; Valeri CR et al, Vox Sang 79:168-174, 2000). Platelets frozen with 6% DMSO can be stored at -80 C for at least 2 years (Melaragno AJ et al, Vox Sang 49:245-258, 1985; Khuri SF et al, J Thorac Cardiovasc Surg 117:172-184, 1999; Valeri CR et al, Transfusion 43:1162-1167, 2003). Pluripotential hematopoietic mononuclear adult stem cells frozen with 10% DMSO can be stored at -80 C for 1 ½ years (Valeri CR and Pivacek LE, Transfusion 36:303-308, 1996). Fresh frozen plasma can be stored at -80 C for at least 14 years (Valeri CR and Ragno G, Transfusion 45(11):1829-1830, 2005).

The technology and logistic support to freeze blood products are more complex and expensive than liquid preservation of RBC, platelets, and pluripotential hematopoietic mononuclear adult stem cells, also referred to as hematopoietic progenitor cells and hematopoietic marrow and blood derived adult stem cells.

The need for cryopreserved blood products can be justified by the availability and quality of these frozen blood products, with regards the recovery of cells associated with the cryopreservation procedure, and the viability and function of the cellular components and plasma proteins.

RBC must circulate to carry oxygen and to remove carbon dioxide from the tissues; RBC affinity for oxygen correlates to their in vivo function to provide oxygen at high oxygen tension to the brain, heart, and kidneys and to exert a hemostatic effect to reduce non-surgical blood loss (Valeri CR et al, Transfusion 47:206S-248S, 2007; Valeri CR and Ragno G, Trans Aph Sci 42:223-233, 2010). Platelets must circulate to the bleeding site to restore hemostasis and pluripotential hematopoietic mononuclear stem cells must circulate to the bone marrow to restore bone marrow function (Valeri CR and Ragno G, Trans Aph Sci 34:271-287 with an editorial on pages 267-269, 2006).

Glycerol must be removed from glycerolized RBC to reduce the glycerol concentration to less than 1% W/V and DMSO should be removed prior to transfusion of platelets and pluripotential adult stem cells to reduce the side effects of the DMSO (Valeri CR, CRC Press, Boca Raton, FL,

1976; Valeri CR and Ragno G, Trans Aph Sci 34:271-287 with an editorial on pages 267-269, 2006).

In May 2001, the FDA approved the automated functionally closed Haemonetics ACP215 instrument to glycerolize and deglycerolize RBC frozen with 40% W/V glycerol at -80 C for at least 10 years and stored as deglycerolized RBC at 4 C in the additive solution AS-3 (Nutricel) for 2 weeks. FDA approval of the Haemonetics ACP215 instrument now makes economically feasible the use of frozen group O Rh-positive and group O Rh-negative RBC to supplement the liquid preserved RBC to control the supply and demand for universal donor RBC for both the military and civilian communities. To maintain a stockpile of universal donor frozen RBC, outdated group O Rh positive and group O Rh negative liquid preserved RBC can be salvaged by biochemical treatment with a solution containing pyruvate, inosine, phosphate and adenine (PIPA) prior to freezing with 40% W/V glycerol and storage at -80 C for at least 10 years (Valeri CR et al, Crit Care Med 7:439-447, 1979; Valeri CR et al, Transfusion 21:138-149, 1981; Valeri CR, Crit Rev Clin Lab Sci 17(4):299-374, 1982; Valeri CR et al , Transfusion 40:1341-1345, 2000; Valeri CR et al, Transfusion 41:933-939, 2001; Bohonek M et al, Transfusion 50:1007-1013, 2010; Ragno G and Valeri CR, Trans Aph Sci35:137-143, 2006).

Automation of RBC glycerolization using the high separation bowl attached to the Haemonetics ACP215 disposable glycerolizing set has simplified glycerolization of both nonrejuvenated and rejuvenated RBC. The procedure to deglycerolize RBC frozen with 40% W/V glycerol using the Haemonetics ACP215 reduces the glycerol concentration to <1%, reduces the level of PIPA solution used to biochemically modify the RBC prior to freezing, reduces the WBC to a mean of 1×10^7 in the unit, and reduces the plasma and non-plasma levels to less than 5%. Leukoreduced and washed deglycerolized RBC stored at 4 C in AS-3 for 2 weeks is the ideal RBC product to prevent transfusion-related acute lung injury (TRALI) and systemic inflammatory response syndrome (SIRS) (Valeri CR et al, Transfusion 45(10):1621-1627, 2005; Valeri CR and Ragno G, Trans Aph Sci 42:223-233, 2010).

NBRL has simplified RBC freezing by removing the supernatant glycerol prior to freezing. NBRL has also simplified freezing of platelets and pluripotential mononuclear adult stem cells by removal of the supernatant DMSO prior to freezing. Removal of the supernatant DMSO removes about 95% of the DMSO and eliminates the need for post-thaw washing prior to transfusion (Valeri CR. at al, Transfusion 21:138-149, 1981, Valeri CR et al, Transfusion 45(12):1890-1898, 2005; Valeri CR, Ragno G, Trans Aph Sci 42:223-233, 2010).

Lelkens and associates have reported the use by the Netherlands military of: a) group O Rh positive and group O Rh negative RBC frozen with 40% W/V glycerol and stored at -80 C; b) group O single donor leukoreduced platelets frozen with 5% DMSO with the removal of the supernatant DMSO solution before freezing and storage at -80 C and, following thawing, resuspended in AB plasma; and c) AB plasma stored at -30 C and then at -80 C (Lelkens CCM et al, Trans Aph Sci 34:289-298, 2006; Noorman F et al, Transfusion 49(35):28A Supplement, 2009; Badloe J and Norman F, Transfusion 51(35) Supplement, 2011). Lelkens CCM and associates experience in the Netherlands military has demonstrated for the first time that frozen RBC, frozen platelets and frozen plasma stored at -80 C can be used to treat patients without the need for the "walking blood banks" which provide fresh whole blood. The method to freeze platelets with 5% or 6% DMSO and removal of the supernatant DMSO prior to freezing and storage at -80 C has not been FDA approved.

Economic assessments are needed to compare the costs of freezing RBC, platelets and plasma at -80 C as described by Lelkens and associates to that of providing "walking blood donors" to provide fresh whole blood, single donor liquid preserved platelets stored at room temperature for 5 days with agitation; red blood cells stored in additive solution (CPD-AS1, CP2D-AS3, and CPD-AS5) at 4 C for 42 days; and FFP and cryoprecipitate stored at -20 C for one year (Lelkens CCM et al, Trans Aph Sc 34:289-298, 2006; Noorman F et al, Transfusion 49(35):28A Supplement, 2009; Badloe J and Noorman F, Transfusion 51(35) Suppl, 2011.; Noorman F and Badloe JF, AABB Oct 2012, Transfusion Practice/Clinical Case Studies, Transfusion Practice in Emergent Settings SP383: Poster Presentation, Transfusion 52(Suppl) SP383, 2012).

Acceptable platelet preservation should ensure that both survival and hemostatic function are maintained. Current guidelines only require satisfactory platelet survival. Studies have demonstrated that platelet survival does not correlate with platelet hemostatic function. Studies of autologous preserved human and baboon platelets demonstrated that platelet survival did not correlate with the hemostatic function to correct an aspirin prolonged bleeding time (Valeri CR, NEJM 290:353-358,

1974; Valeri CR et al, Transfusion 42:1206-1216, 2002; Valeri CR et al, Transfusion 47:206S-248S, 2007; Noorman F et al, Component and Component Processing: Platelets, SP23, Poster Presentation, Transfusion 52(Suppl) SP23, 62A, 2012; Noorman F and JF Badloe AABB Oct 2012, Transfusion Practice/Clinical Case Studies S53:030H; Oral presentation, Transfusion 52(Suppl) S-53-030H; 33A, 2012).

Studies in hemodiluted, anemic, thrombocytopenic patients undergoing CPB surgery demonstrated that allogeneic previously frozen washed platelets had lower survival but increased hemostatic function to reduce non-surgical blood loss and to reduce the need for RBC and FFP during the post-operative period compared to liquid preserved platelets without any adverse events (Khuri SF et al, J Thorac Cardiovasc Surg 117:172-184, 1999).

Experience in freezing human platelets at the Naval Blood Research Laboratory using 6% DMSO and storage at -80 C in mechanical freezers demonstrate that platelets frozen with 6% DMSO at -80 C, thawed, washed and resuspended in autologous plasma have in vivo recovery and lifespan values similar to platelets stored at 22 C for 5 days but improved hemostatic function. Allogeneic platelets frozen with 6% DMSO stored at -80 C for at least 2 years, thawed, washed and resuspended in autologous plasma had freeze-thaw-wash recovery values of about 70% and in vivo recovery values 50% those of fresh allogeneic platelets in stable thrombocytopenic patients (Valeri CR et al, Blood 43:131-136, 1974; Valeri CR, NEJM, 290:353-358, 1974; Melaragno AJ et al, Vox Sang 49:245-258, 1985).

Autologous previously frozen, washed baboon platelets were compared to autologous baboon platelets stored in the liquid, state at 22 C for 5 days. Previously frozen washed baboon platelets had slightly increased in vivo recovery than liquid preserved baboon platelets stored at 22 C for 5 days. Previously frozen washed autologous platelets had significantly better function as assessed by their ability to reduce an aspirin-induced increased bleeding time and to produce shed blood thromboxane at the bleeding time site in normal healthy baboons compared to liquid preserved autologous platelets stored at 22 C for 5 days with agitation (Valeri CR et al, Transfusion 42:1206-1216, 2002; Valeri CR et al, Vox Sang 83:347-351, 2002).

Patients undergoing CPB were transfused with allogeneic previously frozen washed or liquid preserved platelets stored at room temperature for a mean of 3.4 days with agitation. Patients transfused with previously frozen washed platelets had greater reduction in nonsurgical blood loss and required fewer RBC and FFP transfusions than those who received liquid preserved platelets stored at room temperature for a mean of 3.4 days with agitation. Previously frozen washed platelets produced more thromboxane A2 following stimulation in vitro with arachidonic acid and adenosine diphosphate and accumulated more factor V on the platelets than did liquid preserved platelets. The procoagulant activity and increased thromboxane production in previously frozen washed platelets improved hemostasis in patients following transfusion which resulted in reduced nonsurgical blood loss and reduced requirements for allogeneic RBC and FFP in patients following cardiopulmonary bypass surgical procedures (Barnard MR et al, Transfusion 39:880-888, 1989; Khuri SF et al, J Thorac Cardiovasc Surg 117:172-184, 1999; Valeri CR et al, Transfusion 45:596-603, 2005).

The in vivo recovery 1-2 hours after transfusion of autologous human platelets frozen after the removal of the supernatant DMSO, thawed, diluted with 0.9% NaCl and stored at room temperature for 4.hours was 25 to 30% with a lifespan of 7 days compared to that of 35 to 40% and a lifespan of 7 days for the platelets frozen with the supernatant DMSO, thawed, washed and resuspended in plasma (Valeri CR. et al, Transfusion 45(12):1890-1898, 2005).

Platelet surface marker testing was performed on human platelets frozen with supernatant DMSO thawed, washed, and resuspended in plasma had increased numbers of normal GPIb and reduced annexin V binding platelets. Platelets frozen after the removal of the supernatant DMSO, thawed, resuspended in 0.9% NaCl had reduced numbers of GPIb positive and increased annexin binding platelets (Valeri CR et al, Transfusion 45(12):1890-1898, 2005). Autologous human platelets frozen after removal of supernatant DMSO, thawed and diluted with 0.9% NaCl had in vitro recovery of 90% and in vivo recovery 1 to 2 hours after transfusion that was about 10% lower than platelets frozen with the supernatant DMSO, thawed, washed, and resuspended in plasma with in vitro recovery of 70% prior to transfusion (Valeri CR et al, Transfusion 45(12):1890-1898, 2005).

Freezing of human and baboon platelets with 6% DMSO with or without the removal of the supernatant DMSO prior to freezing by storage in a -80 C mechanical freezer following thawing with or without washing and resuspension had a bimodal population of platelets.

Baboon platelets frozen with the supernatant DMSO, thawed, washed and resuspended in plasma had a population of platelets with 35 to 40% in vivo recovery values 1 to 2 hours after transfusion and was associated with platelets with normal GPIb and reduced annexin V binding. The other population of baboon platelets with in vivo recovery values 1 to 2 hours after transfusion of 5% had reduced GPIb and increased annexin V binding and was associated with improved hemostatic function (Barnard MR et al, Transfusion 39:880-888, 1999). The clinical experience by the Netherlands military in combat zones in the Middle East from 2001 to 2012 reported no adverse effects associated with the transfusion of multiple units of single donor leukoreduced group O platelets treated with 5% DMSO; the supernatant DMSO removed prior to freezing and storage at -80 C and following thawing and resuspension in AB plasma. The Netherlands military have reported that frozen platelets were safe and therapeutically effective in the treatment of combat casualties with group O Rh positive and group O Rh negative frozen RBC and frozen AB plasma without the need for fresh whole blood (Noorman F and JF Badloe, AABB Oct 2012, Transfusion Practice/Clinical Case Studies: Transfusion Practice in Emergent Settings, SP383:Poster Presentation, Transfusion 52(Suppl) SP383, 198A, 2012; Noorman F and JF Badloe, AABB Oct 2012, Transfusion Practice/Clinical Case Studies S53-030H, Oral Presentation, Transfusion 52(Suppl) S53-030H, 33A, 2012; Noorman F et al, AABB Oct 2012, Components and Component Processing: Platelets SP23 Poster Presentation, Transfusion 52(Suppl), Supplement SP23, 62A, 2012).

Khuri S et al demonstrated the safety and therapeutic benefit of allogeneic platelets frozen with the supernatant DMSO, thawed, washed and resuspended in plasma had reduced blood loss without producing adverse events in patients following CPB surgery (Khuri S at al, J Thorac Cardiovasc Surg 117:172-184, 1999).

Elimination of the postthaw washing of previously frozen platelets has simplified the procedure by reducing the time for thawing, washing and resuspension of the platelets from 2 hours to 50 minutes for thawing and resuspension of the platelets (Valeri CR et at, Transfusion 45(12):1890-1898, 2005). Noorman and associates have reduced the time to thaw and dilute the frozen platelets using thawed AB plasma stored at 4 C for as long as 7 days to prepare the frozen platelets in 15 minutes (Noorman F et al, AABB Oct 2012, Components and Component Processing: Platelets SP23 Poster Presentation, Transfusion 52(Suppl), Supplement SP23, 62A,

2012). Noorman and associates reported that frozen platelets are activated, clot strength is reduced, onset of clotting and clot development is faster compared to fresh platelets. Frozen platelets can be prepared within 15 minutes without loss of quality compared to the standard 50 minutes procedure and can be used in the early resuscitation of military trauma patients (Noorman F et al, AABB Oct 2012, Components and Component Processing: Platelets SP23 Poster Presentation, Transfusion 52(Suppl), Supplement SP23, 62A, 2012).

Dumont LJ and associates in Transfusion 53:128-137, 2013 have reported a randomized controlled trial evaluating recovery and survival of 6% DMSO-frozen autologous platelets in healthy volunteers. Single donor autologous apheresed platelets treated with 6% DMSO, the supernatant DMSO removed, the platelets frozen at -80 C, thawed, and diluted with 0.9% NaCl had recovery in vivo of 41.6±9.7% and lifespan of 7.0±2.1 days whereas the autologous apheresed fresh platelets had recovery in vivo of 68.4±8.2% and lifespan of 8.4±1.2 days in 24 normal volunteers that were studied.

NBRL has experience freezing pluripotential hematopoietic autologous mononuclear cells (MNC) isolated from peripheral blood to treat dogs subjected to radiation injury (Valeri CR et al, Cryobiology 33:387-394, 1986; Valeri CR et al, Technical Report 85-01). Human mononuclear cells isolated from bone marrow and peripheral blood have been frozen at NBRL. Human peripheral blood MNC (PBMNC) adult stem cells obtained by leukaphersis were purified by ficoll-hypaque density gradient centrifugation to remove the RBC and granulocytes (Valeri CR and G Ragno, Trans Aph Sci 34:271-287, 2006). PBMNC were preserved with 10% DMSO and frozen at 2-3 C per minute by storage in a -80 C mechanical freezer and then stored in polyvinylchloride plastic bags at -80 C or in polyolefin plastic bags at -135 C in mechanical freezers. Following thawing, the MNC were washed and resuspended in autologous plasma prior to testing for in vitro recovery, viability assessed by fluorescein diacetate and ethidium bromide, and growth in the colony forming unit granulocyte-erythroid-monocyte-megakaryocytic (CFU-GEMM) tissue culture assay. PBMNC frozen with 10% DMSO can be stored in PVC bags at -80 C for 1.5 years, PBMNC frozen with 10% DMSO can be stored in polyolefin plastic bags at -135 C for 2.4 years. When these guidelines are followed, in vitro recovery was 90% that of fresh PBMNCs, viability was 90% and growth in the CFU-GEMM tissue culture assay was similar to that of fresh PBMNCs (Valeri CR and LE Pivacek, Transfusion 36:303-308, 1996).

In a procedure similar to that used to freeze RBC with glycerol and platelets with DMSO, NBRL has modified the freezing of PBMNC with 10% DMSO by removal of the supernatant DMSO prior to freezing. Leukapheresed, ficoll-hypaque isolated MNC arc treated with a volume of 27% DMSO in 0.9% NaCl to achieve a final concentration of 10% DMSO in a 300 ml PVC or ethylvinyl acetate plastic bag. MNC treated with 10% DMSO are concentrated by centrifugation at 1250 X g for 10 minutes. All the visible supernatant is removed and the MNC re disaggregated in the presence of 10% DMSO. The 300 ml PVC plastic bag or 300 ml ethylvinyl acetate (EVA) plastic bag is placed in a polyester plastic bag, the air removed from the plasma bag and heat sealed, and then stored in a rigid cardboard box. The MNC concentrate is frozen by storage at the bottom of the -80 C mechanical freezer and stored for 24 hours. MNCs in the EVA plastic bag are transferred to a -135 C mechanical freezer and MNC in the PVC plastic bag are stored at -80 C. Removal of about 95% DMSO prior to freezing allows for the frozen MNC to be thawed, diluted with 0.9% NaCl and infused without the need for post-thaw washing to reduce the DMSO level (Valeri CR, Ann NY Acad Sci 459:353-366, 1985; Carciero R and Valeri CR, Vox Sang 49:373-380, 1985; Valeri CR and Pivacek LE, Transfusion 36:303-308, 1996; Valeri CR and Ragno G, Trans Aph Sci 34:271-287 with an editorial on pgs 267-269, 2006).

The use of mechanical freezers for both freezing and storage reduces the cost and complexity of the cryopreservation procedure of pluripotential mononuclear cells obtained from bone marrow, peripheral blood and umbilical cord blood (Valeri CR and Ragno G, Trans Aph Sci 34:271-287 with an editorial on pgs 267-269, 2006; McCullough. J et al, Transfusion 50:808-819, 2010). The growth of MNC in the CFU-GEMM tissue culture assay is an important in vitro test to help assess the in vivo effect of the MNC to repopulate bone marrow after autologous and allogeneic transplantation. The viability and function of the cryopreserved MNC can only be assessed following infusion to repopulate the bone marrow of the ABO, Rh, and HLA compatible recipients. Unlike frozen RBC which are typed for ABO and Rh antigens and platelets typed for ABO antigens, MNC are typed for ABO, Rh, and HLA antigens to perform compatibility testing prior to allogeneic transplantation. Pluripotential hematopoietic MNC devoid of immunocompetent lymphocytes to prevent graft versus host disease should be ABO, Rh and HLA compatible with the recipient (Valeri CR and Ragno G, Trans Aph Sci 34:371-387, 2006).

Studies performed at NBRL over the past 45 years have demonstrated that the -80 C mechanical freezer can preserve RBC, platelets, pluripotential hematopoietic MNC adult stem cells and plasma to supplement the liquid preserved blood products to treat patients subjected to hemorrhagic shock and radiation injury. The -80 C temperature allows for shipment of these frozen blood products using insulated containers and dry ice. The -80 C mechanical freezer allows for freezing of these blood products without the need for liquid nitrogen and controlled rate freezers to control the rate of freezing at 1 C per minute (Valeri CR and Ragno G, Trans Aph Sci 34:271-287, 2006; McCullough J et al, Transfusion 50:808-819, 2010).

The FDA has approved the storage of RBC frozen with 40% W/V glycerol at -80 C for 10 years. The FDA has approved that RBC glycerolized with 40% W/V glycerol and frozen at -80 C and deglycerolized using the automated functionally closed Haemonetics ACP215 instrument for storage at 4 C in the additive solution AS-3 (Nutricel) for 2 weeks (Valeri CR, Ragno G, Trans Aph Sci 42:223-233, 2010). Single donor non-leukoreduced and leukoreduced group O platelets treated with 6% DMSO, concentrated to remove the supernatant DMSO, frozen at -80 C can be stored at -80 C for at least 2 years (Melaragno AJ et al, Vox Sang 49:245-258, 1985; Valeri CR et al, Transfusion 43:1162-1167, 2003; Valeri CR et al, Transfusion 45(12):1890-1898, 2005).

Leukapheresed ficoll-hypaque isolated pluripotential hematopoietic MNC stem cells treated with 10% DMSO, concentrated to remove the supernatant DMSO, frozen at -80 C can be stored at -80 C for 1 ½ years. With the removal of the supernatant DMSO prior to freezing, thawed DMSO-frozen platelet concentrates and thawed DMSO frozen MNC concentrates can be diluted with a 0.9% NaCl without post-thaw washing prior to transfusion. Thawed fresh frozen plasma can be stored at room temperature (22C ± 2 C) for 8 hours and at 4 C for 24 hours prior to transfusion. Frozen blood products stored at -80 C (i.e. group O Rh positive and group O Rh negative RBC, group O platelets, AB plasma., and ABO, Rh and HLA pluripotential hematopoietic MNC adult stem cells) can be used for contingency and inventory control to supplement liquid preserved blood products to treat both military and civilian patients subjected to hemorrhagic shock and radiation injury. Red blood cells, platelets, plasma and pluripotential mononuclear adult stem cells can all be frozen using a -80 C mechanical freezer without the need for controlled rate freezers and liquid nitrogen (Valeri CR and Ragno G, Trans Aph Sci 34:271-287, 2006) Red blood cells are frozen with 40% W/V glycerol and storage at -80 C. After the addition and removal of glycerol using the Haemonetics ACP215 instrument the RBC are resuspended in AS-3 (Nutricel) and stored at 4 C for 2 weeks (Valeri CR et al, Transfusion 41:933-939, 2001; Valeri CR et al, Transfusion 41:928-932, 2001).

Transfusion associated graft versus host disease (GVHD) results from the presence of viable immunocompetent lymphocytes in transfused allogeneic blood components. Viable immunocompetent lymphocytes have been detected in RBCs that were frozen with glycerol and washed before transfusion. The effect of radiation on human RBC frozen with 40 percent (wt/vol) glycerol and stored at -80 C was studied at the NBRL. In vitro and in vivo testing were done on human RBCs that were frozen with 40 percent (wt/vol) glycerol at -80 C, with some units exposed to 2500 cGy of gamma radiation and others not irradiated, and that after thawing and washing, were stored in a sodium chloride-glucose solution at 4 C for 3 days before autologous transfusion.

The glycerol-frozen RBCs treated with 2500 cGy before deglycerolization had a mean freeze-thaw-wash recovery of 87% and a mean 24-hour posttransfusion survival of 86 percent after storage for 3 days at 4 C in a 0.9% NaCl and 0.2% glucose solution. For the nonirradiated units, the mean freeze-thaw-wash recovery was 85 percent and the mean 24-hour posttransfusion survival was 83 percent.

These data show similar acceptable results for RBCs frozen with 40 percent (wt/vol) glycerol at -80 C and treated in the frozen state with 2500 cGy of gamma radiation and for RBCs that were not irradiated, all of which were washed and then stored in a sodium chloride-glucose solution for 3 days before autologous transfusion (Valeri CR et al, Transfusion 41:545-549, 2001). Gamma radiation of frozen RBC containing 40% W/V glycerol with 2500 cGy inactivated the immunocompetent lymphocytes to prevent GVHD (Valeri CR et al, Transfusion 41:545-549, 2001).

RBC with 40% W/V glycerol in PVC plastic bags and stored in a rigid cardboard box at -80 C for 14-18 years had breakage of 2.4% for nontransported units and 6.7% for transported units (Valeri CR, and Ragno G, Trans Aph Sci 34:271-287, 2006; Valeri CR, Ragno G, Trans Aph Sci 42:223-233, 2010).

Studies at the NBRL have shown that human RBC with 40% W/V glycerol and stored at -80 C tolerate fluctuations in temperature (Valeri CR et al, Transfusion 21:138-149, 1981; Valeri CR et al, Vox Sang 45:25-39, 1983; Valeri CR et al, Transfusion 41;401-405, 2001; Valeri CR et al, Transfusion 43:411-414, 2003).

Platelets are treated with 6% DMSO and the supernatant DMSO removed prior to freezing and storage at -80 C. The thawed previously frozen platelets are diluted with 10 to 20 ml of 0.9% NaCl or a unit of thawed AB plasma prior to transfusion (Valeri CR et al, Transfusion 45:1890-1898, 2005; Lelkens CCM et al, Trans Aph Sci 34:289-298, 2006; Noorman F and Badloe J, Transfusion 52(Suppl) S53-030H, 33A, 2012; Noorman F et al, Transfusion 52(Suppl) SP23, 62A 2012).

The important factors used in the freezing of platelets and mononuclear cells include: DMSO as the FDA approved cryoprotectant, a final concentration of 5 or 6 % DMSO for platelets and 10% DMSO for mononuclear cells (MNC) (Valeri CR et al, Transfusion 47:206S-248S, 2007; Noorman F and Badloe J, Transfusion 52(Suppl) S53-030H, 33A, 2012; Noorman F et al, Transfusion 52(Suppl) SP23, 62A 2012). The supernatant DMSO is removed prior to freezing the platelets and MNC.

The rate of freezing at 2-3 C per minute occurs by storage of the platelets and MNC in 300 ml PVC plastic bags placed in polyester plastic bags and stored in rigid cardboard box with placement on the bottom of a -80 C mechanical freezer (Valeri CR et al, Transfusion 45:1890-1898, 2005; Valeri CR, Ragno G, Trans Aph Sci 34:271-287, 2006; Lelkens CCM et al, Trans Aph Sci 34:289-298, 2006; McCullough J et al, Transfusion 50:808-819, 2010).

Liquid nitrogen is not needed for freezing or storage of platelets and MNC. Platelets with 6% DMSO can be stored at -80 C for at least 2 years and MNC with 10% DMSO can be stored at -80 C for 1 ½ years. Platelets and MNC do not have to be washed prior to transfusion (Valeri CR et al: Frozen Platelets, In: Platelet Therapy: Current Status and Future Trends. Ed. J. Seghatchian, EL, Snyder, and P. Krailadairi, Elsevier Science, Clinical Medicine, Amsterdam, 2000; 105-130; Valeri CR et al, Transfusion 45:1890-1898, 2005; Valeri CR, Ragno G, Trans Aph Sci 34:271-287, 2006; Valeri CR, Ragno G, Trans Aph Sci 42:223-233, 2010; Noorman, F, Badloe J, Transfusion 52(Suppl): S53-030H, 33A, 2012; Noorman F et al, Transfusion 52(Suppl):SP23, 62A, 2012).

Clinical indications for frozen RBC include;

a) To provide preserved RBC for inventory control using O-positive and O- negative frozen RBC
b) To provide RBC that circulate and function immediately or shortly after infusion
c) To provide autologous RBC and rare RBC
d) To provide RBC to prevent transfusion related acute lung injury (TRALI) and systemic inflammatory response syndrome (SIRS)
e) There is no need to leukoreduce autologous red blood cells prior to freezing with 40% W/V glycerol and storage at -80 C which are thawed and washed to remove the glycerol and reduce the WBC count to 1 X 10^7 in the unit. Leukoreduction of autologous RBC prior to freezing removes 10 to 15% of the red blood cells and should not be done to freeze autologous RBC (Ashenden M and Markeberg J, Vox Sang 101:320-326, 2011).

The survival and function of human preserved RBC are maintained by:
a) Storage of liquid preserved RBC in CPDA-l, CPD/AS-1, CP2D/AS-3, and CPD/AS-5 at 4 C for only 2 weeks
b) Storage of RBC frozen with 40% W/V glycerol at -80 C with a range of -65 C to -90 C for 10 years, deglycerolized and stored at 4 C in AS-3 for 2 weeks

The need for cryopreserved blood products can be justified by the availability and quality of these frozen blood products, with regards the recovery of cells associated with the cryopreservation procedure and the viability and function of the red blood cells, platelets, plasma proteins and pluripotential adult stem cells.

In May 2001, the FDA approved the automated functionally closed Haemonetics ACP215 instrument to glycerolize and deglycerolize RBC frozen with 40% W/V glycerol at-80 C and stored as deglycerolized RBC at 4 C in the additive solution AS-3 (Nutricel) for 2 weeks (Valeri CR et al, Transfusion 41:928-932, 2001; Valeri CR et al, Transfusion 41:933-939, 2001).

To maintain a stockpile of universal donor frozen RBC, outdated group O Rh positive and group O Rh negative liquid preserved RBC can be salvaged by biochemical treatment with a solution containing pyruvate, inosine, phosphate and adenine (PIPA) prior to freezing with 40% W/V glycerol and storage at -80 C for at least 10 years (Valeri CR et al, Crit Care Med 7:439-447, 1979; Valeri CR, Crit Rev Clin Lab Sci 17(4):299-374, 1982; Valeri CR et al, Transfusion 40:1341-1345, 2000; Valeri CR,

Ragno G, Trans Aph Sci 34:271-287, 2006; Ragno G, Valeri CR, Trans Aph Sci 35:137-143, 2006).

Automation of RBC glycerolization using the high separation bowl attached to the Haemonetics ACP215 disposable glycerolizing set has simplified glycerolization of both nonrejuvenated and rejuvenated RBC (Valeri CR et al, Transfusion 45(10):1621-1627, 2005). The procedure to deglycerolize RBC frozen with 40% W/V glycerol using the ACP215 reduces the glycerol concentration to <1%, reduces the level of PIPA solution used to biochemically modify the RBC prior to freezing, reduces the WBC to a mean of 1×10^7 in the unit, and reduces the plasma and nonplasma levels to less than 5% (Valeri CR, Ragno G, Trans Aph Sci 34:271-287, 2006).

Leukoreduced and washed deglycerolized RBC stored at 4 C in AS-3 for 2 weeks is the ideal RBC product to prevent transfusion-related acute lung injury (TRALI) and systemic inflammatory response syndrome (SIRS) (Valeri CR Ragno G, Trans Aph Sci 34:271-287, 2006).

NBRL has simplified the RBC freezing by removing the supernatant glycerol prior to freezing. NBRL has simplified freezing of platelet and pluripotential mononuclear cells by removal of the supernatant DMSO prior to freezing. Removal of the supernatant DMSO removes about 95% of the DMSO and eliminates the need for post-thaw washing prior to transfusion (Valeri CR, Ragno G, Trans Aph Sci 34:271-287, 2006). Lelkens CCM and associates have reported the use by the Netherlands military of: Group O Rh positive and group O Rh negative RBC frozen with 40% W/V glycerol and stored at -80 C; group O single donor leukoreduced platelets frozen with 5% DMSO with removal of the supernatant DMSO solution before freezing and storage at -80 C and following thawing resuspended in AB plasma; and AB plasma stored at -30 C and then at -80 C to treat patients who required these blood products (Le1kens CCM et al, Trans Aph Sci 34:289-298, 2006; Noorman F, Badloe J, Transfusion 52(Suppl):S53-030H, 33A, 2012; Noorman F et al, Transfusion 52(Suppl), SP23, 62A, 2012).

Lelkens CCM and associates experience in the Netherlands military has demonstrated for the first time that frozen RBC, frozen platelets and fresh frozen plasma stored at -80 C can be used to treat patients without the need for a "walking blood bank" and fresh whole blood (Lelkens CCM et al, Trans Aph Sci 34:289-298, 2006; Noorman F at al, Transfusion 49(35):28A Suppl 2009; Badloe JF and Noorman F, Transfusion 51(35)

Suppl, 2011). Henkelman S and G Rakhorst in Transfusion (52:2272-2273, 2012) have reported the safety and therapeutic effectiveness of frozen RBC, frozen platelets, and frozen plasma stored in -80 C mechanical freezers to treat combat casualties without the need for fresh whole blood (Noorman F, Badloe JF, AABB Oct 2012, Transfusion Practice/Clinical Case Studies: Transfusion Practice in Emergent Settings SP383: Poster Presentation, Transfusion 52(Suppl) SP383, 198A, 2012).

Economic assessments are needed to compare the costs of:

a) Freezing RBC, platelets and plasma at -80 C as described by CCM Lelkens, F Noorman and JF Badloe et al to that of providing walking blood donors to provide fresh whole blood, liquid preserved red blood cells stored at 4 C in additive solutions for 42 days, apheresed platelets stored at room temperature for 24 hours to 5 days with agitation, and FFP and cryoprecipitate stored at 20 C;

b) Maintaining -20 C .freezers to that of--80 C freezers

c) Transportation of liquid preserved blood products and frozen blood products

Current challenges facing cryopreservation of hematopoietic and nonhematopoeitic adult stem cells are identical to those already investigated for cryopreservation of RBC, platelets and plasma.

Areas needed to be investigated for hematopoietic and non-hematopoietic adult stem cells include the following: source of these cells from bone marrow, peripheral blood, and umbilical vein blood; isolation of these cells free of RBC and granulocytes and devoid of immunocompetent lymphocytes; concentration and total number of cells; cryoprotectant and concentration — DMSO with and without hydroxyethylstarch (HES) and albumin; removal of supernatant DMSO solution prior to freezing; container and controlled and uncontrolled rate of freezing; temperature and length of storage at -80 C, -135 C, - 150 C, and -197 C, post-thaw processing and length of storage at room temperature and at 4 C; and studies to assess repopulation of bone marrow by hematopoietic stem cells and repopulation of heart and liver by non-hematopoietic stem cells in ABO, Rh and HLA-compatible recipients (Valeri CR, Ragno G, Trans Aph Sci 34:271-287 with an editorial on pages 267-268, 2006).

The quality of preserved RBC must ensure that RBC must circulate; RBC should have normal function to carry and release oxygen to tissue; RBC have a critical role in hemostasis to interact with platelets and plasma clotting proteins to reduce nonsurgical blood loss; RBC must maintain not only the RBC volume but also the plasma volume and blood volume; and RBC have an important role in maintaining blood pH. Transfusion trigger for RBC should be set to improve oxygen delivery to tissue; restore hemostasis; and restore the RBC volume, plasma volume, and the blood volume. Platelet transfusion trigger should be dependent upon the hematocrit, platelet count, platelet size, and platelet hemostatic function prior to prophylactic and therapeutic transfusions in anemic thrombocytopenic patients. In anemic and. thrombocytopenic patients, the RBC-platelet interaction in vivo is critical to prevent nonsurgical blood loss for both prophylactic and therapeutic platelet transfusions.

The optimum hematocrit is 35 V% to ensure the function of platelets in nonthrombocytopenic and thrombocytopenic recipients. Prophylactic and therapeutic platelet transfusions should ensure that the platelets circulate and function. The platelet transfusion trigger determined by the platelet counts ranging from 5,000 to 20,000 per ul should also consider the hematocrit and platelet function of the anemic and thrombocytopenic patient. The restoration of the hematocrit of the anemic thrombocytopenic patient to 35 V% by RBC transfusion to restore platelet function may prove to be safer and cheaper than prophylactic platelet transfusions to minimize the bleeding diathesis in these patients (Valeri CR, Ragno G, Trans Aph Sci 42:223-233, 2010).

Clinical indications for use of frozen red blood cells.

Universal donor group O Rh positive and group O Rh negative RBC frozen to supplement the liquid preserved RBC inventory. RBC with improved capacity to deliver oxygen by biochemical treatment prior to freezing for patients with fixed cerebral and coronary blood flow. Autologous RBC, rare RBC and selected RBC lacking antigens that commonly sensitize patients (Valeri CR et al, Crit Care Med 7:439-447, 1979).

RBC with reduced white blood cells, platelets, plasma proteins, microaggregates and free of vasoactive substances, prevent transfusion related acute lung injury (TRALI) and prevent systemic inflammatory response (SIRS) (Valeri CR, Ragno G, Trans Aph Sci 34:271-287, 2006).

Clinical indications for frozen platelets.

A supply of frozen platelets would be valuable for: autologous platelets for patients in remission and previously refractory to platelet transfusions; allogeneic HLA-compatible platelets for patients with refractory thrombocytopenia; single donor plateletpheresed allogeneic platelets for civilian and military contingency planning; cryopreserved PLA-1 negative platelets; and HLA typed platelets for use in compatibility testing (Valeri CR et al, Transfusion 43:1162-1167, 2003).

CHAPTER 9 REFERENCES

1. Ashenden M, Markeberg J. Net haemoglobin increase from reinfusion of refrigerated frozen red blood cells after autologous blood transfusions. Vox Sang 101:320-326, 2011.
2. Badloe J, Noorman F. The Netherlands experience with frozen -80 C red cells, plasma and platelets in combat casualty care S1-0301. Transfusion Suppl 51(35), 2011.
3. Barnard MR, MacGregor H, Ragno G, Pivacek LE, Khuri SF, Michelson AD, Valeri CR. Fresh, liquid preserved, and cryopreserved platelets: adhesive surface receptors and membrane procoagulant activity. Transfusion 39:880-888, 1999.
4. Bohonek M et al. Quality evaluation of frozen apheresis red blood cell storage with 21-day postthaw storage in additive solution 3 and saline-adenine-glucose-mannitol: biochemical and chromium-51 recovery measures. Transfusion 50:1007-1013, 2010.
5. Carciero R, Valeri CR. Isolation of mononuclear leukocytes in a plastic bag system using ficoll-hypaque. Vox Sang 49:373-380, 1985.
6. Dumont LJ, Cancelas JA, Dumont DE, Siegel AH, Szczepiorkowski ZM, Rugg N, Pratt PG, Worsham DN, Hartman EI, Dunn SK, O'Leary M, Ransom JH, Michael RA, Macdonald VW. A randomized controlled trial evaluating recovery and survival of 6% dimethyl sulfoxide-frozen autologous platelets in healthy volunteers. Transfusion 53:128-137, 2013
7. Henkelman S, Rakhorst G. Does modern combat still need fresh whole blood transfusions? Transfusion 52:2272-2273, 2012.
8. Khuri SF, Healey N, MacGregor H, Barnard MR, Szymanski IO, Birjiniuk V, Michelson AD, Gagnon DR. Valeri, CR. Comparison of the effects of transfusions of cryopreserved and liquid preserved platelets on hemostasis and blood loss after cardiopulmonary bypass. J Thorac Cardiovasc Surg. 117:172-184, 1999.
9. Lelkens CCM et al. Experience with frozen blood products in the Netherlands military. Trans Aph Sci 34:289-298, 2006.
10. Melaragno AJ, Carciero R, Feingold H, Talarico L, Weintraub L, Valeri CR. Cryopreservation of human platelets using 6% dimethylsulfoxide and storage at -8OC. Effects of 2 years of frozen

storage at -80C and transportation in dry ice. Vox Sang 49:245-258, 1985.

11. McCullough J, Haley R, Clay M, Hubel A, Lindgren B, Moroff G. Long-term storage of peripheral blood stern cells frozen and stored with a conventional liquid nitrogen technique compared with cells frozen and stored in a mechanical freezer. Transfusion 50:808-819, 2010.
12. Neuhaus S Jr, Wishaw J, Lelkens C. Australian experience with frozen blood products on military operations. MJA 192(4):203-206, 2010.
13. Noorman F, et al. Frozen -80 C red cells, plasma and platelets in combat casualty care. Transfusion 49(35) :28A (Suppl), 2009.
14. Noorman F, Badloe JF. -80 C frozen red blood cells, plasma and platelets: efficient logistics, available, compatible, safe and effective in the treatment of trauma patients with or without massive blood loss in military theatre. AABB Oct 2012, Transfusion practice/clinical case studies: transfusion practices in emergent settings; SP383; poster presentation. Transfusion 52(Suppl) SP383, 198A, 2012.
15. Noorman F, Badloe JF. -80 C frozen platelets, efficient logistics, available, compatible, safe and effective in the treatment of trauma patients with or without massive blood loss in military theatre. AABB Oct 2012, Transfusion practice/clinical case studies, S53-030H oral presentation; Transfusion 52(Suppl) S53-030H, 33A, 2012.
16. Noorman F, Strelitski R. Badloe JF. -80 C frozen platelets are activated compared to 24 hour liquid stored platelets and quality of frozen platelets is unaffected by a quick preparation method (15 min) which can be used to prepare platelets for the early treatment of trauma patients in military theatre. AABB Oct 2012, Components and component processing: platelets SP23. Poster presentation, Transfusion 52(Supp]) SP23, 62A, 2012.
17. Noorman F, Streletski R, Badloe JF. Lyophilized plasma, an alternative to 4 C stored thawed plasma for the early treatment of trauma patients with (massive) blood loss in military theatre. AABB Oct 2012, Components and component processing: plasma. Poster presentation, Transfusion 52(Suppl) SP7, 55A, 2012.
18. Ragno G, Valeri CR: Salvaging of 1iquid-preerved 0-positive and

0-negative red blood cells by rejuvenation said freezing. Trans Aph Sci 35:137-143, 2006.

19. Rosenblatt MS, Hirsch EF, Valeri CR. Frozen red blood cells in combat casualty care: Clinical and logistical considerations. Mil Med 159:392-397, 1994.
20. Valeri CR. Blood Banking and the Use of Frozen Blood Products, Chemica1 Rubber Company Press, Boca Raton, Florida, 1976.
21. Valeri CR. Hemostatic effectiveness of liquid-preserved and previously frozen human platelets. NE.JM 290:353-358 1974.
22. Valeri CR. Cryopreservation of human platelets and bone marrow and peripheral blood totipotential mononuclear stem cells. Ann NY Acad Sci 459:353-366, 1985.
23. Valeri CR. Use of rejuvenation solutions in blood preservation. Crit Rev Clin Lab Sci 17(4):299- 374, 1982.
24. Valeri CR, Ragno G. The effect of storage of fresh frozen plasma at -8OC for as long as 14 years on plasma clotting proteins. Transfusion 45(11): 1829-1830, 2005.
25. Valeri CR, Ragno G. Cryopreservation of human blood products. Trans Aph Sci 34:271-287, with an editorial on pages 267-269, 2006.
26. Valeri CR, Ragno G. An approach to prevent the severe adverse events associated with transfusion of FDA-approved blood products. Trans Aph Sci 42;223-233, 2010.
27. Valeri CR, Pivacek LE. Effects of the temperature, the duration of frozen storage, and the freezing container on in vitro measurements in human peripheral blood mononuclear cells. Transfusion 36:303-303, 1996.
28. Valeri CR, Feingold, H, Marchionni LD. A simple method for freezing human platelets using 6% dimethylsulfoxide and storage at -80 C. Blood 41:131-136. 1974.
29. Valeri CR, Valeri DA, Dennis RC, Vecchione JJ, Emerson CP. Biochemical modification and freeze-preservation of red blood cells: A new method. Crit Care Med 7:439-447, 1979.
30. Valeri CR., Valeri DA, Anastasi J, Vecchione JJ, Dennis RC, Emerson CP. Freezing in the primary polyvinylchloride plastic collection bag: A new system for preparing and freezing nonrejuvenated and rejuvenated red blood cells. Transfusion 21:138-149, 1981.

31. Valeri CR, Sims KL, Bates JF, Reichman D, Lindberg JR, Wilson AC. An integrated liquid frozen blood banking system. Vox Sang 45:25-39, 1983.
32. Valeri CR, Melaragno AJ, Dittmer J, Roy AJ, Vecchione JJ, Cassidy GP, Gray AD and Carciero RE. Bone marrow reconstitution of lethally irradiated beagles by treatment with autologous previously frozen bone marrow or peripheral blood mononuclear cells obtained as a byproduct of plateletpheresis. Technical Report 85-01.
33. Valeri CR, Feingold H, Melaragno AJ, Vecchione JJ. Cryopreservation of dog platelets with dimethylsulfoxide: Therapeutic effectiveness of cryopreserved platelets in the treatment of thrombocytopenic dogs, and the effect of platelet storage at -80 C. Cryobiology 23:387-394, 1986.
34. Valeri CR, Pivacek LE, Gray AD, Cassidy GP, Leavy ME, Dennis RC, Melaragno AJ, Niehoff J Yeston N, Emerson CP, Altschule MD. The safety and therapeutic effectiveness of human red cells stored at -8O C for as long as 21 years. Transfusion 29:429-437, 1989.
35. Valeri CR, Ragno G, Pivacek LE. Frozen Platelets. In: Platelet Therapy: Current Status and Future Trends, ed. J Seghatchian, EL Snyder, and P Krailadairi, Elsevier Science, Clinical Medicine, Amsterdam, 2000:105-130.
36. Valeri CR, Pivacek LE, Cassidy GP, Ragno G. The survival, function, and hemolysis of human RBCs stored at 4 C in additive solution (AS-1., AS-3, or AS-5) for 42 days and then biochemically modified, frozen, thawed, washed, and stored at 4 C in sodium chloride and glucose solution for 24 hours. Transfusion 40:1341-1345, 2000.
37. Valeri CR, Ragno G, Pivacek LE, Cassidy GP, Srey R, Hansson-Wicher M, Leavy ME. An experiment with glycerol-frozen red blood cells stored at -80 C for up to 37 years. Vox Sang 79:168-174, 2000.
38. Valeri CR, Pivacek LE, Cassidy GP, Ragno G. In vitro and in vivo measurements of human RBCs frozen with glycero1 and subjected to various storage temperatures before deglycerolization and storage at 4 C for 3 days. Transfusion 41:401-405, 2001
39. Valeri CR, Pivacek LE, Cassidy GP. Ragno G. In vitro and in vivo measurements of' gamma-radiated, frozen, glycerolized RBCs. Transfusion 41:545-549, 2001.

40. Valeri CR, Ragno G, Pivacek LE, O'Neill EM.. In vivo survival of apheresis RBCs, frozen with 40-percent (wt/vol) glycerol, deglycerolized in the ACP 215 and stored at 4 C in AS-3 for up to 21 days. Transfusion 41:928-932, 2001.
41. Valeri CR. Ragno G, Pivacek LE, Srey R, Hess JR. Lippert TE, Mettille F, Fahie R, O'Neill EM, Szymanski IO. A multicenter study of in vitro and in vivo values in human RBCs frozen with 40-percent (wt/vol) glycerol and stored after deglycerolization for 15 days at 4 C in AS-3: assessment of RBC processing in the ACP 215. Transfusion 41:933-939, 2001.
42. Valeri CR, Giorgio A, MacGregor H, Ragno G. Circulation and distribution of autotransfused fresh, liquid-preserved and cryopreserved baboon platelets. Vox Sang 83:347-351, 2002.
43. Valeri CR, MacGregor H, Giorgio A, Ragno G: Circulation and hemostatic function of autologous fresh, liquid preserved, and cryopreserved baboon platelets transfused to correct an aspirin-induced thrombocytopathy. Transfusion 42:1206-1216, 2002.
44. Valeri CR, Srey R, Lane JP, Ragno G. Effect of WBC reduction and storage temperature on PLTs frozen with 6 percent DMSO for as long as 3 years. Transfusion 43(8):l162-1167, 2003.
45. Valeri CR, Lane J, Srey R, Ragno G. Incidence of breakage of human RBCs frozen with 40 percent wt/vol glycerol using two different methods for storage at -80 C. Transfusion 43(3):411-414, 2003.
46. Valeri CR, MacGregor H, Ragno G. Correlation between in vitro aggregation and thromboxane. A2 production in fresh, liquid-preserved, and. cryopreserved human platelets: Effects of agonists, pH, and plasma and saline resuspension. Transfusion 45:596-603, 2005.
47. Valeri CR, Ragno G, et al. Automation of the glycerolization of the RBC using the high separation bowl in the Haemonetics ACP215 instrument. Transfusion. 45(10):1621-1627, 2005.
48. Valeri CR, Ragno G, Khuri S. Freezing human platelets using 6% DMSO with removal of the supernatant solution prior to freezing and storage at -8O C without post-thaw processing. Transfusion 45(12):1890-1898, 2005.
49. Valeri CR., Khuri S, Ragno G. Non surgica1 bleeding diathesis in anemic thrombocytopenic patients: Role of temperature, RBC, platelets, and p1asma clotting proteins. Transfusion 47:206S-248S, 2007.

CHAPTER 10

AN APPROACH TO PREVENT THE SEVERE ADVERSE EVENTS ASSOCIATED WITH TRANSFUSION OF FDA APPROVED BLOOD PRODUCTS

There have been several retrospective studies reporting severe adverse events of mortality and morbidity associated with blood transfusions. Mortality and morbidity associated with posttransfusion infection, transfusion related acute lung injury (TRALI) and systemic inflammatory response syndrome (SIRS) have been reported in patients undergoing cardiac surgery, after massive transfusions for severe traumatic injuries, and after transfusions for elective and emergency indications. After 35 days of storage at 4 C in additive solutions, RBC have 24-hour posttransfusion survival values of 75% but do not function satisfactorily. For RBC to function satisfactorily shortly after transfusion, they should be stored at 4 C for no more than 2 weeks. Yet while the FDA requires a 24-hour posttransfusion survival value of 75%, there is no requirement for the function of the transfused RBC. It has been shown that red blood cells that circulate and function immediately or shortly after transfusion exert a very important hemostatic effect to reduce the bleeding time and nonsurgical blood loss in anemic thrombocytopenic patients. Greater restoration of hemostasis is seen with viable and functional RBC transfusions than with platelets or plasma even though the platelets and plasma may have satisfactory viability and function (Valeri CR et al, Transfusion 47:206S-248S, 2007).

The length of storage of the blood products affects their survival and function and the transfusion of nonviable compatible RBC, antibodies to granulocytes and WBC HLA antigens and biologically active substances affect the patient's clinical outcome. One of the easiest ways to prevent the severe adverse events that have been observed is to ensure that the transfused blood products survive and function at an optimum level and that the levels of antibodies to granulocytes and WBC HLA antigens and biologically active substances are eliminated or reduced. The best way to ensure this is to store liquid preserved leukoreduced human red blood cells at 4 C in additive solutions for no more than 2 weeks and leukoreduced platelets at room temperature for no more than 2 days. These liquid preserved blood products can be used in conjunction with frozen RBC, frozen platelets, and frozen plasma stored in -80 C mechanical freezers and will avoid the need for fresh whole blood and single donor leukoreduced apheresed platelets stored at room temperature with agitation for 5 days and prevent the severe adverse events associated with the transfusion of blood products (Valeri CR, Ragno G, Trans Aph Sci 42:223-233, 2010; Perkins JG et al, Transfusion 51:242-252, 2011).

Recent publications have reported severe adverse events of mortality and morbidity associated with the transfusion of FDA-approved blood products. Transfusions of RBC, platelets and plasma have been associated with mortality and posttransfusion infections, transfusion related acute lung injury (TRALI) and systemic inflammatory response syndrome (SIRS), leading to increased hospitalization of these patients and increased healthcare costs. Even while reports continue to surface of severe adverse events suffered by recipients of these blood products and even though studies have shown that no such adverse effects have been seen with transfusions of previously frozen RBC and frozen platelets, the reluctance to use freeze preservation procedures continue. Combining freeze-preservation and liquid preservation procedures would prevent the severe adverse events now observed with blood products (Valeri CR, Ragno G, Trans Aph Sci 42:223-233, 2010).

Current methods by which leukoreduced RBC are preserved in the liquid state at 4 C maintain the viability and function of the RBC at 4 C for only two (2) weeks. If the leukoreduced liquid preserved RBC are washed after 4 C storage for 2 weeks it would reduce the antibodies to granulocytes and WBC HLA antigens which activate granulocytes in the recipient's lung and cause transfusion related acute lung injury (TRALI). Washing would also reduce cytokines and other biologically active substances in blood products that activate the recipient's granulocytes to produce oxygen free radicals and the systemic inflammatory response syndrome (SIRS) (Valeri CR, Ragno G, Trans Aph Sci 42:223-233, 2010).

After room temperature storage at 22 $\pm$ 2 C for only 2 days, leukoreduced platelets have acceptable in vivo recovery, lifespan and hemostatic function immediately or shortly following infusion and can reduce the nonsurgical bleeding diathesis in anemic and thrombocytopenic patients (Valeri CR, Ragno G, Trans Aph Sci, 42:223-233, 2010).

Freeze Preservation of RBC, Platelets and Plasma

Frozen blood products have been shown to be safe and therapeutically effective as a supplement to liquid preserved products. Group O Rh positive and group O Rh negative RBC, single donor group O leukoreduced platelets collected by plateletpheresis and AB plasma can all be frozen and stored in a -80 C mechanical freezer. The Naval Blood Research Laboratory for 45 years conducted studies to cryopreserve blood products for military and civilian use. These studies were supported by Naval Medical Research and Development Command (NMRDC) of the U.S. Navy's Bureau of Medicine and Surgery; the Office of Naval Research (ONR); and the Congress of the United States. Frozen blood products stored in -80 C mechanical freezers were deployed to supplement the supply of liquid preserved blood products to treat wounded casualties. During the Vietnam War, the Huggins cytoglomerator was used to deglycerolize RBC frozen with 40% W/V glycerol and stored at -80 C. The Haemonetics Blood Processor 115 was used in the Persian Gulf War and the Haemonetics ACP215 is being used in Iraq, Afghanistan and Bosnia (Valeri CR, Ragno G, Trans Aph Sci 34:271-287, 2006; Lelkens CCM et al Trans Aph Sci 34:289-298, 2006). With the functionally closed Haemonetics ACP215 instrument, group O Rh positive and group O Rh negative RBC are treated with 40% W/V glycerol, the supernatant glycerol is removed, and the RBC containing 40% W/V glycerol are frozen in a -80 C mechanical freezer. These RBC have been stored at -80 C for at least 10 years, and following deglycerolization, leukoreduction and washing, the RBC have been stored at 4 C in the additive solution AS-3 (Nutricel) for 2 weeks (Valeri CR, Ragno G, Trans Aph Sci 34:271-287, 2006.

Single donor leukoreduced group O platelets obtained by plateletpheresis have been treated with 6% DMSO, and after the supernatant DMSO is remove, the platelets are frozen in a -80 C freezer and stored at -80 C for at least 2 years. After dilution of the thawed platelets with 10 to 20 ml of 0.9% NaCl, the platelets can be stored at room temperature without agitation for six hours. Group AB plasma has been frozen in a -80 C mechanical freezer and stored at -80 C for at least 14 years, and after thawing can be stored at room temperature for 8 hours or at 4 C for 24 hours. All these blood products have been shown to be safe and therapeutically effective (Valeri CR, Ragno G, Trans Aph Sci 34:271-287, 2006; Lelkens CCM et al, Trans Aph Sci 34:289-298, 2006; Valeri CR, Ragno G, Trans Aph Sci 42:223-233, 2010).

A -80 C dual cascade air-cooled mechanical freezer maintained at a mean temperature of -80 C with a range from -65 C to -90 C is used for frozen storage of these blood products. There is no need to control the rate of freezing using a special programmed rate freezer and liquid nitrogen. The blood product is frozen in a polyvinylchloride plastic bag inside a polyester plastic bag placed in a rigid cardboard box in a -80 C mechanical freezer. The frozen blood products can be transported on dry ice in polystyrene foam shipping containers which maintain the temperature at -80 C. In our experience, frozen RBC in PVC plastic bags showed an incidence of breakage of about 6% following transportation compared to an incidence of about 3% in PVC plastic bags that were not subjected to transportation (Valeri CR, Ragno G, Trans Aph Sci 34:271-287, 2006).

Severe Adverse Events Related to Liquid Preserved Blood Products

Since 1985 the effects of the length of storage of liquid preserved RBC in the additive solutions CPD AS-1 (ADSOL), CP2D AS-3 (Nutricel) and CPD AS-5 (Optisol) on mortality and morbidity in patients have been debated. Numerous publications have discussed whether or not the length of storage of liquid preserved RBC at 4 C in the current FDA approved additive solutions are responsible for the severe adverse events that have been reported in patients in the articles by Zimrin AB and Hess JR (Vox Sang 96:93-103, 2009), Lelubre C and associates (Transfusion 49:1384-1394, 2009), Bennett-Guerrero E and associates (Transfusion 49:1375-1383, 2009) and the editorial by Steiner ME and Stowell C (Transfusion 49:1286-1290, 2009) report on this controversy. Bennett-Guerrero E and associates report a prospective, double-blind, randomized clinical feasibility trial of controlling the storage age of red blood cells for transfusion in cardiac surgical patients. The report by Bennett-Guerrero E and associates is a limited prospective randomized study in 23 patients who received liquid preserved red blood cells, 12 patients received RBC stored at 4 C for 7 $\pm$ 4 days and 11 patients received RBC stored at 4 C for 21 $\pm$ 4 days. The authors report most (94.4%) of the RBC units were AS-1 or AS-3 units which have a shelf life of 42 days; whereas 12 of the units (5.6%) were CPDA-1 units which have a shelf life of 35 days. The authors do not report the clinical outcome in these 23 patients. This limited prospective randomized study failed to assess liquid preserved RBC stored in additive solutions (AS-1 or AS-3) for 35 to 42 days.

Zimrin AB and Hess JR discussed the quality of liquid preserved RBC stored in three additive solutions (CPD AS-1, CP2D AS-3 and CPD AS-5) approved by FDA for 42 days. They tabulated the adverse events reported in 19 retrospective studies and they concluded there is no need for an immediate change in the current transfusion practice.

There have been retrospective reports showing that the mortality and morbidity of recipients are related to the storage of the RBC in FDA-approved additive solutions for 42 days. Koch CG and associates have reported that red blood cells stored at 4 C for longer than 2 weeks were associated with increased mortality and morbidity in patients subjected to cardiac surgical procedures (N Engl J Med 358:1229-1239, 2008).

Edgren G and associates in the paper "Duration of red blood cell storage and survival of transfused patients" in Transfusion 50:1185-1195, 2010 reported the mortality of less than 5% in patients who received liquid preserved RBC stored in saline, adenine, glucose, mannitol (SAG-M) additive solution at 4 C for longer than 30 days. The mortality reported in this study pooled retrospective data from Sweden which stored RBC in SAGM at 4 C for 42 days and data from Denmark which stored RBC in SAGM at 4 C for 35 days. The authors suggest that the mortality rate of "less than 5% was acceptable" for patients who received long-term stored red blood cells compared to patients who received short-term red blood cells. Unlike the paper of Koch CG and associates in the N Engl J Med (358:1229-1239, 2008) this paper did not report the morbidity in the patients who received short-term and long-term SAGM stored RBC. The authors did report the median length of hospital stay during the first 30 days after transfusion. Of interest, the authors reported that leukoreduction of the preserved RBC did not reduce the mortality rate in the patients, a finding similar to that reported by Koch CG and associates. The length of storage of SAGM RBC at 4 C should be 30 days and not 35 days in Denmark and not 42 days in Sweden to reduce the mortality of less than 5% for RBC stored for longer than 30 days. The retrospective data reported by Edgren G and associates suggest that storage of SAGM RBC for greater than 2 weeks was associated with an increased morbidity. In the 387,130 patients transfused less than 5% of the patients died who received red blood cells preserved in SAGM for longer than 30 days. The reported hazard ratio of greater than 1.0 suggests that severe adverse events were increased when the patients received red blood cells stored at 4 C for longer than 2 weeks. The SAEs in the patients that received red blood cells stored at 4 C for longer than 2 weeks may be similar to that reported by Koch CG and associates. Wang D and associates in Transfusion (52:1184-1195, 2012) have reported a meta-analysis on the transfusion of older stored blood and the risk of death. These authors concluded the use of older stored blood was associated with a significantly increased risk of death.

Articles have reported on the safety and therapeutic effectiveness of liquid preserved red blood cells in the treatment of pediatric and adult patients (Gauvin F et al, Transfusion 50:1902-1913, 2010; Koch CG et al, Crit Care Med 34:1608-1616, 2006; Hajjar LA et al, JAMA 304:1559-1567, 2010; Bennett-Guerrero E et al, JAMA 304:1568-1570, 2010). These articles focus on the red blood cell transfusion trigger assessed by the hematocrit level, the number of units of red blood cells transfused, and the length of storage of the liquid preserved red blood cells at 4 C on morbidity and mortality. Predominantly retrospective studies were reported together with prospective randomized studies related to liberal and restrictive use of preserved red blood cells. These studies report data which focus on the hematocrit value; the number of units transfused; and the liberal and restrictive use of blood products. These studies did not report the limitation of the hematocrit to assess whether the patients were normovolemic or hypovolemic, the falsely elevated hematocrit values in hypovolemic patients, and the quality of the red blood cells with regards their viability and function. In the liberal and restrictive use of red blood cells the quantity of the red blood cells administered was associated with increased severe adverse events (SAEs) but the quality of the preserved red blood cells was not reported in the retrospective and prospective randomized studies. The findings that restrictive use of blood products produced better outcome may reflect the poor quality of the preserved red blood cells transfused.

The current clinical data on the safety and therapeutic effectiveness of red blood cells preserved in the liquid state at 4 C for 42 days; platelets preserved at room temperature for 5 days with agitation; and fresh frozen plasma and cryoprecipitate stored at -20 C should now be compared to the safety and therapeutic effectiveness of frozen blood products; i.e. frozen red blood cells, frozen platelets and frozen plasma all stored in mechanical freezers maintained at -80 C with a range from -65 C to -90 C. The experience of the Netherlands military using only frozen blood products produced excellent clinical outcomes without any adverse events in the treatment of military and civilian casualties (Lelkens CCM et al, Trans Aph Sci 34:289-296, 2006; Neuhaus S et al, MJA 192:203-205, 2010; Noorman F et al, Transfusion 49(35):28A (Suppl), 2009; Noorman F, Badloe J, Transfusion 52(Suppl) SP383, 198A, 2012; Noorman F, Badloe J, Transfusion 52(Suppl) S53-030H, 33A, 2012; Noorman F et al, Transfusion 52(Suppl0 SP23, 62A, 2012; Noorman F et al, Transfusion 52(Suppl) SP7, 55A, 2012).

At the American Society of Hematology meeting in Orlando, Florida held from December 4 to 7, 2010, three (3) papers on Transfusion Medicine: Transfusion Support in Trauma – Military and Civilian Approaches were presented.

1. Trauma Blood Management: Avoiding the Collateral Damage of Trauma Resuscitation Protocols by Timothy Hannon (Hematology 2010, pgs 463-464).
2. Optimal Use of Blood Products in Severely Injured Trauma Patients by John B. Holcomb (Hematology 2010, pgs 465-469).
3. Logistics of Massive Transfusion by Thomas G. DeLoughery (Hematology 2010, pgs 470-473).

These articles suggest that in the treatment of trauma patients blood products at a ratio of 1:1:1 should be utilized. The safety and therapeutic effectiveness of the red blood cells, plasma and platelets at a ratio of 1:1:1 did not report the quality and quantity of the red blood cells, plasma, and platelets that should be transfused (Johansson PI, Stensballe J, Transfusion 50:701-710, 2010). The three articles failed to report and discuss the potential toxicity of the resuscitation solutions like the crystalloid solution Ringer's lactate and the colloid solution albumin (Valeri CR, Veech RL,

Trans Aph Sci 46:121-123, 2012) utilized in the treatment of the patient together with the blood products.

The paper by Thomas DeLoughery suggests that the thromboelastogram testing provides information regarding the blood products that should be transfused to restore hemostasis in the patients. Blood products to treat the parameters recorded in the thromboelastogram tracing of blood obtained from the patient were reported in the DeLoughery paper in Hematology 2010, pgs 470-473. Specific blood products and antifibrinolytic agents were reported to treat the R time, K time, alpha angle, maximum amplitude (MA) and lysis index recorded in the thromboelastogram tracing. Fresh frozen plasma, cryoprecipitate, platelets and antifibrinolytic agents were recommended for specific parameters recorded in the thromboelastogram tracing. However, the quality and quantity of the blood products and the quantity of the antifibrinolytic agents recommended are not reported. The thromboelastogram testing does not recommend the need for red blood cells. The treatment of the tracing produced by the thromboelastogram with specific blood products and antifibrinolytic agents needs to be supported by data to document that the in vitro testing provided by the thromboelastogram correlates to the restoration of hemostasis in patients. The experience at the NBRL has shown that the R time of the thromboelastogram suggests a platelet dysfunction (Valeri CR et al, J Trauma 57:S22-S25, 2004). In our experience the increase in the maximum amplitude (MA) correlated to an increase in the bleeding time and a decrease in hematocrit and hemoglobin concentration (Valeri CR, Ragno G, Trans Aph Sci 35:33-41, 2006). In our studies, the thromboelastogram testing did not correlate to the bleeding time and nonsurgical blood loss. The recommendation that the parameters recorded in the thromboelastogram tracing should be treated by specific blood products is not supported by our studies. Viable and functional red blood cells reduce the bleeding time and reduce nonsurgical blood loss in patients and reduce the maximum amplitude (MA) (Crowley JP et al, Am J Clin Pathol 108:579-584, 1997; Valeri CR et al, Transfusion 41:977-983, 2001; Valeri CR, Ragno G, Trans Aph Sci, 35:33-41, 2006). Viable and functional RBC are more effective than viable and functional platelets and plasma clotting proteins to reduce the bleeding time and reduce nonsurgical blood loss. The suggestion that the thromboelastogram should be used to indicate what blood products should be infused into the patient cannot be supported by our data (Khuri SF et al, J Thorac Cardiovasc Surg 104:94-107, 1992; Valeri CR et al, J Trauma 57:S22-S25,

2004; Valeri CR and Ragno G, Trans Aph Sci 35:33-41, 2006; Fischer TH et al, J Trauma 71:S176-S182, 2011) and data reported by Roeloffzen WWH and associates (Transfusion 50:1536-1544, 2010).

The thromboelastogram cannot detect the effect of viable and functional RBC which reduce the bleeding time and reduce nonsurgical blood loss. The maximum amplitude of the thromboelastogram is increased in blood with reduced hematocrit and hemoglobin concentration values. Bolliger D and associates (Trans Med Rev 26:1-13, 2012) reported that low hematocrit values increase the bleeding time in vivo but improves the thromboelastogram maximum amplitude (MA) in vitro which support our data that the thromboelastogram does not detect RBC to restore hemostasis in patients.

The multicenter study by Carson JL and associates in N Engl J Med (365:2453-2462, 2011) performed at 47 clinical sites in the USA and Canada from July 14, 2004 to Feb. 28, 2009 reported on 1007 orthopedic patients who received liberal RBC transfusions and 1009 orthopedic patients who received restrictive RBC transfusions. The mortality and physical performance of these orthopedic patients were evaluated using the hemoglobin concentration to assess the need for red blood cell transfusion and the therapeutic response to the RBC transfusions in the liberal and restrictive groups of elderly patients. The length of storage of the RBC at 4 C transfused after randomization was 22.0 $\pm$ 9.5 days for the liberal group and 22.1 $\pm$ 9.9 days for the restrictive group. The patients subjected to the orthopedic surgical procedures were 81.8 $\pm$ 8.8 years old for the liberal group and 81.5 $\pm$ 9.0 years old for the restrictive group. The authors did not report that the hemoglobin concentrations in the hypovolemic anemic patients are falsely increased. The hemoglobin concentration does not correlate to the red blood cell volume deficit and the need for RBC transfusions (Biron PE, J Bone Joint Surg 54-A:1001-1014, 1972; Valeri CR and Altschule MD, Chemical Rubber Company Press, Boca Raton, FL, 1981; Cordts PR, Surg Gynec Obstet 175:243-248, 1992; Valeri CR, et al, Transfusion 46:365-371, 2006; Valeri CR, et al, Transfusion 47:206S-248S, 2007). No significant difference in mortality was observed between the liberal and restrictive groups. The physical performance was assessed by the ability of the patients to walk independently on day 60 following the orthopedic surgery. The measurement of the temperature of the big toes of the extremities would have provided clinical information on the therapeutic benefit of the transfused red blood cells to increase perfusion to the extremities in the patients in the restrictive and liberal groups.

The Carson JL and associates report on liberal and restrictive transfusion in patients after orthopedic surgery did not emphasize to the readership of the N Engl J Med 365:2453-2462, 2011 the limitations of the hemoglobin concentration to assess the need for red blood cell transfusion in patients subjected to orthopedic surgical procedures. The signs and symptoms of the patients and not the hemoglobin concentrations should determine the need and the quality and quantity of the red blood cells that should be provided. The hemoglobin concentration does not assess whether the patient is normovolemic and anemic or hypovolemic and anemic. The hemoglobin concentration does not assess the red blood cell volume, the red blood cell deficit, and the therapeutic effect of the transfused preserved red blood cells (Biron PE, J Bone Joint Surg 54-A:1001-1014, 1972; Valeri CR and Altschule MD, Chemical Rubber Company Press, Boca Raton, FL, 1981; Cordts PR, Surg Gynec Obstet 175:243-248, 1992; Valeri CR, et al, Transfusion 46:365-371, 2006; Valeri CR, et al, Transfusion 47:206S-248S, 2007).

The orthopedic patients transfused in the Carson JL and associates paper were hypovolemic and anemic and the hemoglobin concentrations were falsely elevated. The signs and symptoms reported in the retrospective study show that the patients in the restrictive group had increased symptoms related to rapid bleeding, chest pain, congestive heart failure, tachycardia, and hypotension compared to the patients in the liberal group. The authors report the length of storage of the transfused red blood cells but the anticoagulant-preservative solution utilized and the hematocrit of the red blood cell concentrates that were transfused were not reported. The quality of the preserved red blood cells should report the survival values and the red blood cell function to provide high pO2 levels to the brain, heart, and kidneys and to prevent nonsurgical bleeding. This retrospective study does not provide information to determine the need, the quality, and the quantity of preserved red blood cells that should be administered to patients. The retrospective data reported indicate that the signs and symptoms of the patient and not the hemoglobin concentration should be treated by the physician. The ambulation of the patients 60 days after the surgical procedures to assess the therapeutic effectiveness of the red blood cell transfusion strategy was reported. The measurement of the temperature of the big toes in these patients may have provided clinical data to document the improvement in the blood flow to the extremities in the patients in the liberal and restrictive groups.

In a review article on red cell storage and prognosis L van de Watering in Vox Sang (100:36-45, 2011) concluded that there was no consensus on the possible associations between the length of storage of RBC at 4 C for 35 days in saline, adenine, glucose and mannitol (SAGM) and morbidity and mortality. L van de Watering and associates have reported in a retrospective study in Transfusion (46:1712-1718, 2006) on the mortality and the length of stay in the intensive care unit (ICU) for patients subjected to coronary artery bypass surgery between 1993 to 1999 who received buffy coat depleted SAGM preserved RBC. The retrospective study assessed short term SAGM RBC stored at 4 C for less than 18 days with a mean of 13 days and long term SAGM RBC stored at 4 C for greater than 18 days with a mean of 24 days showed no difference in mortality and length of stay in the ICU. The length of stay in the ICU was the only indicator in the retrospective review of the records to assess the occurrence of adverse events in the patients. The authors report the storage time in days and the number of red blood cell units transfused. The authors provide no information on the number of patients who received greater than 10 units of RBC during the perioperative period including three (3) days before until one (1) day after surgery regarding mortality, length of ICU stay, and severe adverse events. The outcome in the patients who received greater than 10 units of RBC stored at 4 C in SAGM for 28 to 35 days were not reported to support the conclusion of the authors that "In their analyses, pertaining to 2732 CABG patients, no justification could be found for use of a particular maximum storage time for RBC transfusion."

The paper by Perkins, JG and associates (Transfusion 51:242-252, 2011) reported a retrospective review of the medical records of casualties at the military hospital in Baghdad, Iraq between January 2004 and December 2006. Patients requiring massive transfusion of greater than 10 units of red blood cells in a 24-hour period for resuscitation were divided into two groups: those receiving fresh whole blood and additional blood products (n=85) or apheresed platelets and additional blood products (n=284) during their resuscitation. This report selected 369 charts for review from 2024 patients who received blood transfusions. In the 369 charts, 284 patients received 2 units (with a range of 1 to 9) of apheresis platelets and 19 units (with a range of 10 to 58) of liquid preserved red blood cells stored at 4 C for 33 days; 12 units (with a range of 2 to 42) of fresh frozen plasma; 10 units (with a range of 0 to 52) of cryoprecipitate and 7.2 mg of factor VIIa (with a range of 2.4 to 36). Eighty-five (85) patients received 5 units of fresh whole blood (with a range of 1 to 21) and 15 units (with a range of 4 to 42) of liquid preserved red blood cells stored at 4 C for 33 days, 8 units (with a range of 0 to 28) fresh frozen plasma, and 9.0 mg (with a range of 2.4 to 21.6) of factor VIIa. The authors concluded that the 24 hour and 30 days survival rate of massively transfused trauma patients receiving fresh whole blood appeared to be similar to patients resuscitated with apheresed platelets. This paper raises numerous questions regarding the methods used to test the fresh whole blood and the apheresed platelets prior to transfusion and screening of the donors of the fresh whole blood and the apheresed platelets for infectious disease like Leishmaniasis, malaria, Chagas diseases, schistosomiasis, brucellosis, and tuberculosis. The logistics of collection and storage of 3.0 to 3.5 X 10^{11} platelets stored at 20-24 C for 24 hours to 5 days with agitation and the outdating of the apheresed platelets and the liquid preserved RBC air transported to the Middle East from the USA should be reported together with the methods used to perform the compatibility testing of the RBC and platelets that were transfused. The major concern with the use of fresh whole blood and apheresed platelets is the transmission of infectious disease associated with bacterial contamination of these blood products at the time of collection in addition to the transmission of hepatitis, HIV, malaria, Chagas disease and other infectious disease agents in these blood products.

The logistics to collect and test fresh whole blood and the apheresed platelets in combat zones for infectious diseases were not discussed in this retrospective review of the records. Both the donors and the recipients of the fresh whole blood and the apheresed platelets should now be tested to determine the possible transmission of infectious disease by these blood products that were administered in this study.

The Netherlands military observed the transmission of hepatitis following the use of non-tested fresh whole blood at the time of transfusion and now deploy -80 C mechanical freezers to freeze universal donor group O Rh positive and group O Rh negative red blood cells, freeze Group O single donor leukoreduced platelets, and freeze AB plasma to treat wounded casualties with excellent clinical outcome in patients who received greater than 10 units of RBC in a 24-hour period with an increase in survival rate from 44% to 84%. The patients received 4 units of deglycerolized group O Rh positive or Group O Rh negative RBC, 3 units of AB plasma; one unit of single donor leukoreduced group O frozen 2.5 - 3.0 $X10^{11}$platelets thawed and resuspended in a unit of AB plasma with no adverse events associated with this transfusion schedule. The frozen blood products are tested prior to freezing and the blood donor screened to ensure the safety of the frozen blood products.

A study is needed to compare the use of fresh whole blood and apheresed platelets to frozen RBC, frozen plasma, and frozen platelets to treat wounded servicemen and servicewomen who required more than 10 units of blood in a 24 hour period for resuscitation.

This study needs to assess the safety and therapeutic effectiveness to treat wounded patients with fresh whole blood and apheresed platelet stored at 22 $\pm$ 2 C with agitation for 5 days, liquid preserved red blood cells stored at 4 C for 42 days in the additive solution AS-5 (Optisol), cryoprecipitate and fresh frozen plasma stored at -20 C for one year compared to universal donor group O Rh positive and group O Rh negative RBC frozen at -80 C for at least 10 years, group O frozen single donor leukoreduced platelets frozen at -80 C for at least 2 years; and frozen AB plasma stored at -80 C for at least 10 years in mechanical freezers. The safety and therapeutic effectiveness of the blood products, the logistic support to provide the blood products and the cost to provide the blood products need to be reported. The safety and therapeutic effectiveness together with the logistic requirement and cost can now be compared to assess the deployment of a -20 C mechanical freezer to store FFP and cryoprecipitate to a -80 C mechanical freezer to freeze RBC, freeze platelets and freeze plasma. The frozen blood products stored at -80 C can be quarantined for 6 months and the donors retested for the mandated infectious disease markers. Transfusion medicine has been preoccupied since 1981 with the testing of the blood and the donor screened for potential diseases that may be transmitted by blood products. The use of fresh whole blood and apheresed platelets to treat wounded patients need to test the blood products and the screened donors to ensure that the blood products do not transmit diseases as mandated by the FDA. The study reported by Perkins and associates did not provide information on the testing of the fresh whole blood and apheresed platelets prior to transfusion and the screening of the blood donors who provided these blood products.

The 369 records that were reviewed by Perkins and associates should report on the current health of the 284 patients who received apheresed platelets and 85 patients who received fresh whole blood. The data reported on the results 24 hours and 30 days after transfusion indicated that massively transfused patient survived but the morbidity associated with the blood products transfused was not reported. The difference between the two groups of patients may be due to the crystalloid and colloid solutions administered following injury that were not reported.

The data reported in the Perkins paper question the need for the testing of the blood prior to transfusion and the selection criteria for blood donors now mandated by FDA. The limited data recovered from the records of 8618 trauma patients who received 2024 blood transfusions should now examine the 284 patients who received apheresed platelets and the 85 patients who received fresh whole blood between January 2004 to December 2006 to determine whether these patients now have evidence that the blood transfusion transmitted diseases in these recipients. The Perkins et al report provide data that raise several questions concerning the quality and quantity of blood products that were administered and whether the blood products produced adverse events beyond the 30 day posttransfusion period. The toxicity of crystalloid solution Ringer's D,L lactate and the poor therapeutic effect of albumin as the colloid to increase the blood volume may have affected the outcomes. The failure to report the crystalloid solutions and colloid solutions used is a major limitation to interpret the data regarding the safety and therapeutic effectiveness of the blood products transfused to the patients (Valeri CR, Veech RL, Trans Aph Sci 46:121-123, 2012). The patients transfused must now be evaluated to assess whether or not the selection of the blood donors and failure to test the blood products prior to transfusion as mandated by the FDA may have produced long-term adverse events, related to the transmissible diseases they received in the blood products that were utilized during the period January 2004 to December 2006 at the military hospital in Baghdad, Iraq.

The Perkins JG and associates study in Transfusion 51:242-252, 2011 underscores the limitation of the retrospective review of records to obtain information to resolve important issues related to the need to select blood donors for collection of blood products, the need for testing of the blood products prior to transfusion, the quality and quantity of the blood products with regards safety and therapeutic effectiveness to treat patients. The retrospective review of the records failed to provide the information required to assess safety and therapeutic effectiveness of fresh whole blood and apheresed platelets in the treatment of patients requiring large volumes of blood products for resuscitation in a combat zone.

Hakre S and associates reported in Transfusion (51:473-485, 2011) the testing of sera obtained from wounded military personnel in war zones in Iraq and Afghanistan from March 1, 2002 to September 30, 2007 prior to and following transfusion of fresh whole blood and apheresed platelets collected from military pre-screened personnel as the "walking blood bank". The sera were frozen and transported from the Middle East to Washington, D.C. and stored at the suboptimal temperature of -30 C, prior to testing for hepatitis C virus (HCV), human immunodeficiency virus (HIV), and hepatitis B virus (HBV). Of the 761 military personnel who were transfused only 475 were tested for HCV, 472 for HIV and 469 for HBV.

Spinella PC and associates reported in Transfusion (52:1146-1153, 2012) that the U.S. Army transfused 47,825 units of blood products to U.S. casualties from October 2001 to Sept 2007. Of these, 4,856 (10%) were freshly collected blood products including 3,384 units of fresh whole blood and 1,472 units of apheresed platelets that were not FDA compliant with regards testing for transfusion transmissible infections in accordance with FDA regulations. Apheresed platelets stored at room temperature with agitation for as long as 5 days were available at a combat support hospital in Iraq in November 2004. Spinella PC and associates have reported in Transfusion (52:1146-1153, 2012) the total number of patients, the total number of units transfused and the mean number of units of blood products, i.e. fresh whole blood, liquid preserved RBCs, apheresed platelets, fresh frozen plasma and cryoprecipitate transfused per patient from November 2001 to December 2010: 1,364 patients received 8,259 units of fresh whole blood with a mean of 6.1 units per patient and 3,311 patients received 7,772 units of apheresed platelets with a mean of 2.3 units per patient. A total of 4,675 patients received 16,041 units of non-FDA compliant fresh whole blood and apheresed platelets stored at room temperature for 24 hours to 5 days with agitation not tested for FDA mandated transfusion transmissible infections. Fresh whole blood, apheresed platelets, liquid preserved red blood cells stored at 4 C in the additive solution (AS-5) for 42 days, and fresh frozen plasma and cryoprecipitate stored at -20 C for one year were provided to treat the wounded casualties. The authors reported on the ratios of red blood cells, plasma and platelets administered to patients who required at least 10 units of red blood cells in a 24-hour period for resuscitation. However, the crystalloid and colloid resuscitation fluids administered to the patients were not reported. The reported ratios of 1:1:1 – RBC, plasma and platelets administered to the patients did not define the quality and the quantity of the red blood cells, plasma and platelets that were transfused. Hakre S and associates in Transfusion (51:473-485, 2011) did not report the mortality and morbidity associated with the transfusion transmissible viral agents that were detected in the sera stored at the suboptimal temperature of -30 C. In a similar manner, Spinella PC and associates paper in Transfusion (52:1146-1153, 2012) did not report the mortality and morbidity associated with the blood products administered to the patients transfused using specific ratios of red blood cells, plasma and platelets as reported by Johansson PI and Stensballe J (Transfusion 50:701-710, 2010).

The Survival and Function of Preserved Red Cells and Platelets

One should not assume that because the survival of preserved RBC and preserved platelets is satisfactory, their ability to function following transfusion will also be satisfactory. Yet, while the FDA requires that preserved RBC have 24-hour posttransfusion survival of 75%, they make no regulations requiring that the RBC function satisfactorily immediately or shortly after transfusion. The FDA also assumes that when a sick patient received 25% of nonviable compatible RBC with a transfusion, the recipient will suffer no adverse effects. Although there are no data to show how many nonviable compatible RBC can be safely administered to sick patients, it is know that if a patient receives 10 units of RBC with a 24-hour posttransfusion survival value of 75%, then 2.5 units of the RBC will be nonviable. The removal of these compatible nonviable RBC by the reticuloendothelial (RE) system may interfere with the removal of infectious disease agents, tumor cells, particulate matter present in the stored RBC, and affect cellular and humoral immunity.

The FDA considers preserved platelets to be acceptable if the in vivo recovery is 66% that of fresh platelets and the lifespan is 50% that of fresh platelets but again offers no guidelines with regard to hemostatic function. Although in vitro levels of RBC ATP, DPG and p50 levels give an indication of the in vivo function of the preserved RBC, there are no in vitro tests that can predict the in vivo function of fresh and preserved platelets. Studies at the NBRL have shown that when preserved platelets function satisfactorily they will correct an aspirin induced prolonged bleeding time in normal volunteers and in healthy baboons (Valeri CR et al, Transfusion 47:206S-248S, 2007).

The survival and function of preserved red blood cells are critical especially for patients with cerebral, myocardial or renal insufficiency and for patients with a nonsurgical bleeding diathesis. Studies at the NBRL have shown that preserved red blood cells stored in CPDA-1 or in any one of the additive solutions CPD/AS-1, CP2D/AS-3 or CPD/AS-5 for only 2 weeks have acceptable 24-hour posttransfusion survival values of greater than 75% and only moderately impaired oxygen transport function at the time of infusion (Valeri CR, Ragno G, Trans Aph Sci 42:223-233, 2010).

Liquid preserved autologous platelets that were stored at room temperature (22$\pm$2C) with agitation for 48 hours were found to have acceptable in vivo recovery and survival and were able to function to reduce an aspirin-induced prolonged bleeding time in normal volunteers and healthy baboons. On the other hand, when autologous baboon platelets were stored at room temperature for 3 or 5 days, they did circulate but did not reduce the prolonged bleeding time in aspirin treated healthy baboons (Handin RI, Valeri CR, NEJM 285:538-543, 1971; Khuri S et al, J Thorac Cardiovasc Surg 117:172-184, 1999; Valeri CR et al, Transfusion 42:1206-1216, 2002; Valeri CR et al, Transfusion 47:206S-248S, 2007).

Between 1988 and 1992, the NBRL collaborated with Shukri Khuri, Chief of Surgery and Chief of Cardiothoracic Surgery at the West Roxbury Veteran's Administration Hospital, to evaluate liquid preserved and washed previously frozen platelets transfused to cardiopulmonary bypass (CPB) patients. The previously frozen platelets had been frozen with 6% DMSO at 2-3 C per minute and stored at -80 C for a mean of 289 $\pm$ 193 days (SD) and for as long as 2 years and were washed prior to transfusion. A prospective randomized study was conducted in 73 patients undergoing CPB surgery. The study was designed to measure nonsurgical blood loss, i.e. blood loss not related to the surgical procedure and not controlled by surgical intervention, after neutralization of the heparin with protamine sulfate. Nonsurgical blood loss was collected intraoperatively and during the 24-hour postoperative period.

The allogeneic single donor washed previously frozen platelets transfused to these patients had been processed in the following manner. The platelets were frozen at the University of Massachusetts in Worcester, Massachusetts, and transported in the frozen state with dry ice to the NBRL where they were stored in -80 C mechanical freezers for at least 3 months. They were then transported in the frozen state in insulated containers with dry ice to West Roxbury Veteran's Administration Hospital blood bank where they were stored at -80 C in mechanical freezers. After thawing, the platelets were washed and stored in ACD plasma at room temperature without agitation for as long as 5 hours. The platelets were transfused to the patients after the CPB surgery.

The prospective randomized study compared the need for allogeneic RBCs and FFP to treat the nonsurgical blood loss in two groups of patients. One group received previously frozen washed platelets and the other group received liquid preserved platelets. The patients who received previously frozen washed platelets showed a reduction in the nonsurgical blood loss and required fewer units of allogeneic RBCs and FFP than the patients who received the liquid preserved platelets stored with agitation at 22 C for a mean of 3.4 days. The platelet survival 2 hours after transfusion was 37% for patients who received the liquid preserved platelets compared to 24% for the patients who received the previously frozen washed platelets (Khuri SF et al, J Thorac Cardiovasc Surg 117:172-184, 1999).

The total number of liquid preserved platelets infused was $6.9 \times 10^{11} \pm 3.9 \times 10^{11}$ per patient which was significantly greater than the $4.5 \times 10^{11} \pm 2.1 \times 10^{11}$ per patient for the previously frozen washed platelets. The difference was the result of the freeze-thaw-wash (FTW) recovery value of 70% for the previously frozen washed platelets. The in vivo recovery and function of the liquid preserved and cryopreserved platelets in these patients were similar to values seen in studies in which liquid preserved and cryopreserved platelets were transfused to aspirin-treated human volunteers and baboons. Although in the patients in this study, the in vivo recovery values were higher for the liquid preserved allogeneic platelets than for the washed previously frozen platelets, nonsurgical blood loss was lower in the patients who received the washed previously frozen platelets and they required fewer units of allogeneic RBCs and FFP (Khuri SF et al, J Thorac Cardiovasc Surg 117:172-184, 1999).

In the paper in Transfusion (51:2603-2610, 2011) Bilgin YM and associates reported the postoperative complications associated with transfusion of platelets and plasma in cardiac surgery. The authors reported that to improve hemostasis, platelets and fresh frozen plasma (FFP) are often transfused in the perioperative and postoperative period, however, neither the efficacy nor the safety of platelets and plasma transfusions have been demonstrated. The authors failed to report that a paper written by Khuri SF and associates was published in J Thorac Cardiovasc Surg 117:172-184, 1999 compared the effects of cryopreserved and liquid preserved platelets on hemostasis and blood loss after cardiopulmonary bypass. The cryopreserved platelets produced a significantly greater reduction in nonsurgical blood loss and the reduced need for liquid preserved red blood cells and fresh frozen plasma. Khuri SF and associates studied the in vivo recovery and function of cryopreserved platelets compared to liquid preserved platelets in a prospective randomized study in 73 patients undergoing cardiopulmonary bypass surgery. The liquid preserved platelets were stored with agitation at 22 C for as long as 5 days and single donor platelets were frozen with 6% DMSO at 2 to 3 C per minute by storage in a -80 C mechanical freezer for as long as 2 years, thawed, and washed prior to transfusion. The study was designed to measure nonsurgical blood loss in the patients after neutralization of the heparin with protamine sulphate. Nonsurgical blood loss was collected intraoperatively and during the 24-hour postoperative period. The prospective randomized study compared the need for liquid preserved red blood cells stored at 4 C and fresh frozen plasma stored at -20 C to treat the nonsurgical blood loss in the two groups of patients. One group received previously frozen single donor washed platelets and the other group receive liquid preserved single donor or pooled platelets stored at 22 C for a mean of 3.4 days with agitation. Although the in vivo recovery values of the liquid preserved platelets were higher than those of the previously frozen washed platelets, the patients who received the washed previously frozen platelets had significantly reduced nonsurgical blood loss and required fewer units of liquid preserved red blood cells and fresh frozen plasma than the patients who received liquid preserved platelets stored at room temperature for a mean of 3.4 days with agitation. No adverse effects were observed in either group (Bilgin YM et al, Transfusion 51:2603-2610, 2011; Khuri SF et al, J Thorac Cardiovasc Surg 117:172-184, 1999).

In 2000 the NBRL modified the method of freezing platelets with 6% DMSO at 2-3 C per minute by storage in a -80 C mechanical freezer. With the modified method the supernatant DMSO was removed before the platelets were frozen and the previously frozen platelets were not washed prior to transfusion (Valeri CR et al, Transfusion 45:1890-1898, 2005). When platelets are washed the in vitro recovery is reduced by 25%. With the modified method, the thawed platelets are not washed but are diluted with 10 ml to 20 ml of 0.9% NaCl and stored without agitation at room temperature for 4 hours. The freeze-thaw (FT) recovery is approximately 90%, with 8% platelet microparticles, in vivo recovery is 25-30%, and the linear lifespan is 7 days. The diluted platelets have a bimodal population: one population is GPIb-normal and annexin V-reduced and the other is GPIb-reduced with increased annexin V binding.

Mechanical freezers maintained at -80 C are needed for the storage of these safe and therapeutically effective frozen blood products for use by both the military and civilian communities to supplement the liquid preservation of RBC stored at 4 C, platelets stored at 22 C, and fresh frozen plasma and cryoprecipitate stored at -20 C.

Studies at the NBRL have shown that previously frozen washed RBC have not been associated with the severe adverse events observed with the current FDA-approved blood products. Frozen RBC stored at -80 C have been successfully deployed during the Vietnam War and the Persian Gulf War by the U.S. Navy and the Department of Defense (Valeri CR, Ragno G, Trans Aph Sci 34:271-287, 2006).

The past 45 years have seen many significant improvements in cryopreservation procedures. Removing supernatant glycerol from RBC and DMSO from platelets prior to freezing have simplified the postthaw processing. Platelets diluted with 10 ml to 20 ml of 0.9% NaCl after thawing can be stored at room temperature without agitation for 6 hours prior to use. With the functionally closed Haemonetics ACP215 instrument, deglycerolized RBC can be stored at 4 C in AS-3 (Nutricel) for 2 weeks. With this instrument, the volume of solution needed to deglycerolize RBC is now only 1.6 liters compared to 3.2 liters with the Haemonetics Blood Processor 115 and 6.8 liters with the Huggins cytoglomerator (Valeri CR, Ragno G, Trans Aph Sci 34:271-287, 2006).

The increased mortality and morbidity, and the occurrence of posttransfusion infectious diseases, TRALI, and SIRS observed after transfusion of red blood cells stored in the liquid state at 4 C for 42 days or platelets stored at room temperature for 5 days can be prevented by using previously frozen blood products stored in a -80 C mechanical freezer.

Some have suggested that if RBC, platelets, and plasma were disinfected, this would eliminate the risk of transmission of infectious disease agents that are not detected with the current FDA mandated testing. Unfortunately, the disinfection of red blood cells, platelets, and plasma may alter their antigenicity so that repeated infusion of allogeneic disinfected blood products could reduce their safety and therapeutic effectiveness. Studies over the past 45 years using glycerol to freeze RBC and DMSO to freeze platelets have produced no alteration of the antigenicity of the RBC and platelets. Likewise, attempts at enzymatic conversion of group A and group B RBC to universal donor O red blood cells have not yet been successful (Valeri CR, Ragno G, Trans Aph Sci 42:223-233, 2010).

Extensive funding has been provided to companies trying to produce a hemoglobin based oxygen carrier as a replacement for RBC. Natanson C and associates in JAMA 299:2304-2312, 2008 have reported the mortality, myocardial infarction, and morbidity associated with the clinical use of HBOCs. Moreover, to compare the safety and therapeutic effectiveness of HBOCs to red blood cells is disingenuous since HBOCs circulate for only 24 hours and RBC have a 24-hour posttransfusion survival value of at least 75% and normal lifespan of 3 months. Even if the HBOCs were proven to be safe and therapeutically effective, they would provide oxygen to the brain, heart and kidney and exert a hemostatic effect for only 24 hours and would not reduce the need for safe and therapeutically effective allogeneic RBC. The severe adverse events that have been reported with HBOCs have been related to their vasoconstriction to impair blood flow to the brain, heart and kidneys, to activate the hemostatic mechanism and to generate oxygen free radicals (Valeri CR, Ragno G, Trans Aph Sci 42:223-233, 2010).

Prevention of TRALI and SIRS

Severe adverse events associated with transfusions of RBC, platelets and plasma include transfusion related acute lung injury (TRALI) and systemic inflammatory response syndrome (SIRS). TRALI is caused by the presence in RBC, platelets and plasma of antibodies to granulocytes and to WBC HLA antigens which occur at a higher incidence in female blood donors. The cytokines and the biologically active substances in blood products following infusion activate the recipient's granulocytes that produce oxygen free radicals, and the severe adverse events related to the systemic inflammatory response syndrome (SIRS).

When leukoreduced liquid preserved RBC are washed, the level of antibodies to granulocytes and WBC HLA antigens which activate granulocytes in the recipient's is reduced. Washing the RBC also reduces the level of cytokines and biologically active substances involved in the activation of the recipient's granulocytes that generate oxygen free radicals and produce the systemic inflammatory response syndrome (SIRS). The incidence of both TRALI and SIRS could be reduced by washing the leukoreduced RBC stored at 4 C.

The procedure to deglycerolize the RBC produces leukoreduced and washed RBC which is the ideal RBC product to reduce or eliminate TRALI and SIRS. The enzymatic conversion of group A and group B red blood cells to group O RBC could be supplemented simply by freezing group O Rh-positive and group O Rh-negative RBC. Nonrejuvenated group O Rh positive and group O Rh negative RBC can be frozen without prior biochemical modification. Indated and outdated group O Rh positive and group O Rh negative RBC can be biochemically modified to increase the RBC 2,3 DPG, ATP, and p50 levels before freezing. Biochemically modified universal donor RBC have acceptable 24-hour posttransfusion survival and normal or improved oxygen transport function. RBC that are frozen do not have to be leukoreduced because the addition and removal of 40% W/V glycerol from the RBC produces a leukoreduced RBC which after washing have a total number of 1 X 10^7 WBC per unit (Valeri CR, Ragno G, Trans Aph Sci 42:223-233, 2010).

At the international symposium on the diagnosis and prevention of TRALI, attendees from Poland suggested that TRALI is associated with storage of liquid preserved RBC at 4 C for longer than 2 weeks and the storage of platelets at room temperature (22 ± 2 C) for longer than 2 days. Participants at the International Forum on TRALI recommended that females should be excluded as plasma donors because they have a higher incidence of antibodies against granulocytes and WBC HLA antigens. They now recommend that group AB plasma from male donors should be frozen and stored at -20 C. With all due respect to the investigators making these suggestions, what is needed are studies comparing the incidence of TRALI and SIRS associated with transfusions of liquid preserved leukoreduced RBCs, previously frozen washed leukoreduced RBCs and washed leukoreduced liquid preserved RBCs. The NBRL, over a 45-year period, transfused multiple units of washed previously frozen leukoreduced RBC from predominantly male donors and never observed any incidence of either TRALI or SIRS.

The Netherlands military has been actively freezing universal donor group O Rh positive and group O Rh negative RBC, single donor leukoreduced frozen group O platelets and frozen AB plasma in -80 C mechanical freezers. These frozen blood products have been collected from donors meeting FDA regulations that were safely transported and stored in -80 C mechanical freezers without breakage. These blood products have been obtained from screened blood donors and tested for the mandated infectious markers and have eliminated the need for fresh whole blood or apheresed leukoreduced platelets stored at room temperature for 5 days with agitation to treat patients suffering traumatic injuries with therapeutically effective outcomes and without adverse events.

The utilization of the -80 C mechanical freezers to freeze RBC, platelets, and plasma by the Netherlands military has demonstrated the safety and therapeutic effectiveness of these frozen blood products without the need for fresh whole blood or apheresed platelets to treat military and civilian casualties requiring more than 10 units of red blood cells within a 24-hour period in combat zones in Afghanistan and Iraq. It is of interest that the Netherlands investigators pioneered the use of liquid nitrogen to freeze blood products but now utilize -80 C mechanical freezers to freeze red blood cells, platelets and plasma to treat combat casualties. The use of the -80 C mechanical freezer which is the temperature of dry ice and alcohol was used with high concentration of glycerol (40% W/V) to salvage fresh RBC that were discarded by Dr. Edwin J. Cohn who isolated plasma from fresh whole blood to prepare albumin using the Cohn Blood Fractionator. The fresh RBC that were routinely discarded were frozen with high concentration of glycerol (40% W/V) and frozen in dry ice and alcohol at -79 C using the protocol reported by Audrey Smith, a veterinarian who initially froze bull spermatozoa and then human red blood cells. The Cohn Blood Fractionator procedure was funded by the Office of Naval Research (ONR) to produce human albumin concentrates as the first blood substitute which has been shown to be a poor plasma volume expander and should not be used to resuscitate patients in hemorrhagic shock. The salvaging of routinely discarded fresh red blood cells in the Cohn Blood Fractionator with 40% W/V glycerol were stored successfully at -80 C for at least 21 years. The deployment of -80 C mechanical freezers to freeze RBC for at least 10 years, platelets for at least 2 years, and plasma for at least 10 years to treat wounded patients in a combat zone by the Netherlands military demonstrated that chaotic observations, serendipity and patience have provided blood products for clinical use that are safe and therapeutically effective to treat wounded military and civilian patients in a combat zone without the need for fresh whole blood and apheresed platelets.

Dr. John Badloe at the ATACCC meeting on August 16, 2010 at St. Pete, Florida reported that in Afghanistan from 2000 to 2010, 859 patients received 6,335 blood products which include 1918 units of frozen red blood cells, 841 units of liquid preserved red blood cells, 2560 units of frozen plasma and 1,016 units of frozen platelets with no transfusion reactions. Fresh whole blood was not used by the Netherlands military because the fresh whole blood could not be tested prior to transfusion for the mandated infectious disease markers whereas all the frozen blood products were tested for the mandated infectious disease markers prior to freezing from the donors who were screened prior to donation.

Dr. John Badloe reported the Netherlands military experience in the Middle East war zones using frozen blood products, i.e. frozen group O Rh positive and group O Rh negative RBC, frozen AB plasma, and frozen group O single donor leukoreduced 2.5 - 3.0 X 10^{11} platelets with removal of supernatant DMSO, all frozen and stored at -80 C in mechanical freezers at ratio of 1:1:1 increased survival of patients from 44% to 84%. No adverse events were reported and only frozen blood products without the need for fresh whole blood were safe, available, effective and efficient in the treatment of patients requiring at least 10 units of red blood cells in a 24-hour period for resuscitation.

The abstract which was reported by Dr. John Badloe and Dr. Femke Noorman from the Ministry of Defense, Military Blood Bank, Leiden, Netherland at the Annual meeting of the AABB, San Diego, CA October 22-25, 2011 confirm the procedures provided by the NBRL, Boston, MA to freeze human RBC, plasma, and platelets in -80 C mechanical freezers. This report by the Netherlands military demonstrates that fresh whole blood advocated by Colonel William Crosby and the U.S. Army is no longer used by the Netherlands military and the fresh whole blood can be replaced by frozen RBC, frozen plasma, and frozen platelets significantly improved survival of the massively transfused patients (Badloe J, Noorman F, Transfusion 51(35)(Suppl), 2011).

Fully tested, frozen blood products, readily available after thawing proved to be safe, available, effective and efficient blood support for combat casualty care and together with the use of a 1:1:1 ratio increased survival significantly in massively transfused patients.

The recent paper by Henkelman S and Rakhorst G in Transfusion 52:2272-2273, 2012 "Does modern combat still need fresh whole blood transfusions?" reported that the Dutch military blood bank eliminated the use of fresh whole blood (FWB) on site and implemented the routine use of frozen group O Rh positive and group O Rh negative deglycerolized RBC, frozen single donor leukoreduced group O platelets, and fresh frozen AB plasma to treat wounded casualties in war zones. The RBC, platelets and plasma are frozen and stored at -80 C in mechanical freezers. Following thawing, the deglycerolized RBC are stored at 4 C in the additive solution AS-3 for 14 days, the thawed AB plasma stored at 4 C for 7 days, and the thawed single donor leukoreduced platelets frozen with 5% DMSO after the removal of the supernatant DMSO at -80 C, thawed, and resuspended containing 2.5 to 3.0 X 10^{11} platelets in AB plasma stored at 4 C for as long as 7 days and then stored at room temperature without agitation for 6 hours prior to transfusion.

The authors reported that fresh whole blood (FWB) needs to be tested for infectious transmissible disease and bacterial contamination. In addition, the transfusion of FWB has been associated with a high incidence of febrile non-hemolytic transfusion reactions, human leukocyte antigen alloimmunization and transfusion associated graft versus host disease (TA-GVHD). They reported that Gilstad C and associates in Transfusion (52:930-935, 2012) described clinical symptoms of TA-GVHD in a trauma patient who was resuscitated with FWB. The Dutch army has demonstrated that frozen RBC, frozen platelets, and frozen plasma are safe, effective, and provide an adequate blood supply. The authors reported the infusion of FWB in military settings is not recommended because of the possible risk of disease transmission, burden to the military personnel, and is not justified when frozen blood products are utilized. The authors were concerned with the potential that FWB can produce graft versus host disease in recipients. The frozen RBC containing 40% W/V glycerol and stored at -80 C can be treated with 2500 cGy of gamma radiation to inactive the immunocompetent lymphocytes in the deglycerolized RBC as reported by Valeri CR and associates (Transfusion 41:545-549, 2001). The data show similar acceptable results for RBC frozen with 40% W/V glycerol at -80 C and treated in the frozen state with 2500 cGy gamma radiation and for RBC that were not irradiated, all of which were washed and then stored in a sodium chloride glucose solution for 3 days before autologous transfusion.

Universal donor group O Rh positive and group O Rh negative RBC, group O single donor leukoreduced platelets resuspended in AB plasma, and AB plasma eliminate the need to provide fresh whole blood and apheresed platelets to treat patients subjected to traumatic injuries in war zones. The current method to freeze single donor leukoreduced platelets containing 2.5 – 3.0 X 10^{11} platelets equivalent to the number of platelets isolated from 4 to 6 units of fresh whole blood treated with 5% DMSO in 0.9% NaCl, the supernatant DMSO removed prior to freezing and storage in a -80 C mechanical freezer, and following thawing resuspended in a unit of AB plasma stored at 4 C for as long as 7 days eliminate the need to provide the platelets in fresh whole blood and single donor apheresed platelets stored at room temperature for 5 days with agitation.

Spinella PC and associates in Transfusion (52:1146-1153, 2012) reported that the Netherlands military under the direction of Lelken CCM and associates in Transfusion and Apheresis Science (34:289-296, 2006) used frozen red blood cells, frozen platelets and frozen plasma all stored at -80 C in mechanical freezers to treat wounded casualties in combat zones. The frozen blood products were collected from screened blood donors and tested for FDA mandated transfusion transmissible infections prior to freezing at -80 C. The logistic requirements and technical procedures utilized by the U.S. Army to collect fresh whole blood, single donor apheresed platelets stored at room temperature with agitation for 24 hours to 5 days, red blood cells stored in the additive solutions (AS-5) at 4 C for 42 days collected in the U.S. and air transported to the Middle East, and the storage of fresh frozen plasma and cryoprecipitate at -20 C for one year which were collected in the U.S. and air transported to the Middle East should be compared to the logistic requirements and technical procedures utilized by Lelkens CCM, Noorman F and Badloe J and associates to provide the Netherlands military frozen blood products (Noorman F, Strelitski R, Badloe JF, Transfusion 52(Suppl) SP23, 62A, 2012; Noorman F and Badloe JF, Transfusion 52(Suppl) S-53-030H; 33A, 2012; Noorman F and Badloe J, Transfusion 52(Suppl) SP383, 198A, 2012; Noorman F, Strelitski R, Badloe JF, Transfusion 52(Suppl) SP7, 55A, 2012).

Universal donor group O Rh positive and group O Rh negative RBC frozen with 40% W/V glycerol, the supernatant glycerol removed prior to freezing, and stored at -80 C for at least 10 years, thawed, deglycerolized in the Haemonetics Blood Processor ACP215 and stored in the additive solution AS-3 for 2 weeks; group O leukoreduced single donor platelets with the removal of the supernatant DMSO and the 2.5 to 3.0 X 1011 platelets frozen with 5% DMSO and stored at -80 C for at least 2 years, thawed, diluted with a unit of AB plasma stored at 4 C for as long as 7 days and then stored at room temperature without agitation for 6 hours; and AB plasma frozen at -30 C and then at -80 C, thawed and stored at 4 C for as long as 7 days are now utilized by the Netherlands military. All the frozen blood products were collected from pre-screened donors and tested for FDA mandated infectious disease markers prior to freezing. All frozen blood products were stored in insulated containers with dry ice to maintain a temperature of -65 C to -80 C during transportation from the Netherlands to Iraq and Afghanistan.

A comparison between the U.S. Army procedures and the Netherlands procedures to provide blood products to treat combat casualties needs to be performed regarding the logistic requirements, outdating of blood products, compatibility, cost of transportation, availability, safety and therapeutic effectiveness. The quality and quantity of the blood products to resuscitate casualties requiring at least 10 units of red blood cells in a 24-hour period and the mortality and morbidity utilizing the U.S. Army procedures to the Netherlands procedures in combat zones need to be compared. The Netherlands military has documented that -80 C frozen red blood cells, plasma and platelets were deployable, available, compatible, safe and effective in the treatment of trauma patients with or without massive blood loss in military theatre.

Dr. Femke Noorman and Dr. John Badloe and associates presented one oral presentation and three (3) poster presentations at the AABB annual meeting from October 6-9, 2012 in Boston, MA. Studies can now be done to compare the safety and therapeutic effectiveness of universal donor RBC frozen for at least 10 years, group O platelets frozen for at least 2 years, and AB plasma frozen for at least 10 years (all blood products frozen and stored at -80 C with a range of -65 C to -90 C) without the need for fresh whole blood to the current use of fresh whole blood, liquid preserved RBC stored at 4 C in additive solutions for as long as 42 days, single donor leukoreduced apheresed platelets stored with agitation at room temperature for 5 days, and fresh frozen plasma and cryoprecipitate stored at -20 C for one year on mortality and morbidity in the recipients.

In the study to compare the procedures utilized by the Netherlands military to the current FDA approved procedures to provide blood products; the volume and composition of the resuscitation fluids used with the blood products need to be reported. The morbidity and mortality associated with the current FDA approved blood products can be compared to the use of frozen RBC, frozen platelets, and frozen plasma without the need for fresh whole blood and apheresed platelets which Netherlands military has successfully utilized to treat combat casualties in war zones from August 2006 to February 2012; 2,175 units of frozen RBC, 1,070 units of frozen platelets, 3,001 units of frozen plasma and 879 units of liquid preserved RBC stored at 4 C were transfused to 1,011 casualties without any transfusion reactions observed. The blood products were obtained from pre-screened donors and the blood products were tested for infectious diseases prior to freezing. The frozen red blood cells, plasma and platelets were available, compatible, safe and effective in the treatment of trauma patients with or without massive blood loss in military theatre (Noorman F and Badloe J, Transfusion 52(Suppl) SP383, 198A, 2012).

In the response to the paper "Transfusion of older stored blood and risk of death" published by Wang D and associates in Transfusion (52:1184-1195, 2012) an editorial was written by Warkentin TE and Eikelboom JW "Old blood bad? Either the biggest issue in transfusion medicine or a nonevent" in Transfusion (52:1165-1167, 2012). These authors stated "the demonstration of an association (if indeed there is one) between the age of transfused blood and outcome would be useless if one could not do anything about it". This editorial failed to acknowledge that freeze preservation of red blood cells, platelets and plasma has been reported to provide safe and therapeutically effective blood products from 2006 to 2012 to successfully treat 1,011 civilian and military casualties in war zones in Iraq and Afghanistan by the Netherlands military using -80 C mechanical freezers to freeze and store universal donor group O Rh positive and group O Rh negative red blood cells, single donor leukoreduced group O platelets, and group AB plasma without the need for fresh whole blood and apheresed platelets.

It is imperative that civilian and military blood banking communities change their methods of collection and preservation of blood products if they are really interested in providing patients with the safest and most therapeutically effective blood products and in avoiding risks associated with transfusion.

The primary concern of the blood banking community appears to be the extension of the length of storage of the blood products. This practice may make inventory management easier, but it certainly will not improve patient outcomes or provide patients with the safest and most therapeutically effective blood products. Of course, cost must be a consideration but, in the long run, the recipient's health should be paramount. With regard to the mortality and morbidity associated with the severe adverse reactions to transfused blood products, it is important to note that Medicare has informed hospitals that they will no longer pay for long stays in hospitals when proper care could have prevented the increased hospitalization. A concerted effort must be made to determine why the transfused blood products are causing mortality and morbidity associated with posttransfusion infection, TRALI and SIRS, and what can be done to eliminate the causes. Our studies have shown that the severe adverse events observed after the transfusion of blood products stored under current FDA guidelines can be prevented by using washed liquid preserved RBC that have been stored at 4 C for no more than 2 weeks in combination with washed previously frozen red blood cells that have satisfactory survival and function. Washing liquid preserved red blood cells with 0.9% NaCl and 0.2 gm% glucose does not adversely affect their posttransfusion survival and function. The health of the patient must be of greater importance than making blood collection and inventory management easier.

There have been several retrospective studies reporting severe adverse events of mortality and morbidity associated with posttransfusion infection, TRALI, and SIRS have been reported in patients undergoing cardiac surgery, after massive transfusions for severe traumatic injuries, and after transfusions for elective and emergency indications. After 35 days of storage at 4 C in additive solutions, RBC have 24-hour posttransfusion survival values of 75% but do not function satisfactorily. For RBC to function satisfactorily shortly after transfusion, they should be stored at 4 C for no more than 2 weeks. Yet while the FDA requires a 24-hour posttransfusion survival value of 75%, they make no requirement for the function of the transfused RBC. It has been shown that red blood cells that circulate and function immediately or shortly after transfusion exert a very important hemostatic effect to reduce the bleeding time and nonsurgical blood loss in anemic and thrombocytopenic patients. Greater restoration of hemostasis is seen with viable and functional RBC transfusions than with platelets or plasma even through the platelets and plasma may have satisfactory viability and function.

The Food and Drug Administration (FDA), Health and Human Services (HHS), the American Red Cross (ARC) and the blood banking community have focused on further testing of blood products to reduce the rate of disease transmission. Another consideration is the disinfection of red blood cells, platelets, and plasma. What is more important, they should be looking at the quality of the blood products being transfused. We know that the length of storage of blood products affects their survival and function and that transfusion of nonviable compatible RBC, antibodies to granulocytes and WBC HLA antigens, and biologically active substances affect the patient's clinical outcome. One of the easiest ways to prevent the severe adverse effects that have been observed is to ensure that the transfused blood products survive and function at an optimum level and that the levels of antibodies to granulocytes and WBC HLA antigens and biologically active substances are reduced. The best way to ensure this is to store liquid preserved human red blood cells at 4 C in additive solutions for no more than 2 weeks and leukoreduced platelets at room temperature for no more than 2 days. These liquid preserved blood products can be used in conjunction with frozen RBC, frozen platelets and frozen plasma all stored in -80 C mechanical freezer.

The reluctance of the blood banking community to consider any changes that could improve the safety and therapeutic effectiveness of blood products brings to mind two very relevant questions: one by Maurice Maeterlinck is that "At every crossway on the road that leads to the future, every progressive spirit is opposed by a thousand men appointed to guard the past". The other by Winston Churchill is that "Occasionally man will stumble on the truth but will manage to pick himself up and continue on as if nothing had happened." It is time now to investigate the current blood banking procedures and to seek ways to improve them.

CHAPTER 10 REFERENCES

1. Badloe J, Noorman F. The Netherlands experience with frozen -80 C red cells, plasma and platelets in combat casualty care S1-0301. Transfusion 51(35) (Suppl). 2011.
2. Bennett-Guerrero E, Stafford-Smith M, Wawera PM, Bredehoeft S, Campbell ML, Haley NR et al. A prospective, double-blind randomized clinical feasibility trial of controlling the storage age of red blood cells for transfusion in cardiac surgical patients. Transfusion 49:1375-1383, 2009.
3. Bennett-Guerrero E, Zhao Y, O'Brien M, et al. Variation in use of blood transfusion in coronary artery bypass graft surgery. JAMA 304:1568-1575, 2010.
4. Bilgin YM, van de Watering LMG, Versteegh MIM, van Oers MHJ, Vamvakas EC, Brand A. Postoperative complications associated with transfusion of platelets and plasma in cardiac surgery. Transfusion 51:2603-2610, 2011.
5. Biron PE, Howard J, Altschule MD, Valeri CR. Chronic deficits in red-cell mass in patients with orthopaedic injuries (stress anemia). J Bone Joint Surg 54-A:1001-1014, 1972.
6. Bolliger D, Seebergen MD, Tanake KA. Principles and practice of thromboelastography in clinical coagulation management and transfusion practice. Trans Med Rev 26:1-13, 2012.
7. Carson JL, et al. Liberal or restrictive transfusion in high risk patients after hip surgery. N Engl J Med 365:2453-2462, 2011.
8. Cordts PR, LaMorte WW, Fisher JB, DelGuercio C, Niehoff J, Pivacek LE, Dennis RC, Siebens H, Giorgio A, Valeri CR, Menzoian JO. Poor predictive value of hematocrit and hemodynamic parameters for erythrocyte deficits after extensive elective vascular operations. Surg Gynec Obstet 175:243-248, 1992.
9. Crowley JP, Metzger JB, Valeri CR. The volume of blood shed during the bleeding time correlates with the peripheral venous hematocrit. Am J Clin Pathol 108:579-584, 1997.
10. DeLoughery TG. Logistics of massive transfusions. Transfusion medicine: transfusion support in trauma – military and civilian approaches. Hematology 2010; pages 470-473.

11. Edgren G, Kamper-Jorgensen M, Eloranta S, Rostgaard K, Custer B, Ullum H, Murphy EL, Busch MP, Reilly M, Melbye M, Hjalgrin H, Nyren O. Duration of red blood cell storage and survival in transfused patients. Transfusion 50:1185-1195, 2010.
12. Fischer TH, Hays WE, Valeri CR. Poly-N-acetyl glucosamine fibers accelerate hemostasis in patients treated with antiplatelet drugs. J Trauma 71:S176-S182, 2011.
13. Gauvin F, Spinella PC, Lacroix J, Choker G, Ducruet T, Karant O, Hebert PC, Hutchison JS, Hume HA, Tucci M on behalf of the Canadian Critical Care Trials Group and the Pediatric Acute Lung Injury and Sepsis Investigators (PALISI) Network. Association between length of storage of transfused red blood cells and multiple organ dysfunction syndrome in pediatric intensive care patients. Transfusion 50:1902-1913, 2010.
14. Gilstad C, Raschewski M, Wells J, Delmas A, Lackey J, Urive P, Popa C, Jardeleza T, Roop S. Fatal transfusion-associated graft-versus-host disease with concomitant immune hemolysis in a group A combat trauma patient resuscitated with group O fresh whole blood. Transfusion 52:930-935, 2012.
15. Hajjar LA, Vincent J-L, Galas FRBG, et al. Transfusion requirements after cardiac surgery: the TRACS randomized controlled trial. JAMA 304:1559-1567, 2010.
16. Handin RI. Valeri CR. Hemostatic effectiveness of platelets stored at 22 C. NEJM 285:538-543, 1971.
17. Hakre SH, Peel SA, O'Connell RJ, Sanders-Buell EE, Jagodzinsk LL, Eggleston JC, Myles O, Waterman PE, McBride RH, Eader SA, Davis KW, Rentas FJ, Sateren WB, Naito NA, Tobler SK, Tovanabutra S, Petruccelli BP, McCutchan FE, Michael NL, Cersovsky SB, Scott PT. Transfusion-transmissible viral infections among US military recipients of whole blood and platelets during Operation Enduring Freedom and Operation Iraqi Freedom. Transfusion 51:473-485, 2011.
18. Hannon T. Trauma blood management: avoiding the collateral damage of trauma resuscitation protocols. Transfusion Medicine: Transfusion Support in Trauma – Military and Civilian Approaches. Hematology, 2010, pgs 463-464.
19. Henkelman S, Rakhorst G. Does modern combat still need fresh whole blood transfusions? Transfusion 52:2272-2273, 2012.

20. Holcomb JB. Optimal use of blood products in severely injured trauma patients. Transfusion Medicine: Transfusion Support in Trauma – Military and Civilian Approaches. Hematology, 2010, pgs 465-469.
21. Johansson PI and Stensballe J. Hemostatic resuscitation for massive bleeding: the paradigm of plasma and platelets – a review of the current literature. Transfusion 50:701-710, 2010.
22. Khuri SF, Wolfe JA, Josa M, Axford TC, Szymanski I, Assousa S, Ragno G, Patel M, Silverman A, Park M, Valeri CR. Hematologic changes during and after cardiopulmonary bypass and their relationship to the bleeding time and nonsurgical blood loss. J Thorac Cardiovasc Surg 104:94-107, 1992.
23. Khuri SF, Healey N, MacGregor H, Barnard MR, Szymanski IO, Birjiniuk V, Michelson AD, Gagnon DR, Valeri CR. Comparison of the effects of transfusions of cryopreserved and liquid-preserved platelets on hemostasis and blood loss after cardiopulmonary bypass. J Thorac Cardiovasc Surg. 117:172-184, 1999.
24. Koch CG, Li L, Duncan AI, Mihaljevla T, Cosgrove DM, Loop FD, Starr NJ, Blackstone EH. Morbidity and mortality risk associated with red blood cell and blood-component transfusion in isolated coronary artery bypass grafting. Crit Care Med 34:1608-1616, 2006.
25. Koch CG, Li L, Sessler DL et al. Duration of red cell storage and complications after cardiac surgery. N Engl J Med 358:1229-1239, 2008.
26. Lelkens CCM, Koning JR, deKort B, Floot IBG, Noorman F. Experience with frozen blood products in the Netherlands military. Trans Aph Sci 34:289-296, 2006.
27. Lelubre C, Piagnerelli M, Vincent JL. Association between duration of storage of transfused red blood cells and morbidity and mortality in adult patients: myth or reality? Transfusion 49:1384-1394, 2009.
28. Natanson C, Kern SH, Lurie P, et al. Cell-free hemoglobin blood substitute and risk of myocardial infarction and death: a metaanalysis. JAMA 299:2304-2312, 2008.
29. Neuhaus S Jr, Wishaw K, Lelkens C. Australian experience with frozen blood products on military operations. MJA Volume 192:203-206, 2010.

30. Noorman F, et al. Frozen -80 C red cells, plasma and platelets in combat casualty care. Transfusion 49(35) :28A (Suppl), 2009.
31. Noorman F, Badloe JF. -80 C frozen red blood cells, plasma and platelets : efficient logistics, available, compatible, safe and effective in the treatment of trauma patients with or without massive blood loss in military theatre. AABB Oct 2012, Transfusion practice/ clinical case studies : transfusion practices in emergent settings ; SP383 ; poster presentation. Transfusion 52(Suppl) SP383, 198A, 2012.
32. Noorman F, Badloe JF. -80 C frozen platelets, efficient logistics, available, compatible, safe and effective in the treatment of trauma patients with or without massive blood loss in military theatre. AABB Oct 2012, Transfusion practice/clinical case studies, S53-030H, oral presentation ; Transfusion 52(Suppl) S53-030H, 33A, 2012.
33. Noorman F, Strelitsid R, Badloe JF. -80 C frozen platelets are activated compared to 24 hour liquid stored platelets and quality of frozen platelets is unaffected by a quick preparation method (15 min) which can be used to prepare platelets for the early treatment of trauma patients in military theatre. AABB Oct 2012, Components and component processing: platelets; SP23, Poster presentation, Transfusion 52(Suppl) SP23, 62A, 2012.
34. Noorman F, Strelitski R, Badloe JF. Lyophilized plasma, an alternative to 4 C stored thawed plasma for the early treatment of trauma patients with (massive) blood loss in military theatre. AABB Oct 2012, Components and component processing: plasma, Transfusion 52(Suppl) SP7, 55A, 2012.
35. Perkins JG, Cap AP, Spinella PC, et al. Comparison of platelet transfusion as fresh whole blood versus apheresis platelets for massively transfused combat patients. Transfusion 51 :242-252, 2011.
36. Roeloffzen WWH, Kluin-Nelemans HC, Bosman L, de Wolf JTM. Effects of red blood cells on hemostasis. Transfusion 50 :1536-1544, 2010.
37. Spinella PC, Dunne J, Beilman GJ, O'Connell RJ, Borgman MA, Cap AF, Rentas F. Constant challenges and evolution of US military transfusion medicine and blood operations in combat. Transfusion 52 :1146-1153, 2012.

38. Steiner ME, Stowell C. Does red blood cell storage affect clinical outcome ? When in doubt, do the experiment. Transfusion 49 :1286-1290, 2009.
39. Valeri CR, Altschule MD. Hypovolemic Anemia of Trauma. The Missing Blood Syndrome. Chemical Rubber Company Press, Boca Raton, Florida, 1981.
40. Valeri CR, Pivacek LE, Cassidy GP, Ragno G. In vitro and in vivo measurements of gamma-radiated, frozen, glycerolized RBCs. Transfusion 41:545-549, 2001.
41. Valeri CR, Cassidy G, Pivacek LE, Ragno G, Lieberthal W, Crowley JP, Khuri SF, Loscalzo J. Anemia-induced increase in the bleeding time: implications for treatment of nonsurgical blood loss. Transfusion 41:977-983, 2001.
42. Valeri CR, MacGregor H, Giorgio A, Ragno G: Circulation and hemostatic function of autologous fresh, liquid preserved, and cryopreserved baboon platelets transfused to correct an aspirin-induced thrombocytopathy. Transfusion 42:1206-16, 2002.
43. Valeri CR, Srey R, Tilahun D, Ragno G. In vitro effects of polymerized N-acetyl glucosamine (NAG) on the activation of platelets in platelet rich plasma with and without red blood cells. J Trauma (suppl) 57:S22-S25, 2004.
44. Valeri CR, Ragno G, Khuri S. Freezing human platelets using 6% DMSO with removal of the supernatant solution prior to freezing and storage at -80 C without post-thaw processing. Transfusion 45(12):1890-1898, 2005.
45. Valeri CR, Ragno G. In vitro testing of platelets using the thromboelastogram, platelets function analyzer, and the clot signature analyzer to predict the bleeding time. Trans Aph Sci 35(1) :33-41, 2006.
46. Valeri CR, Dennis RC, Ragno G, MacGregor H, Menzoian JO, Khuri SF. Limitations of the hematocrit to assess the need for RBC transfusion in hypovolemic anemic patients. Transfusion 46:365-371, 2006.
47. Valeri CR, Ragno G. Cryopreservation of human blood products. Trans Apher Sci 34:271-287, with an editorial on pages 267-269, 2006.
48. Valeri CR, Khuri S, Ragno G. Non-surgical bleeding diathesis in anemic thrombocytopenic patients: Role of temperature, RBC,

platelets, and plasma clotting proteins. Transfusion 47:206S-248S, 2007.

49. Valeri CR, Ragno G. An approach to prevent the severe adverse events associated with transfusion of FDA-approved blood products. Trans Aph Sci 42:223-233, 2010.
50. Valeri CR, Veech RL. The unrecognized effects of the volume and composition of the resuscitation fluid used during the administration of blood products. Trans Aph Sci 46:121-123, 2012.
51. van de Watering L, Lorinser J, Versteegh M, Westendorp R, Brand A. Effects of storage time of red blood cell transfusions on the prognosis of coronary artery bypass graft patients. Transfusion 46:1712-1718, 2006.
52. van de Watering L. Red cell storage and prognosis. Vox Sang 100:36-45, 2011.
53. Wang D, Sun J, Solomon SB, Klein HG, Natanson C. Transfusion of older stored blood and risk of death: a meta-analysis. Transfusion 52:1184-1195, 2012.
54. Warkentin TE, Eikelboom JW. Old blood bad? Either the biggest issue in transfusion medicine or a nonevent. Transfusion 52:1165-1167, 2012.
55. Zimrin AB, Hess JR. Current issues relating to the transfusion of stored red blood cells. Vox Sang 96 :93-103, 2009.

CHAPTER 11

OPTIMUM TREATMENT OF COMBAT CASUALTIES

1. Hemostatic agents to treat blood loss, stimulate wound healing, and exert an antimicrobial effect.
2. Resuscitation fluids to provide small volume of hypertonic NaD betahydroxybutyrate solution.
3. Transfusion of universal donor group O Rh positive and group O Rh negative RBC that circulate and function; transfusion of AB plasma clotting proteins from male donors that circulate and function, and transfusion of single donor leukoreduced group O platelets that circulate and function.
4. Optimum volume and composition of resuscitation fluids and optimum ratio of viable and functional RBC, functional plasma clotting proteins, and viable and functional platelets have to be determined. Both the quality and quantity of the blood products need to be specified.
5. Deployment of -80 C mechanical freezers maintained at a mean temperature of -80 C with a range of -65 C to -90 C with a tank of liquid CO2 to be triggered to be added to the mechanical freezer when the temperature is less than -65 C. Transportation of frozen blood products in insulated containers with dry ice.
6. Current FDA requirements for testing of the blood products for infectious disease markers and the history of blood donors regarding where the donor lives and what the donor eats require the testing of fresh whole blood for 24 to 48 hours prior to transfusion.

7. Current requirements for testing of blood products for infectious disease markers and the donor history now support the use of frozen RBC, platelets and plasma which can be tested and quarantined for six months to retest the donor to ensure the safety of the frozen blood products.
8. Adverse effects of hypothermia and the need to warm the wounded individual and to administer blood products maintained at 40 C at the time of transfusion with warming devices to ensure optimum hemostasis in the recipient.
9. Need to provide resuscitation fluids, irrigation fluids, and solutions to process frozen RBC from potable water using membrane technology with the need for final heat sterilization.

General Approach to Optimum Treatment of Combat Casualties

1. Control surgical blood loss and use of hemostatic agents with external pressure.
2. Resuscitation solution: low volume hypertonic ketone solution containing NaD betahydroxybutyrate instead of Ringer's lactate solution and human albumin.
3. Hypothermia produces adverse effects on hemostasis.
4. Nonsurgical bleeding diathesis is associated with hemodilution, anemia, thrombocytopenia, and reduction in plasma clotting proteins.
5. Quality and quantity of blood products should ensure the survival and function of compatible RBC, platelets, and plasma clotting proteins.

Resuscitation Fluid

1. Low volume substrate resuscitation using hypertonic NaD betahydroxybutyrate ketone solution instead of high volume resuscitation using Ringer's lactate solution which produces hemodilution and an anticoagulant effect. Human albumin should not be used because albumin is a poor oncotic protein and does not increase the plasma volume and blood volume in patients.
2. Hemodilution and hypothermia produce adverse effects on hemostasis and increase nonsurgical blood loss.

3. Production of parenteral solution for resuscitation, irrigation solutions, and solutions to process blood products at the site of need from potable water and concentrates.

Hemostatic Agents to Reduce Blood Loss

1. Safety and therapeutic effectiveness of the Marine Polymer hemostatic agents applied with external pressure.
2. Microalgae produce a highly polymerized N-acetylglucosamine (NAG) nanofiber. The long NAG nanofiber required that it has to be removed from the bleeding site. Gamma radiation procedures produce a short NAG nanofiber that does not have to be removed from the bleeding site.
3. The sNAG like the lNAG activates the platelets, RBC and clotting proteins to exert a hemostatic effect, stimulates wound healing and exerts an antimicrobial effect.
4. The subendothelial glycocalyx may release derivatives of N-acetylglucosamine which the microalgae in the bioreactor produces as an endotoxin free, sterile, highly purified substance.

Mechanism of action of the sNAG nanofiber

1. Activates the platelets, RBC and plasma clotting proteins to produce thrombin.
2. Activates wound healing by stimulation of the keratinocytes, fibroblasts and endothelial cells.
3. Provides a surface to stimulate the production of the alpha and beta defensin peptides to exert an antimicrobial effect.
4. Function as a platelet substitute.

Safety and therapeutic effectiveness of the N-acetylglucosamine (NAG) hemostatic agent with external pressure

1. Maintains patency of the catheters used to perform hemodialysis.
2. Restores hemostasis and accelerates healing at the arterial site used to catheterize patients to perform diagnostic and therapeutic interventions.
3. Accelerates wound healing in patients with chronic ulcers used

with and without the vacuum assisted closure (VAC) device (Maus E, BMJ Case Reports doic:10.1138, 2012).

Insulin resistance and therapeutic effect of D-B-hydroxybutyrate

Severe injury, infection and hemorrhage all cause insulin resistance. Insulin resistance blocks glucose entry into cells and conversion of pyruvate to acetyl CoA impairing the Krebs cycle and energy production. Lactate fluids decrease energy production by inhibiting the production of pyruvate and pyruvate entry into mitochondria. D-B-hydroxybutyrate containing fluids increase metabolic efficiency and bypass blocks caused by insulin resistance and lactate.

Quality and quantity of RBC, platelets, plasma to treat patients with surgical and nonsurgical blood loss using liquid preserved red blood cell products stored at 4 C, platelets stored at 22 C with agitation, and frozen plasma and cryoprecipitate stored at -20 C.

1. Acceptable survival and function of liquid preserved RBC stored at 4 C in CPDA-1, CPD/AS-1, CP2D/AS-3, and CPD/AS-5 for only 2 weeks.
2. Acceptable survival and function of platelets stored at room temperature (22 $\pm$ 2 C) with agitation for only 2 days.
3. Fresh frozen plasma and cryoprecipitate containing functional plasma clotting proteins stored at -20 C for one year, thawed and stored at 4 C for 24 hours.

Supplementation of liquid preserved platelets stored at room temperature with agitation, red blood cells stored at 4 C, and fresh frozen plasma and cryoprecipitate stored in the frozen state at -20 C with RBC frozen for at least 10 years, platelets frozen for at least 2 years, plasma frozen for at least 10 years all stored at -80 C in mechanical freezers.

Compatible fresh whole blood was advocated by Colonel William Crosby, the U.S. Army and Department of Defense to treat wounded casualties in Korea, Vietnam, Persian Gulf War, and now in the Middle East in combat zones. In 1981 the need to test fresh whole blood for autoimmune deficiency syndrome (AIDS), hepatitis and other infectious disease agents and the history of the blood donor related to where the donor lives and what the donor eats limits the availability of fresh whole blood which now requires 24 to 48 hours to test the blood products prior to transfusion. The therapeutic benefit of fresh whole blood observed by Colonel Crosby is due primarily to the survival and function of the compatible red blood cells which provide oxygen to the brain, heart, and kidneys and provide an important hemostatic effect to reduce nonsurgical blood loss.

Safe and Therapeutically Effective Frozen Blood Products

1. Blood products frozen and stored in -80 C mechanical freezers provide universal donor red blood cells, platelets and plasma clotting proteins that circulate and function immediately or shortly after transfusion prevent the severe adverse events associated with the transfusion of FDA approved blood products.
2. Universal donor group O Rh positive and group O Rh negative RBC frozen with 40% W/V glycerol and stored at -80 C for at least 10 years, thawed, washed in the Haemonetics Blood Processor ACP215 instrument and stored at 4 C in the additive solution AS-3 (Nutricel) for 2 weeks.
3. Universal donor AB plasma collected from male donors stored at -80 C for at least 14 years thawed and stored at 4 C for 24 hours maintain the function of the clotting proteins factors V and VIII and fibrinogen.
4. Universal donor group O single donor leukoreduced platelets treated with 6% DMSO, the supernatant DMSO removed, and DMSO treated 2.5 to 3.0 X 10^{11} platelets frozen and stored at -80 C for at least 2 years, thawed, diluted with 10 to 20 ml of 0.9% NaCl or a unit of AB plasma and stored at room temperature without agitation for 6 hours.
5. Universal donor group O Rh positive and group O Rh negative RBC stored in the liquid state at 4 C for 3 to 6 days treated with 40% W/V glycerol, the supernatant glycerol removed, frozen at -80 C in mechanical freezers for at least 10 years as nonrejuvenated glycerolized RBC.
6. Universal donor group O Rh positive and group O Rh negative RBC stored in the liquid stored in the liquid state at 4 C for 3 to 6 days, biochemically treated with pyruvate, inosine, phosphate, and adenine (PIPA) solution and treated with 40% glycerol, the supernatant glycerol removed can be stored at -80 C in mechanical freezers for at least 10 years as indated rejuvenated glycerolized RBC.

7. Universal donor group O Rh positive and group O Rh negative RBC stored in the liquid state at 4 C at the time of outdating salvaged by biochemical treatment with pyruvate, inosine, phosphate and adenine solution (PIPA), treated with 40% W/V glycerol and the supernatant glycerol remove, can be frozen at -80 C in mechanical freezers for at least 10 years as outdated rejuvenated glycerolized RBC.
8. Using the functional closed Haemonetics Blood Processor ACP215 instrument, deglycerolized nonrejuvenated and rejuvenated indated RBC can be stored at 4 C in the additive solution (AS-3) for 2 weeks and the rejuvenated outdated RBC can be stored at 4 C in the additive solution (AS-3) for at least 24 hours.

Summary

Hemostatic agents are needed to treat blood loss, stimulate wound healing and exert an antimicrobial effect. Surgical and nonsurgical blood loss should be treated at the site of injury with hemostatic agents and external pressure.

Resuscitation fluids should be administered in the field or as soon as possible by infusion of small volume of the hypertonic 600 mM Na-D-B-hydroxybutyrate solution to treat traumatic injury and head injury to reduce the incidence of posttraumatic stress disorder.

Low volume of hypertonic solution containing 600 mM Na-D-B-hydroxybutyrate solution which provide a ketone substrate to restore the metabolic state and function of the cells following hemorrhage rather than hypertonic NaCl which produces hyperchloremic acidosis or high volume Ringer's lactate which produces hemodilution, inhibits hemostasis, and produces oxygen free radicals. Human albumin should not be used because human albumin does not increase the plasma volume and blood volume in patients.

Hemodilution and hypothermia produce adverse effects on hemostasis and increase nonsurgical blood loss. The need to warm the wounded individual and to administer blood products maintained at 40 C at the time of transfusion using warming devices to ensure optimum hemostasis in the recipient.

Current FDA requirements for testing of the blood products for infectious disease markers and the history of blood donors regarding where the donor lives and what the donor eats require the testing of fresh whole blood for 24 to 48 hours prior to transfusion.

Current requirements for testing of blood products for infectious disease markers and the donor history now support the use of frozen RBC, platelets and plasma which can be tested and quarantined for six months at -80 C in the mechanical freezer to retest the blood donor to ensure the safety of the blood products.

Frozen blood products can be stored in -80 C mechanical freezers to provide universal donor red blood cells, platelets and plasma clotting proteins that circulate and function immediately or shortly after transfusion.

Universal donor group O Rh positive and group O Rh negative RBC frozen with 40% W/V glycerol can be stored at -80 C for at least 10 years, thawed, washed using the Haemonetics Blood Processor (ACP) 215 and stored at 4 C in the additive solution AS-3 (Nutricel) for 2 weeks.

Universal donor AB plasma collected from male donors can be frozen and stored at -80 C for at least 14 years, thawed and stored at 4 C for 24 hours maintain the function of the clotting proteins factors V and VIII and fibrinogen.

Universal single donor group O leukoreduced 2.5 to 3.0 X 10^{11} platelets treated with 6% DMSO, the supernatant DMSO removed, and DMSO platelets frozen at -80 C for at least 2 years, thawed, diluted with 10 to 20 ml of 0.9% NaCl or with a unit of AB plasma and stored at room temperature without agitation for 6 hours.

Transfusion of universal donor frozen group O Rh positive and group O Rh negative RBC that circulate and function; transfusion of frozen AB plasma clotting proteins from male donors that circulate and function, and transfusion of single donor frozen, leukoreduced 2.5 to 3.0 X 10^{11} group O platelets that circulate and function have provided safe and therapeutically effective blood products without the need for fresh whole blood and apheresed platelets to treat military and civilian casualties by the Netherlands military in combat zones in Iraq and Afghanistan.

Optimum volume and composition of resuscitation fluids and optimum ratio of RBC, plasma and platelets to treat patients subjected to trauma with massive blood loss are being evaluated.

Deployment of -80 C mechanical freezers maintained at a mean temperature of -80 C with a range of -65 C to -90 C with a tank of liquid CO2 to be added to the mechanical freezer when the temperature is less than -65 C. Transportation of frozen blood products in insulated containers with dry ice have been demonstrated by the U.S. Navy in Vietnam in 1968 to 1974 and by the Netherlands military in the Middle East war zones in 2001 to 2012.

Need to provide resuscitation fluids, irrigation fluids, and solutions to process frozen RBC from potable water using membrane technology at the site of need to prepare these solutions from sterile water for injection and concentrates of the solutions.

An approach to prevent the severe adverse events associated with FDA approved blood products like transfusion related acute lung injury (TRALI) and systemic inflammatory response syndrome (SIRS) with the use of liquid methods and freezing procedures to provide safe and therapeutically effective blood products to treat combat casualties. The experience by the Netherlands military under the direction of Dr. Charles Lelkens, Dr. Femke Noorman and Dr. John Badloe and associates have demonstrated that frozen RBC, frozen platelets, frozen plasma all stored at -80 C in mechanical freezers provided safe and therapeutically effective blood products to treat wounded casualties without the need for fresh whole blood and apheresed platelets in combat zones (Badloe J and Noorman F, Transfusion 51(35) (Suppl), 2011; Henkelman S and Rakhorst G, Transfusion 52:2272-2273, 2012; Lelkens CCM et al, Trans Aph Sci 34:289-296, 2006; Noorman F et al, Transfusion 49(35) 28A(Suppl), 2009; Noorman F and Badloe JF, Transfusion 52(Suppl): SP383, 198A, 2012).

Freezing provides red blood cells with normal and improved function to release oxygen to tissue at high pO2 tensions. Indated RBC and outdated RBC can be treated with a solution containing pyruvate, inosine, phosphate and adenine (PIPA) to increase the RBC ATP, DPG and p50 levels prior to glycerolization and freezing. The procedures designed, developed and tested at the NBRL provided outdated rejuvenated universal donor group O Rh positive and group O Rh negative RBC with normal or moderately increased affinity for oxygen and indated rejuvenated RBC with decreased affinity for oxygen and improved function to deliver oxygen to tissue.

Liquid preserved group O Rh positive and group O Rh negative RBC stored in ACD, CPD, CPDA-1 and in the additive solutions CPD/AS-1, CP2D/AS-3, and CPD/AS-5 for 3 to 6 days can be glycerolized, the supernatant glycerol removed, and frozen as nonrejuvenated RBC and stored at -80 C. Liquid preserved RBC stored in ACD, CPD, CPDA-1, and in the additive solutions CPD/AS-1, CP2D/AS-3, and CPD/AS-5 for 3 to

6 days can be biochemically treated with pyruvate, inosine, phosphate, and adenine (PIPA) and then glycerolized, the supernatant glycerol removed, and frozen at -80 C as indated rejuvenated RBC. Liquid preserved RBC stored in the additive solutions CPDA-1 for 36 to 38 days and in CPD/AS-1, CP2D/AS-3, and CPD/AS-5 for 42 days can be biochemically treated with PIPA then glycerolized, the supernatant glycerol removed, and frozen as outdated rejuvenated RBC. The nonrejuvenated glycerolized RBC frozen at -80 C, thawed and deglycerolized in the Haemonetics Blood Processor ACP215 and stored in the additive solution AS-3 for 2 weeks have 24 hour posttransfusion survival of 75%, red cell oxygen affinity which is moderately increased and less than 1% hemolysis. The indated rejuvenated glycerolized frozen RBC following thawing, deglycerolization in the Haemonetics Blood Processor ACP215 and stored in AS-3 for 2 weeks have 24 hour posttransfusion survival of 75%, red cell oxygen affinity which is normal or slightly decreased and less than 1% hemolysis. The outdated rejuvenated glycerolized frozen RBC following thawing, deglycerolization in the Haemonetics Blood Processor ACP215 and stored at 4 C in AS-3 for at least 24 hours have 24 hour posttransfusion survival of 75%, normal or slightly increased red blood cell oxygen affinity and less than 1% hemolysis. The biochemical treatment of universal donor indated RBC and outdated RBC produce RBC with ATP, DPG and P50 levels similar to those of patients with chronic hypovolemic anemia of trauma and patients with cardiopulmonary disorders. The serendipitous observations made in patients with chronic hypovolemic anemia of trauma and in patients with cardiopulmonary insufficiency provided the levels of RBC ATP, DPG and P50 that can be achieved to biochemically modify indated and outdated RBC prior to glycerolization and freezing. The biochemical treatment produced RBC with function that patients which chronic hypovolemic anemia of trauma and with cardiopulmonary insufficiency produce to provide optimum oxygen delivery to the brain, heart, and kidneys without the need to increase blood flow.

The unrecognized effects of the volume and composition of the resuscitation fluid used during the administration of blood products have been reported in Transfusion and Apheresis Science (46:121-123, 2012) by Valeri CR and RL Veech.

Recent publications have reported the severe adverse events associated with blood products but have not considered the effect of the volume and composition of the resuscitation fluids infused with the blood products. Injured tissue leads to cellular reaction characterized by insulin resistance which prevents glucose to enter muscle and fat cells. In all cells, mitochondrial pyruvate dehydrogenase (PDH) activity is decreased during insulin deficiency leaving cells deficient in substrates needed to power the Krebs cycle and make ATP.

D-B-hydroxybutyrate, a normal ketone body metabolite, enters cells on the monocarboxylate transport channel mimicking the action of insulin and bypassing the enzymatic block at PDH. Metabolism of ketone bodies increases efficiency of mitochondrial energy production and cellular ATP level. Infusion of 250 ml of 600 mM Na-D-B-hydroxybutyrate solution, with the same osmotic strength as the hypertonic NaCl solution currently being used would correct insulin resistance, provide energy substrates for cells to produce ATP, correct the tendency of injured tissue to swell due to decreased energy of ionic gradients and correct acidosis observed in hemorrhage (Table 1).

Substrate Resuscitation

Carl W. Walter in 1937 established the first blood bank in Boston. He later developed the plastic blood bag for collection of blood thus facilitating the separation of blood components. In 1986, he wrote a commentary on a paper which outlined the toxicity of current parenteral fluids by RL Veech published in Am J Clin Nutri 44:519-551, 1986. "The prescribing of parenteral fluids has become so routine that most physicians have become oblivious to the toxic impact of current practices on the cellular metabolism of their patients. Few physicians recognize the iatrogenic threat of replacement of body fluids based solely on volumetric and caloric needs. Understanding the metabolic and ionic organization of cells can provide the physician means to use parenteral fluids to control the inherent metabolic energy of cells. Application of new insight into physical chemistry and metabolic properties of the cell can enhance the physician's therapy in critically ill patients".

Injury of any sort leads to a cellular reaction which is characterized by insulin resistance (Li L and Messina JL, Trends Endocrinol Metab 20:429-435, 2009). During insulin resistance, glucose cannot enter muscle and fat cells nor can the cell metabolize the lactate given in lactated Ringer's solutions. More importantly in all cells the mitochondrial pyruvate dehydrogenase (PDH) activity is decreased during insulin deficiency leaving the cell deficient in substrates needed to power the Krebs cycle to make ATP. The normal ketone body metabolite D-B-hydroxybutyrate enters cells via the monocarboxylate transporter channel mimicking the action of insulin (Kashiwaya Y, et al, Am J Cardiol 80:50A-64A, 1997) and bypassing the enzymatic block at PDH. Even more importantly, the metabolism of ketone bodies increases the efficiency of mitochondrial energy production and increases the energy contained within the ATP molecule that is the delta GATP (Sato K, et al, FASEB J 9:651-658, 1995).

In order to prevent dilutional coagulopathy, one needs to use low volume substrate based resuscitation fluids which are capable of correcting the metabolic and physical chemical abnormalities and the energy deficit of the injured cell. The most important metabolic defect needing correction in injured patients receiving blood products and parenteral fluids is insulin resistance (Li L and Messina JL, Trends Endocrinol Metab 20:429-435, 2009). Insulin resistance prevents the injured cell from metabolizing the lactate produced by glycolysis as well as the lactate administered in Ringer's lactate fluids. This metabolic defect of insulin resistance in the cells of injured patients can be overcome by the administration of Na D-B-hydroxybutyrate containing solutions which bypass the metabolic block at PDH and increase the energy of the ATP molecule (Kashiwaya Y, et al, Am J Cardiol 80:50A-64A, 1997; Sato K et al, FASEB J 9:651-658, 1995). The six most important metabolic effects of ketone bodies can be summarized as follows (Sato K et al, FASEB J 9:651-658, 1995).

1. Increases the concentration of Krebs cycle substrates depleted by insulin resistance
2. Production of mitochondrial NADH required as a substrate for electron transport
3. Oxidation of coenzyme Q thus lowering the production of free radicals
4. Increases the energy of ATP hydrolysis, delta GATP
5. Mimics the action of insulin
6. Overcomes the blockade of insulin resistance

Such a solution would be one comprised of 600 mM Na D-B-hydroxybutyrate which would have the same osmotic strength as the current hypertonic NaCl solutions being used. It could be provided in 250 ml volumes. Infusions of Na D-B-hydroxybutyrate containing solutions would correct the insulin resistance of injury (Li L and Messina JL, Trends

Endocrinol Metab 20:429-435, 2009), provide energy substrates for the cell to produce ATP (Sato K et al, FASEBJ 9:651-658, 1995), correct the tendency of injured tissue to swell due to the decrease in the energy of ionic gradients (Veech RL, et al, IUBMB Life 54:241-252, 2002) as well as correcting the acidosis often accompanying hemorrhage.

Table 1*

High volume resuscitation and low volume resuscitation fluids.

Component (mM)	Ringer's lactate	Ringer's ketone	Normal saline	Hypertonic saline	Hypertonic ketone
Na+	130	130	155	603	600
K+	4	4			
Ca++	3	3			
Cl-	109	109	155	603	
Lactate	28				
D-B-hydroxy butyrate		28			600
pH	6.5	6.5	6.5	6.5	6.5
Osmolarity (mOsm/l)	275	275	310	1200	1200
Sterilization	Heat	Heat	Heat	Heat	Heat

*with permission from Elsevier, Inc., 360 Park Avenue South, New York, NY 10010-1710, USA

CHAPTER 11 REFERENCES

1. Badloe J, Noorman F. The Netherlands experience with frozen -80 C red cells, plasma and platelets in combat casualty care S10-301. Transfusion 51(35) (Suppl). 2011.
2. Henkelman S, Rakhorst G. Does modern combat still need fresh whole blood transfusions? Transfusion 52:2272-2273, 2012.
3. Kashiyawa Y, King MT, Veech RL. Substrate signaling by insulin: a ketone bodies ratio mimics insulin action in heart. Am J Cardiol 80:50A-64A, 1997.
4. Lelkens CCM, Koning JR, deKort B, Floot IBG, Noorman F. Experience with frozen blood products in the Netherlands military. Trans Aph Sci 34:289-296, 2006.
5. Li L, Messina JL. Acute insulin resistance following injury. Trends Endocrinol Metab 20:429-435, 2009.
6. Maus EA. Successful treatment of two refractory venous stasis ulcers treated with a novel poly-N-acetyl glucosamine-derived membrane. BMJ Case Reports doic:10.1138, 2012.
7. Noorman F, et al. Frozen -80 C red cells, plasma and platelets in combat casualty care. Transfusion 49(35) :28A (Suppl), 2009.
8. Noorman F, Badloe JF. -80 C frozen red blood cells, plasma and platelets : efficient logistics, available, compatible, safe and effective in the treatment of trauma patients with or without massive blood loss in military theatre. AABB Oct 2012, Transfusion practice/ clinical case studies : transfusion practices in emergent settings ; SP383 ; poster presentation. Transfusion 52(Suppl) SP383, 198A, 2012.
9. Sato K, Kashiwaya Y, Keon CA, Tsuchiya N, King MT, Radda GK, et al. Insulin, ketone bodies, and mitochondrial energy transduction. FASEB J 9:651-658, 1995.
10. Valeri CR, Ragno G. An approach to prevent the severe adverse events associated with transfusion of FDA-approved blood products. Trans Aph Sci 42:223-233, 2010.
11. Valeri CR, Veech RL. The unrecognized effects of the volume and composition of the resuscitation fluid used during the administration of blood products. Trans Aph Sci 46:121-123, 2012.

12. Veech RL. The toxic impact of parenteral solutions on the metabolism of cells: a hypothesis for physiological parenteral therapy. Am J Clin Nutri 44:519-551, 1986.
13. Veech RL, Kashiwaya Y, Gates DN, King MT, Clarke K. The energetics of ion distribution: the origin of the resting electric potential of cells. IUBMB Life 54:241-252, 2002.

CHAPTER 12

PEER REVIEWED PUBLICATIONS, BOOKS AND MONOGRAPHS PUBLISHED BY THE NBRL

A. PEER REVIEWED PUBLICATIONS

1. Lionetti FJ, Valeri CR, Bond JC, Fortier NL. Measurement of hemoglobin binding capacity of human serum or plasma by means of dextran gels. J Lab Clin Med 64:519-528, 1964.
2. Valeri CR, Henderson ME. Recent difficulties with frozen glycerolized blood. JAMA 188:1125-1131, 1964.
3. Valeri CR, Bond JC, Fowler K, Sobucki J. Quantitation of serum hemoglobin binding capacity using cellulose acetate membrane electrophoresis. Clin Chem 11:581-588, 1965.
4. Valeri CR, McCallum LE. Relationship between glutathione stability and in vivo survival of autologous, deglycerolized, resuspended red blood cells. Nature 205:561-563, 1965.
5. Valeri CR. Effect of resuspension medium on in vivo survival and supernatant hemoglobin of erythrocytes preserved with glycerol. Transfusion 5:25-35, 1965.
6. Valeri CR. Observations on recipient plasma hemoglobin concentration after transfusion with glycerolized frozen blood. Transfusion 5:36-53, 1965.
7. Valeri CR, Mercado-Lugo R, Danon D. Relationship between osmotic fragility and in vivo survival of autologous deglycerolized resuspended red blood cells. Transfusion 5:267-272, 1965.

8. Valeri CR. The in vivo survival, mode of removal of the non-viable cells, and the total amount of supernatant hemoglobin in deglycerolized, resuspended erythrocytes. I. The effect of the period of storage in ACD at 4C prior to glycerolization. II. The effect of washing deglycerolized, resuspended erythrocytes after a period of storage at 4C. Transfusion 5:273-285, 1965.
9. Valeri CR, McCallum L. The age of human erythrocytes lost during freezing and thawing with glycerol using the Cohn Fractionator. Transfusion 5:421-426, 1965.
10. Valeri CR. In vivo survival and supernatant hemoglobin of autologous, deglycerolized, resuspended erythrocytes processed using centrifugation. III. The effect of the length of storage at -80 C. Transfusion 6:112-115, 1966.
11. Lionetti FJ, Valeri CR, Bond JC, Kivowitz C, Weinman E. Nucleotides in frozen glycerolized erythrocytes. Transfusion 6:116-123, 1966.
12. Valeri CR. The in vivo survival of Coombs positive autologous erythrocytes produced by agglomeration. Transfusion 6:247-253, 1966.
13. Valeri CR, Bond JC. Observations on the preservation of autologous human erythrocytes using glycerol, slow-freeze technic and agglomeration. Transfusion 6:254-262, 1966.
14. Valeri CR, Bond JC, McCallum LE. Relationships between metabolic state and (1) in vivo survival and (2) density distribution of previously frozen human erythrocytes. Transfusion 6:543-553, 1966.
15. Valeri CR, McCallum LE, Danon D. Relationships between in vivo survival and (1) density distribution, (2) osmotic fragility of previously frozen, autologous, agglomerated, deglycerolized erythrocytes. Transfusion 6:554-564, 1966.
16. Valeri CR. Clinical evaluation of erythrocytes preserved for prolonged periods. Military Med 131:705-710, 1966.
17. Valeri CR. Frozen blood. NEJM 275:365-373, 1966.
18. Valeri CR. Frozen blood (concluded). NEJM 275:425-431, 1966.
19. Almond DV, Valeri CR. Relationship between lipid fractions of erythrocytes and their in vivo survival following preservation with glycerol and the slow freeze technic. Transfusion 7:10-16, 1967.

20. Szymanski IO, Valeri CR, Almond DV, Emerson CP, Rosenfield RE. Automated differential agglutination for measurement of red cell survival. Brit J Haemat 13 (Suppl):50-53, 1967.
21. Almond DV, Valeri CR. The in vivo effects of deglycerolized agglomerated erythrocytes transfused in multiple units to stable anemic patients. Transfusion 7:95-104, 1967.
22. Brodine CE, Sell KW, Moss GS, Valeri CR. Navy surgical research programs. J Surg Res 7:545-548, 1967.
23. Valeri CR, Runck AH, McCallum LE. Observations on autologous, previously frozen, deglycerolized, agglomerated, resuspended red cells. I. Effect of storage temperatures. II. Effect of adenine supplementation of glycerolized red cells prior to freezing. Transfusion 7(2) : 105-16, 1967 Mar-Apr.
24. Szymanski IO, Valeri CR, McCallum LE, Emerson CP, Rosenfield RE. Automated differential agglutination technic to measure red cell survival. I. Methodology. Transfusion 8:65-73, 1968.
25. Szymanski IO, Valeri CR. Automated differential agglutination technic to measure red cell survival. II. Survival in vivo of preserved red cells. Transfusion 8:74-83, 1968.
26. Valeri CR. Observations on the chromium labelling of ACD-stored and previously frozen red cells. Transfusion 8:210-219, 1968.
27. Szymanski IO, Valeri CR. Evaluation of double 51Cr technique. Vox Sang 15:287-292, 1968.
28. Valeri CR. Preservation of human red blood cells. Bull NY Acad Med 44:3-17, 1968.
29. Moss GS, Valeri CR, Brodine CE. Clinical experience with the use of frozen blood in combat casualties. NEJM 278:748-752, 1968.
30. Valeri CR. Clinical effectiveness of concentrated liquid stored red cells and previously frozen red cells. Maryland State Med J 17:59-64, 1968.
31. Valeri CR, Brodine CE. Current methods for processing frozen red cells. Cryobiology 5:129-135, 1968.
32. Runck AH, Valeri CR, Sampson WT. Comparison of the effects of ionic and non-ionic solutions on the volume and intracellular potassium of frozen and non-frozen human red cells. Transfusion 8(1):9-18, 1968.

33. Valeri CR, Brodine CE, Moss GE. Use of frozen blood in Vietnam. Bibl Haematol 29:735-8 1968.
34. Szymanski IO, Valeri CR. Clinical evaluation of concentrated red cells. NEJM 280:281-287, 1969.
35. Valeri CR, Fortier NL. Red-cell 2,3-diphosphoglycerate and creatine levels in patients with red-cell mass deficits or with cardiopulmonary insufficiency. NEJM 81:1452-1455, 1969.
36. Valeri CR, Runck AH. Long term frozen storage of human red blood cells: Studies in vivo and in vitro of autologous red blood cells preserved up to six years with high concentrations of glycerol. Transfusion 9:5-14, 1969.
37. Valeri CR, Runck AH, Sampson WT. Effects of agglomeration on human red blood cells. Transfusion 9:120-134, 1969.
38. Runck AH, Valeri CR. Recovery of glycerolized red blood cells frozen in liquid nitrogen. Transfusion 9:297-305, 1969.
39. Valeri CR, Runck AH. Viability of glycerolized red blood cells frozen in liquid nitrogen. Transfusion 9:306-313, 1969.
40. Valeri CR, Fortier NL. Red-cell mass deficits and erythrocyte 2,3 DPG levels. Forsvarsmedicin 5:212-218, 1969.
41. Fortier NL, Hirsch NM, Valeri CR. Restoration of 2,3 DPG and ATP in ACD-stored red blood cells. Forsvarsmedicin 5:250-257, 1969.
42. Daane TA, Valeri CR, Barton RK. Autotransfusions of previously frozen blood in elective gynecologic surgery. Am J Obstet Gynecol 105:394-399, 1969.
43. Valeri CR, Hirsch NM. Restoration in vivo of erythrocyte adenosine triphosphate, 2,3-diphosphoglycerate, potassium ion, and sodium ion concentrations following the transfusion of acid-citrate-dextrose-stored human red blood cells. J Lab Clin Med 73:722-733, 1969.
44. Valeri CR. Can current blood-bank practices be improved? Medical Counterpoint 1:25-28, 1969.
45. Valeri CR, Runck AH, Brodine CE. Recent advances in freeze-preservation of red blood cells. JAMA 208:489-492, 1969.
46. Valeri CR. International Forum - In routine blood transfusion, should ACD anticoagulants be replaced by CPD or ACD adenine solutions? If so, by which one? Vox Sang 19:560-561, 1970.
47. Szymanski IO, Valeri CR. Factors influencing chromium elution

from labelled red cells in vivo and the effect of elution on red-cell survival measurements. Br J Haemat 19:397-409, 1970.

48. Daane TA, Valeri CR, Barton RK. Autologous transfusions of previously frozen blood in elective surgery. Lab Med 1:34-35, 1970.
49. Valeri CR, Szymanski IO, Runck AH. Therapeutic effectiveness of homologous erythrocyte transfusions following frozen storage at -80C for up to seven years. Transfusion 10:102-112, 1970.
50. Szymanski IO, Dean HM, Valeri CR, Bougas JA, Desforges JF. Measurement of erythrocyte survival during open-heart surgery. Transfusion 10:163-170, 1970.
51. Valeri CR, Bougas JA, Talarico L, Emerson CP, Didimizio T, Pivacek, L. Behavior of previously frozen erythrocytes used during open-heart surgery. Transfusion 10:238-246, 1970.
52. Szymanski IO, Valeri CR. Analysis of erythrocyte survival curves obtained simultaneously by 51Cr and an automated differential agglutination technic. Transfusion 10:287-298, 1970.
53. Handin RI, Fortier NL, Valeri CR. Platelet response to hypotonic stress after storage at 4 C or 22 C. Transfusion 10:305-309, 1970.
54. Valeri CR, Fortier NL. Is red-cell creatine metabolic garbage? NEJM 282(17):979-980, 1970.
55. Szymanski IO, Valeri CR. Lifespan of preserved red cells. Vox Sang 21:97-108, 1971.
56. Valeri CR, Collins FB. The physiologic effect of transfusing preserved red cells with low 2,3-diphosphoglycerate and high affinity for oxygen. Vox Sang 20:397-403, 1971.
57. Valeri CR. Viability and function of preserved red cells. NEJM 284:81-88, 1971.
58. Handin RI, Valeri CR. Hemostatic effectiveness of platelets stored at 22C. NEJM 285:538-543, 1971.
59. Handin RI, Lawler KC, Valeri CR. Automated platelet counting. Am J Clin Path 56:661-664, 1971.
60. Valeri CR, Szymanski IO, Pivacek LE. Effects of the host on transfused preserved red blood cells. J Med 2:228-247, 1971.
61. Valeri CR, Collins FB. Physiologic effects of 2,3-DPG-depleted red cells with high affinity for oxygen. J Appl Physiol 31:823-827, 1971.

62. Valeri CR, Hirsch NM, Runck AH, Szymanski IO, Johnson LH. An automated low ionic antiglobulin test. Transfusion 11:110-116, 1971.

63. Herman CM, Rodkey FL, Valeri CR, Fortier NL. Changes in the oxyhemoglobin dissociation curve and peripheral blood after acute red cell mass depletion and subsequent red cell mass restoration in baboons. Ann Surg 74:734-743, 1971.

64. Valeri CR, Kopriva CJ. Oxyhemoglobin dissociation curve in hemorrhagic and septic shock. Adv Exp Med Biol 23(0):177- 93, 1971.

65. Kopriva CJ, Ratliff JL, Fletcher JR, Fortier NL, Valeri CR. Biochemical and hematological changes associated with massive transfusion of ACD-stored blood in severely injured combat casualties. Ann Surg 176:585-589, 1972.

66. Valeri CR, Zaroulis CG. Rejuvenation and freezing of outdated stored human red cells. NEJM 287:1307-1313, 1972.

67. Handin RI, Valeri CR. Improved viability of previously frozen platelets. Blood 40:509-513, 1972.

68. Runck AH, Valeri CR. Continuous-flow centrifugation washing of red blood cells. Transfusion 12:237-244, 1972.

69. Valeri CR, Szymanski IO, Zaroulis CG. 24-hour survival of ACD- and CPD-stored red cells. I. Evaluation of nonwashed and washed stored red cells. Vox Sang 2:289-308, 1972.

70. Valeri CR, Altschule MD, Pivacek LE. The hemolytic action of adrenochrome, an epinephrine metabolite. J Med 3:20-40, 1972.

71. Biron PE, Howard J, Altschule MD, Valeri CR. Chronic deficits in red-cell mass in patients with orthopaedic injuries (stress anemia). J Bone Joint Surg 54-A:1001-1014, 1972.

72. Valeri CR, Zaroulis CG, Handin RI, Marchionni LD. Prostaglandins in the preparation of blood components. Science 175:539-542, 1972.

73. Valeri CR, Zaroulis CG, Marchionni LD, Patti KJ. A simple method for measuring oxygen content in blood. J Lab Clin Med 79:1035-1040, 1972.

74. Valeri CR, Zaroulis CG. Rejuvenation and freezing of outdated human red cells. Adv Exp Med Biol 28:457-77, 1972.

75. Rogers JC, Boldt D, Kornfeld S, Skinner SA, Valeri CR. Excretion of deoxyribonucleic acid by lymphocytes stimulated with

phytohemmagglutinin or antigen. Proc Natl Acad Sci 69(7):1685-1689, 1972.

76. Szymanski IO, Lipson CS, Valeri CR. Elution of chromium label from preserved red blood cells transfused during surgery. Transfusion 13:13-18, 1973.
77. Valeri CR. Status report: blood preservation. Contemporary Surgery 2:21-31, 1973.
78. Kingsley JR, Valeri CR, Peters H, Cole BC, Fouty WJ, Herman CM. Citrate anticoagulation and on-line cell washing in intraoperative autotransfusion in the baboon. Surg Forum 24:258-260, 1973.
79. Miller ME, Rorth M, Parving HH, Howard D, Reddington I, Valeri CR, Stohlman F Jr. pH effect on erythropoietin response to hypoxia. NEJM 288:706-710, 1973.
80. Valeri CR, Contreras TJ, Feingold H, Sheibley RH, Jaeger RJ. Accumulation of di-2-ethylhexyl phthalate (DEHP) in whole blood, platelet concentrates, and platelet-poor-plasma. I. Effect of DEHP on platelet survival and function. Environmental Health Perspectives, Jan 1973, pp 103-118.
81. Valeri CR, Cooper AG, Pivacek LE. Limitations of measuring blood volume with iodinated I 125 serum albumin. Arch Int Med 132:534-538, 1973.
82. Howarth HC, Valeri CR. Autotransfusion of frozen red blood cells in elective oral surgery. J Oral Surg 31:40-43, 1973.
83. Valeri CR, Szymanski IO. Further studies on the rapid and slow components of chromium elution in vivo from preserved erythrocytes. Vox Sang 24:502-514, 1973.
84. Valeri CR, Altschule MD. Hemolysis in vitro of blood obtained from patients with traumatic injuries. J Trauma 13:678-686, 1973.
85. Miller ME, Zaroulis CG, Valeri CR, Stohlman F Jr. Oxygen transport by the red cell: Effects of chronic hemodialysis. Blood 43:49-56, 1974.
86. Valeri CR, Feingold, H, Marchionni LD. A simple method for freezing human platelets using 6% dimethylsulfoxide and storage at -80 C. Blood 43:131-136, 1974.
87. Crowley JP, Rene A, Valeri CR. Changes in platelet shape and structure after freeze preservation. Blood 44:599-603, 1974.

88. Valeri CR. Factors influencing the 24-hour posttransfusion survival and the oxygen transport function of previously frozen red cells preserved with 40 percent W/V glycerol and frozen at -80 C. Transfusion 14:1-15, 1974.

89. Contreras TJ, Sheibley RH, Valeri CR. Accumulation of DI-2-ethylhexyl phthalate (DEHP) in whole blood, platelet concentrates, and platelet poor plasma. Transfusion 14:34-46, 1974.

90. Crowley JP, Valeri CR. The purification of red cells for transfusion by freeze preservation and washing. I. The mechanism of leukocyte removal from washed, freeze-preserved red cells. Transfusion 14:188-195, 1974

91. Crowley JP, Valeri CR. The purification of red cells for transfusion by freeze preservation and washing. II. The residual leukocytes, platelets, and plasma in washed, freeze-preserved red cells. Transfusion 14:196-202, 1974.

92. Valeri CR, Feingold H, Marchionni LD. The relation between response to hypotonic stress and the ^{51}Cr recovery in vivo of preserved platelets. Transfusion 14:331-337, 1974.

93. Crowley JP, Valeri CR. Recovery and function of granulocytes stored in plasma at 4 C for one week. Transfusion 14:574-580, 1974.

94. Crowley JP, Valeri CR. The purification of red cells for transfusion by freeze-preservation and washing. III. Leukocyte removal and red cell recovery after red cell freeze-preservation by the high or low glycerol concentration method. Transfusion 14:590-594, 1974.

95. Valeri CR, Feingold H, Zaroulis CG, Sphar RL, Adams GM. Effects of hyperbaric exposure on human platelets. Aerospace Medicine 45:610-616, 1974.

96. Crowley JP, Rene A, Valeri CR. The recovery, structure, and function of human blood leukocytes after freeze-preservation. Cryobiology 11:395-409, 1974.

97. Valeri CR. Hemostatic effectiveness of liquid-preserved and previously frozen human platelets. NEJM 290:353-358, 1974.

98. Allen JE, Valeri CR. Prostaglandins in hematology. Arch Int Med 133:86-96, 1974.

99. Valeri CR. Oxygen transport and viability of preserved red blood cells. J Med 5:278-291, 1974.

100. Crowley JP, Skrabut EM, Valeri CR. Immunocompetent lymphocytes in previously frozen washed red cells. Vox Sang 26:513-517, 1974.
101. Valeri CR. Prostaglandins and their role in clinical medicine. Chicago Clin Chem 7:32-34, 1974.
102. Valeri CR. Quality of red cell transfusion. Giornale di Medicina Militare 124:7-18, 1974.
103. Valeri CR. Viability and function of preserved red cells. Wadley Med Bull 5:205-238, 1975.
104. Dennis RC, Hechtman HB, Berger RL, Vito L, Weisel RD, Valeri CR. Transfusion of 2,3 DPG enriched red blood cells to improve cardiac function. Hematologia 17:7-18, 1975.
105. Valeri CR. Blood components in the treatment of acute blood loss: Use of freeze preserved red cells, platelets, and plasma proteins. Anesthesia and Analgesia-Current Researches 54:1-14, 1975.
106. Rice CL, Herman CM, Kiesow LA, Homer LD, John DA, Valeri CR. Benefits from improved oxygen delivery of blood in shock therapy. J Surg Res 19:193-198, 1975.
107. Dennis RC, Vito L, Weisel RD, Valeri CR, Berger RL, Hechtman HB. Improved myocardial performance following high 2,3 diphosphoglycerate red cell transfusions. Surgery 77:741-747, 1975.
108. Crowley JP, Skrabut EM, Valeri CR. The determination of leukocyte phagocytic oxidase activity by measurement of the initial rate of stimulated oxygen consumption. J Lab Clin Med 86:586-594, 1975.
109. Crowley JP, O'Donnell M, Sell KW, Valeri CR. The purification of red cells for transfusion by freeze-preservation and washing. IV. The use of micropore filtration to reduce the residual HL-A antigenicity of previously frozen, washed red cells. Transfusion 15:34-38, 1975.
110. Valeri CR. Simplification of the methods for adding and removing glycerol during freeze-preservation of human red blood cells with the high or low glycerol methods: Biochemical modification prior to freezing. Transfusion 5:195-218, 1975.
111. Walker R, Wilder M, Valeri CR. The effects of Staphylococcus Aureus and Klebsiella Pneumoniae peritonitis in mice exposed

to normal and hypoxic conditions on red cell oxygen transport function. J Med 6:113-120, 1975.

112. Valeri CR, Rorth M, Zaroulis CG, Jakubowski MS, Vescera S. Physiologic effects of hyperventilation and phlebotomy in baboons: Systemic and cerebral oxygen extraction. Ann Surg 181:99-105, 1975.

113. Valeri CR, Rorth M, Zaroulis CG, Jakubowski MS, Vescera S. Physiologic effects of transfusing red blood cells with high or low affinity for oxygen to passively hyperventilated, anemic baboons: Systemic and cerebral oxygen extraction. Ann Surg 181:106-113, 1975.

114. Skrabut EM, Crowley JP, Catsimpoolas N, Valeri CR. The effect of cryogenic storage on human erythrocyte membrane proteins as determined by polyacrylamide-gel electrophoresis. Cryobiology 13:395-403, 1976.

115. Miller ME, Rorth M, Stohlman F Jr, Valeri CR, Lowrie G, Howard D, McGilvray N. The effects of acute bleeding on acid-base balance, erythropoietin (Ep) production and in vivo p50 in the rat. Brit J Haemat 33:379-385, 1976.

116. Catsimpoolas N, Griffith AL, Skrabut EM, Platsoucas CD, Valeri CR. Differential ^{51}Cr uptake of human peripheral lymphocytes separated by density gradient electrophoresis. Cell Immunol 25:317-321, 1976.

117. Skrabut EM, Catsimpoolas N, Crowley JP, Valeri CR. (^{51}Cr) sodium chromate incorporation into the soluble protein fraction of the human erythrocyte: Binding not associated with the hemoglobin monomeric subunit. BBRC 69:672-677, 1976.

118. Valeri CR. Circulation and hemostatic effectiveness of platelets stored at 4 C or 22 C: Studies in aspirin-treated normal volunteers. Transfusion 16:20-23, 1976.

119. Contreras TJ, Valeri CR. A comparison of methods to wash liquid-stored red blood cells and red blood cells frozen with high or low concentrations of glycerol. Transfusion 16:539-565, 1976.

120. Contreras TJ, Valeri CR. Removal of HB_sAg from blood in vitro. I. Effects of washing alone, glycerol addition and removal, and glycerolization, freezing, and washing. Transfusion 16:594-609, 1976.

121. Kingsley JR, Valeri CR, Peters H, Cole BC, Fouty WJ, Sears

HF, Herman CM. Citrate anticoagulation and cell washing for intraoperative autotransfusion in the baboon. Am J Surg 131:717-721, 1976.

122. Valeri CR. Cryobiology overview of red cell preservation: achievements and prospective. Prog Clin Biol Res 11:55-87, 1976.

123. Weisel RD, Vito L, Dennis RC, Valeri CR, Hechtman HB. Myocardial depression during sepsis. Am J Surg 133:512-521, 1977.

124. Lionetti FJ, Hunt SM, Lin PS, Kurtz SR, Valeri CR. Preservation of human granulocytes. II. Characteristics of granulocytes obtained by counterflow centrifugation. Transfusion 17:465-472, 1977.

125. Foster ED, Spector JI, Talarico L, Umlas J, Valeri CR, Dobnick DB, Berger RL. Polybrene neutralization as a means of monitoring heparin therapy for extracorporeal circulation. Ann Thoracic Surg 23:514-519, 1977.

126. Valeri CR. Current concepts of blood transfusion. J Oral Surg 35:707-712, 1977.

127. Crowley JP, Wade PH, Wish C, Valeri CR. The purification of red cells for transfusion by freeze-preservation and washing. V. Red cell recovery and residual leukocytes after freeze-preservation with high concentrations of glycerol and washing in various systems. Transfusion 17:1-7, 1977.

128. Spector JI, Yarmala JA, Marchionni LD, Emerson CP, Valeri CR. Viability and function of platelets frozen at 2 to 3 C per minute with 4 or 5 percent DMSO and stored at -80C for 8 months. Transfusion 17:8-15, 1977.

129. Spector JI, Skrabut EM, Valeri CR. Oxygen consumption, platelet aggregation and release reactions in platelets freeze-preserved with dimethylsulfoxide. Transfusion 17:99-109, 1977.

130. Wade PH. Skrabut EM, Vinciguerra L, Valeri CR. In vitro function of granulocytes isolated from blood of normal volunteers using continuous-flow centrifugation in the IBM-Aminco Celltrifuge and adhesion-filtration leukapheresis using nylon fiber. Transfusion 17:136-140, 1977.

131. Spector JI, Zaroulis CG, Pivacek LE, Emerson CP, Valeri CR. Physiologic effects of normal-or-low-oxygen affinity red cells in hypoxic baboons. Am J Physiol 232(1):H79-84, 1977.

132. Lang DJ, Valeri CR. Hazards of blood transfusion. Adv Pediatr 24:311-38, 1977.

133. Weisel RD, Dennis RC, Manny J, Mannick JA, Valeri CR, Hechtman HB. Adverse effects of transfusion therapy during abdominal aortic aneurysectomy. Surgery 83:682-690, 1978.

134. Roy AJ, Ramirez M, Valeri CR. Stability of neutrophil-releasing activity of plasma obtained from leukapheresed rats and stored at 4 C. Transfusion 18:734-737, 1978.

135. Zaroulis CG, Schweitzer RL, Barchet S, Valeri CR. Enhanced red cell oxygen transport in a pregnant woman with hemoglobin SD disease. J Obstet Gynec 52:358-360, 1978.

136. Kurtz SR, Van Deinse WH, Valeri CR. The immunocompetence of residual lymphocytes at various stages of red cell cryopreservation with 40% W/V glycerol in an ionic medium at -80 C. Transfusion 18:441-447, 1978.

137. Catsimpoolas N, Griffith AL, Skrabut EM, Valeri CR. An alternate method for the preparative velocity sedimentation of cells at unit gravity. Analyt Biochem 87:243-248, 1978.

138. Contreras TJ, Hunt SM, Lionetti FJ, Valeri CR. Preservation of human granulocytes. III. Liquid preservation studied by electronic sizing. Transfusion 18:46-53, 1978.

139. Valeri CR. What is the clinical importance of alterations of the hemoglobin oxygen affinity in preserved blood- especially as produced by variations of red cell 2,3 DPG content? International Forum. Vox Sang 34:111-127, 1978.

140. Zaroulis CG, Kourides IA, Valeri CR. Red cell 2,3 diphosphoglycerate and oxygen affinity of hemoglobin in patients with thyroid disorders. Blood 52:181-185, 1978.

141. Lionetti FJ, Hunt SM, Mattaliano RJ, Valeri CR. In vitro studies of cryopreserved baboon granulocytes. Transfusion 18(6):685-692, 1978.

142. Dennis RC, Hechtman HB, Berger RL, Vito L, Weisel RD, Valeri CR. Transfusion of 2,3 DPG-enriched red blood cells to improve cardiac function. Ann Thoracic Surg 26(1):17-26, 1978.

143. Valeri CR, Weisel RD, Dennis RC, Mannick JA, Berger RL, Hechtman HB. Oxygen transport function of preserved red blood cells and myocardial performance. Prog Clin Biol Res 21:597-616, 1978.

144. Robblee LS, Shepro D, Vecchione JJ, Valeri CR. Increased thrombin sensitivity of human platelets after storage at 4C. Transfusion 19:45-52, 1979.

145. Contreras TJ, Lang DJ, Pivacek LE, Valeri CR. Occurrence of HB_sAg, anti-HB_s, and anti-CMV following the transfusion of blood products. Transfusion 19:129-136, 1979.

146. Contreras TJ, Lindberg JR, Lowrie GB, Pivacek LE, Austin RM, Vecchione JJ, Valeri CR. Liquid and freeze-preservation of dog red blood cells. Transfusion 19:279-292, 1979.

147. Kurtz SR, Carciero R, Valeri CR. Factors that influence the process of ^{51}chromium labeling of human granulocytes isolated from blood by counterflow centrifugation. Transfusion 19:398-403, 1979.

148. Spector JI, Flor WJ, Valeri CR. Ultrastructural alterations and phagocytic function of cryopreserved platelets. Transfusion 19:307-312, 1979.

149. Zaroulis CG, Spector JI, Emerson CP, Valeri CR. Therapeutic transfusions of previously frozen washed human platelets. Transfusion 19:371-378, 1979.

150. Valeri CR, Valeri DA, Dennis RC, Vecchione JJ, Emerson CP. Human red blood cells with normal or improved oxygen transport function prepared and frozen in the primary polyvinyl chloride plastic blood collection container. Blood Transfusions and Immunohematology 22:467-486, 1979.

151. McLoughlin GA, Grindlinger GA, Manny J, Valeri CR, Lipinski B, Mannick JA, Hechtman HB. Intrapulmonary clotting and fibrinolysis during abdominal aortic aneurysm surgery. Ann Surg 190:623-630, 1979.

152. Dennis RC, Bechthold D, Valeri CR. In vitro measurement of p50--the pH correction, and use of frozen red blood cells as controls. Crit Care Med 7:385-390, 1979.

153. Hechtman HB, Grindlinger GA, Vegas AM, Manny J, Valeri CR. Importance of oxygen transport in clinical medicine. Crit Care Med 7:419-423, 1979.

154. Valeri CR, Valeri DA, Dennis RC, Vecchione JJ, Emerson CP. Biochemical modification and freeze-preservation of red blood cells: A new method. Crit Care Med 7:439-447, 1979.

155. Costa JL, Dobson CM, Kirk KL, Poulsen FM, Valeri CR,

Vecchione JJ. Studies of human platelets by ^{19}F and ^{31}P NMR. FEBS Letters 99:141-146, 1979.

156. Catsimpoolas N, Kurtz SR, Skrabut EM, Griffith AL, Valeri CR. Cytotoxins alter the sedimentation behavior of human granulocytes. Science 205:936-937, 1979.

157. Zaroulis CG, Pivacek LE, Lowrie GB, Valeri CR. Lactic acidemia in baboons after transfusion of red blood cells with improved oxygen transport function and exposure to severe arterial hypoxemia. Transfusion 19:420-425, 1979.

158. Hechtman HB, Manny J, Valeri CR, Shepro D. Optimization of oxygen transport in acute respiratory failure. Bibl Anat 18:164-5, 1979.

159. Bryan-Brown CW, Valeri CR, Altschule MD. "The colouring substance of blood". Crit Care Med 7(9):358-9, 1979.

160. Valeri CR, Vecchione JJ, Pivacek LE, Lowrie GB, Austin RM, Emerson CP. Viability and function of outdated human red blood cells after biochemical modification to improve oxygen transport function, freezing, thawing, washing, postthaw storage at 4 C, perfusion in vitro through a bubble oxygenator, and autotransfusion. Transfusion 20:39-46, 1980.

161. Valeri CR, Zaroulis CG, Vecchione JJ, Valeri DA, Anastasi J, Pivacek LE, Emerson CP. Therapeutic effectiveness and safety of outdated human red blood cells rejuvenated to restore oxygen transport function to normal, frozen for 3 to 4 years at -80 C, washed, and stored at 4 C for 24 hours prior to rapid infusion. Transfusion 20:159-170, 1980.

162. Valeri CR, Zaroulis CG, Vecchione JJ, Valeri DA, Anastasi J, Pivacek LE, Emerson CP. Therapeutic effectiveness and safety of outdated human red blood cells rejuvenated to improve oxygen transport function, frozen for about 1.5 years at -80 C, washed, and stored at 4 C for 24 hours prior to rapid infusion. Transfusion 20:263-276, 1980.

163. Vecchione JJ, Chomicz SM, Emerson CP, Valeri CR. Cryopreservation of human platelets isolated by discontinuous-flow centrifugation using the Haemonetics Model 30 Blood Processor. Transfusion 20:393-400, 1980.

164. Valeri CR. Status report on red blood cell and platelet freezing. Weekly Anesthesiology Update 3:1-7, 1980.

165. Lionetti FJ, Hunt SM, Schepis JP, Roy AJ, Liss RH, Valeri CR. In vivo distribution of cryogenically preserved guinea pig granulocytes. Cryobiology 17:1-11, 1980.
166. Lionetti FJ, Luscinskas FW, Hunt SM, Valeri CR, Callahan AB. Factors affecting the stability of cryogenically preserved granulocytes. Cryobiology 17:297-310, 1980.
167. Crowley JP, Valeri CR. Antibody inhibition of polymorphonuclear phagocytosis: Dissociation of bacterial attachment and bacterial killing. J Lab Clin Med 95:868-876, 1980.
168. Hechtman HB, Krausz MM, Utsunomiya T, Valeri CR. Preoperative assessment of the high risk surgical patient. Surg Clin NA 60:1349-1358, 1980.
169. Utsunomiya T, Krausz MM, Valeri CR, Shepro D, Hechtman HB. Treatment of pulmonary embolism with prostacyclin. Surgery 8:25-30, 1980.
170. Costa JL, Dobson CM, Kirk KL, Poulsen FM, Valeri CR, Vecchione JJ. Nuclear magnetic resonance studies of blood platelets. Phil Trans R Soc Lond 289:413-423, 1980.
171. Dennis RC, Valeri CR. Measuring percent oxygen saturation of hemoglobin, percent carboxyhemoglobin and methemoglobin, and concentrations of total hemoglobin and oxygen in blood of man, dog, and baboon. Clin Chem 26:1304-1308, 1980.
172. Valeri CR, Yarnoz M, Vecchione JJ, Dennis RC, Anastasi J, Valeri DA, Pivacek LE, Hechtman HB, Emerson CP, Berger RL. Improved oxygen delivery to the myocardium during hypothermia by perfusion with 2,3 DPG-enriched red blood cells. Ann Thoracic Surg 30:527-535, 1980.
173. Lionetti FJ, Lin PS, Mattaliano RJ, Hunt SM, Valeri CR. Temperature effects on shape and function of human granulocytes. Exp Hemat 8:304-317, 1980.
174. Grindlinger GA, Vegas AM, Churchill WH Jr, Valeri CR, Hechtman HB. Is respiratory failure a consequence of blood transfusion? J Trauma 20:627-631, 1980.
175. Callow AD, Ledig CB, O'Donnell TF, Kelly JJ, Rosenthal D, Korwin S, Hotte C, Kahn PC, Vecchione JJ, Valeri CR. A primate model for the study of the interaction of ^{111}IN-labeled baboon platelets with dacron® arterial prostheses. Ann Surg 191:362-366, 1980.

176. Cullen DJ, Kunsman J, Caldera D, Dennis RC, Valeri CR. Comparative evaluation of new fine-screen filters: Effects on blood flow rate and microaggregate removal. Anesthesiology 53:3-8, 1980.

177. Catsimpoolas N, Skrabut EM, Griffith AL, Valeri CR. Gravity sedimentation analysis of human blood leukocytes. J Biochem Biophys Methods 2:147-153, 1980.

178. Crawshaw HM, Quist WC, Serrallach E, Valeri CR, LoGerfo FW. Flow disturbance at the distal end-to-side anastomosis: Effect of patency of the proximal outflow segment and angle of anastomosis. Arch Surg 115:1280-1284, 1980.

179. Sweetman HE, Costa JL, Vecchione JJ, Valeri CR, Shepro D. Dense bodies and total calcium in human platelets following aspirin ingestion for a two-week period. Thrombosis Res 17:55-61, 1980.

180. Valeri CR, Valeri DA, Anastasi J, Vecchione JJ, Dennis RC, Emerson CP. Freezing in the primary polyvinylchloride plastic collection bag: A new system for preparing and freezing nonrejuvenated and rejuvenated red blood cells. Transfusion 21:138-149, 1981.

181. Kurtz SR, Valeri DA, Melaragno AJ, Gray A, Vecchione JJ, Emerson CP, Valeri CR. Leukocyte-poor red blood cells prepared by the addition and removal of glycerol from red blood cell concentrates stored at 4 C. Transfusion 21:435-442, 1981.

182. Smith DA, Monaghan WP, Orcutt RM, Barracchini AE, Valeri CR. A pre-sealed overwrap method of protecting frozen blood components during water immersion thawing. Transfusion 21:447-449, 1981.

183. Szymanski IO, Harper JM, Odgren PR, Valeri CR. Freezing red blood cells prepared for quality control of antiglobulin sera. Transfusion 21:498-501, 1981.

184. Vecchione JJ, Chomicz SM, Emerson CP, Valeri CR. Enumeration of previously frozen platelets using the Coulter counter, phase microscopy, and the Technicon optical system. Transfusion 21:511-516, 1981.

185. Valeri CR, Valeri DA, Gray A, Melaragno AJ, Vecchione JJ, Dennis RC, Emerson CP. Cryopreserved red blood cells for pediatric transfusion: Frozen storage of small aliquots in polyvinyl chloride (PVC) plastic bags. Transfusion 21:527-536, 1981.

186. Kurtz SR, McMican A, Carciero R, Melaragno AJ, Abdu W, Katchis R, Valeri CR. Plateletpheresis experience with the Haemonetics Blood Processor 30, the IBM Blood Processor 2997, and the Fenwal CS-3000 Blood Processor. Vox Sang 41:212-218. 1981.

187. Kuehl GV, Harkness DR, Skrabut EM, Bechthold DA, Emerson CP, Valeri CR. In vitro interactions of 51Cr in human red blood cells and hemolysates. Vox Sang 40:260-272, 1981.

188. Valeri CR, Kuehl GV, Skrabut EM, Bechthold DA, Vecchione JJ, Harkness DR, Emerson CP. Studies on the in vivo elution of ^{51}Cr from baboon red blood cells. Vox Sang 40:338-345, 1981.

189. Valeri CR, Lindberg JR, Contreras TJ, Pivacek LE, Austin RM, Valeri DA, Gray A, Emerson CP. Liquid preservation of baboon red blood cells in acid-citrate-dextrose or citrate-phosphate-dextrose anticoagulant: Effects of washing liquid-stored red blood cells. Am J Vet Res 42:1011-1013, 1981.

190. Valeri CR, Lindberg JR, Contreras TJ, Pivacek LE, Austin RM, Valeri DA, Gray A, Emerson CP. Measurement of red blood cell volume, plasma volume, and total blood volume in baboons. Am J Vet Res 42:1025-1029, 1981.

191. Valeri CR, Lindberg JR, Contreras TJ, Lowrie GB, Pivacek LE, Gray A, Valeri DA, Emerson CP. Freeze-preserved baboon red blood cells: Effects of biochemical modification and perfusion in vitro. Am J Vet Res 42:1590-1594, 1981.

192. Valeri CR. Optimal use of blood products in the treatment of hemorrhagic shock. Surgical Rounds 4:38-46, 1981.

193. Valeri CR. Blood products and resuscitative solutions to treat hemorrhagic shock. Crit Care Monitor, 1981.

194. Costa JL, Dobson CM, Fay DD, Kirk KL, Poulsen FM, Valeri CR, Vecchione JJ. Nuclear magnetic resonance studies of amine storage in pig platelets. FEBS Letters 136:325-328, 1981.

195. LoGerfo FW, Crawshaw HM, Nowak M, Serrallach E, Quist WC, Valeri CR. Effect of flow split on separation and stagnation in a model vascular bifurcation. Stroke 12:660-665, 1981.

196. Melaragno AJ, Abdu W, Katchis R, Doty A, Valeri CR. Liquid and freeze preservation of baboon platelets. Cryobiology 18:445-452, 1981.

197. Arnaout MA, Luscinskas FW, Lionetti FJ, Alper CA, Valeri CR.

Alternative complement pathway-dependent ingestion of fluolite particles by human granulocytes. J Immunol 127:278-281, 1981.

198. Valeri CR. Status report on blood products and resuscitative solutions to treat hemorrhagic shock. Med Bull US Army, Europe 38:13-17, 1981.

199. Hunt SM, Lionetti FJ, Valeri CR, Callahan AB. Cryogenic preservation of monocytes from human blood and plateletpheresis cellular residues. Blood 57:592-598, 1981.

200. Krausz MM, Dennis RC, Utsunomiya T, Grindlinger GA, Vegas AM, Churchill WH Jr, Mannick JA, Valeri CR, Hechtman HB. Cardiopulmonary function following transfusion of three red blood cell products in elective abdominal aortic aneurysmectomy. Ann Surg 194:616-624, 1981.

201. Thompson CB, Quinn PG, Valeri CR, Catsimpoolas N. Sedimentation behavior of activated human granulocytes: Aggregation and volume effects. Cell Biophysics 3:279-288, 1981.

202. Utsunomiya T, Krausz MM, Valeri CR, Shepro D, Hechtman HB. Treatment of aspiration pneumonia with ibuprofen and prostacyclin. Surgery 90:170-176, 1981.

203. Utsunomiya T, Krausz MM, Valeri CR, Levine L, Shepro D, Hechtman HB. Treatment of pulmonary embolism with positive end-expiratory pressure and prostaglandin E_1. Surg Gynec Obstet 153:161-168, 1981.

204. Valeri CR. Status report on rejuvenation and freezing of red blood cells. Plasma Therapy 2:155-170, 1981.

205. Kurtz SR, Skrabut EM, Catsimpoolas N, Valeri CR. 51Cr and $DF^{32}P$ labeling of human blood cells in leukocyte-rich plasma. Exp Hemat 9:957-965, 1981.

206. Valeri CR. The current state of platelet and granulocyte cryopreservation. Crit Rev Clin Lab Sci 14(1):21-74, 1981.

207. Valeri CR. New developments in red blood cell preservation using liquid and freezing procedures. J Clin Lab Automation 2:425-439, 1982.

208. Kurtz SR, Kuszaj T, Ouellet R, Valeri CR. Survival of homozygous Co^{a}(Colton) red cells in a patient with anti-Co^{a1}. Vox Sang 43:28-30, 1982.

209. Kurtz SR, Valeri DA, Gray A, Lindberg JR, McMican A,

Blumberg N, Valeri CR. A new approach to washing red blood cells frozen with a high concentration of glycerol in a special freezing container. Vox Sang 43:132-137, 1982.

210. Melaragno AJ, Abdu WA, Katchis RJ, Vecchione JJ, Valeri Cr. Cryopreservation of platelets isolated with the IBM 2997 Blood Cell Separator: A rapid and simplified approach. Vox Sang 43:321-326, 1982.

211. Justice RE, Utsunomiya T, Krausz MM, Valeri CR, Shepro D, Hechtman HB. A miniaturized chamber for the measure of oxygen consumption. J Appl Physiol 52:488-490, 1982.

212. Thompson CB, Eaton KA, Princiotta SM, Rushin CA, Valeri CR. Size dependent platelet subpopulations: relationship of platelet volume to ultrastructure, enzymatic activity, and function. Brit J Haemat 50:509-519, 1982.

213. Utsunomiya T, Krausz MM, Dunham B, Valeri CR, Levine L, Shepro D, Hechtman HB. Modification of inflammatory response to aspiration with ibuprofen. Am J Physiol 243:H903-H910, 1982.

214. Callow AD, Connolly R, O'Donnell TF Jr, Gembarowicz R, Keough E, Ramberg-Laskaris K, Valeri CR. Platelet-arterial synthetic graft interaction and its modification. Arch Surg 117:1447-1455, 1982.

215. Krausz MM, Utsunomiya T, Dunham B, Valeri CR, Shepro D, Hechtman HB. Inhibition of permeability edema with imidazole. Surgery 92:299-308, 1982.

216. Vecchione JJ, Melaragno AJ, Weiblen BJ, Halkett JAE, Callow AD, Valeri CR. Repeated intravenous administrations of human albumin and human fibrinogen in the baboon: Survival measurements. J Med Primatol 11:91-105, 1982.

217. Valeri CR, Valeri DA, Gray A, Melaragno AJ, Dennis RC, Emerson CP. Red blood cell concentrates stored at 4 C for 35 days in CPDA-1, CPDA-2, or CPDA-3 anticoagulant-preservative, biochemically modified, and frozen and stored in the polyvinyl chloride plastic primary collection bag with 40% W/V glycerol at -80 C. Transfusion 22:102-106, 1982.

218. Vecchione JJ, Melaragno AJ, Hollander A, Defina S, Emerson CP, Valeri CR. Circulation and function of human platelets isolated from units of CPDA-1, CPDA-2, and CPDA-3 anticoagulated blood and frozen with DMSO. Transfusion 22:206-209, 1982.

219. Valeri CR, Valeri DA, Gray A, Melaragno AJ, Dennis RC, Emerson CP. Viability and function of red blood cell concentrates stored at 4 C for 35 days in CPDA-1, CPDA-2, or CPDA-3. Transfusion 22:210-216, 1982.

220. Valeri CR. Use of low affinity red cells in the treatment of anemic hypoxic patients with normal or reduced arterial oxygen tensions at normothermic or hypothermic temperatures. The Physiologist 25:90, 1982.

221. Gembarowicz R, Connolly R, Callow AD, O'Donnell TF, Keough E, Schultz M, Ramberg-Laskaris K, Melaragno AJ, Valeri CR. Effect of PGI2 on the interaction of platelets with small caliber dacron grafts. Surg Forum 33:466-468, 1982.

222. Valeri CR. Use of rejuvenation solutions in blood preservation. Crit Rev Clin Lab Sci 17(4):299-374, 1982.

223. Luscinskas FW, Lionetti FJ, Melaragno AJ, Valeri CR. Long-term cryopreservation of dog granulocytes. Cryobiology 20:1-6, 1983.

224. Immerman KL, Melaragno AJ, Ouellet RP, Weinstein R, Valeri CR. Morphology of glutaraldehyde-fixed preserved red blood cells and 24-hr post-transfusion survival. Cryobiology 20:30-35, 1983.

225. Thompson CB, Jakubowski JA, Quinn PG, Deykin D, Valeri CR. Platelet size as a determinant of platelet function. J Lab Clin Med 101:205-213, 1983.

226. Weiblen BJ, Melaragno AJ, Catsimpoolas N, Valeri CR. Measurement of the distribution of Indium-111 on human plasma proteins using immunoprecipitation. J Immunol Methods 58:73-81, 1983.

227. Kurtz SR, Ouellet R, McMican A, Valeri CR. Survival of MM red blood cells during hypothermia in two patients with anti-M. Transfusion 23:37-39, 1983.

228. Valeri CR, Ouellet R, Mostacci J, Harmening-Pittiglio D, Vecchione JJ. Phosphate ion exchange resin used in the liquid preservation of baboon red blood cells. Transfusion 23:215-220, 1983.

229. Hunt SM, Lionetti FJ, Valeri CR. Isolation and cryopreservation of monocytes from plateletapheresis cellular residues. Transfusion 23:387-390, 1983.

230. Valeri CR, Sims KL, Bates JF, Reichman D, Lindberg JR, Wilson AC. An integrated liquid-frozen blood banking system. Vox Sang 45:25-39, 1983.
231. Bernhard WF, Clay W, Shoen FJ, Gernes D, Sherman C, Carr JG, Melaragno AJ, Burke D, Poirier VL, Valeri CR. Clinical and laboratory investigations related to temporary and permanent ventricular bypass. Heart Transplantation III:16-25, 1983.
232. Jakubowski JA, Thompson CB, Vaillancourt R, Valeri CR, Deykin D. Arachidonic acid metabolism by platelets of differing size. Brit J Haemat 53:503-511, 1983.
233. Valeri CR, Valeri DA, Gray A, Leavy PD, Contreras TJ, Lindberg JR. Rhesus macaque red blood cells frozen with 40% glycerol and stored at -80C. Am J Vet Res 44:1786-1788, 1983.
234. Huval WV, Dunham BM, Lelcuk S, Valeri CR, Shepro D, Hechtman HB. Thromboxane mediation of cardiovascular dysfunction following aspiration. Surgery 94:259-266, 1983.
235. Mathieson MA, Dunham BM, Huval WV, Lelcuk S, Stemp LI, Valeri CR, Shepro D, Hechtman HB. Ischemia of the limb stimulates thromboxane production and myocardial depression. Surg Gynec Obstet 157:500-504, 1983.
236. Hechtman HB, Huval WV, Mathieson MA, Stemp LI, Valeri CR, Shepro D. Prostaglandin and thromboxane mediation of cardiopulmonary failure. Surg Clin NA 63:263-283, 1983.
237. Skrabut EM, Catsimpoolas N, Kurtz SR, Griffith AL, Valeri CR. Alteration of the electrophoretic mobility of human peripheral blood mononuclear cells following treatment with dimethyl sulfoxide. Cryobiology 20:652-656, 1983.
238. Thompson CB, Diaz DD, Quinn PG, Lapins M, Kurtz SR, Valeri CR. The role of anticoagulation in the measurement of platelet volumes. Am J Clin Path 80:327-332, 1983.
239. Weiblen BJ, DeBell K, Valeri CR. "Acquired immunodeficiency" of blood stored overnight. NEJM 309:793, 1983.
240. Thompson CB, Love DG, Quinn PG, Valeri CR. Platelet size does not correlate with platelet age. Blood 62:487-494, 1983.
241. Valeri CR, Valeri DA, Gray A, Contreras TJ, Lindberg JR. Horse red blood cells frozen with 20% (W/V) glycerol and stored at -150C for five years. Am J Vet Res 44:2200-2202, 1983.
242. Krausz MM, Utsunomiya T, Valeri CR, Hechtman HB.

One- stage technique for chronic hemodynamic and lung lymphatic cannulation in sheep. Israeli J Med Sci 19:199-201, 1983.

243. Hesketh PJ, Sullivan R, Valeri CR, McCarroll LA. The production of granulocyte-monocyte colony-stimulating activity by isolated human T lymphocyte subpopulations. Blood 63:1141-1146, 1984.

244. Valeri CR, Gray AD, Cassidy GP, Riordan W, Pivacek LE. The 24-hour posttransfusion survival, oxygen transport function, and residual hemolysis of human outdated-rejuvenated red cell concentrates after washing and storage at 4 C for 24 to 72 hours. Transfusion 24:323-326, 1984.

245. Thompson CB, Jakubowski JA, Quinn PG, Deykin D, Valeri CR. Platelet size and age determine platelet function independently. Blood 63:1372-1375, 1984.

246. Alexander F, Mathieson M, Teoh KHT, Huval WV, Lelcuk S, Valeri CR, Shepro D, Hechtman HB. Arachidonic acid metabolites mediate early burn edema. J Trauma 24:709-712, 1984.

247. Lelcuk S, Huval WV, Valeri CR, Shepro D, Hechtman HB. Inhibition of ischemia-induced thromboxane synthesis in man. J Trauma 24:393-396, 1984.

248. Valeri CR, Pivacek LE, Ouellet R, Gray A. A comparison of methods of determining the 100 percent survival of preserved red cells. Transfusion 24:105-108, 1984.

249. Weiblen BJ, Debell K, Giorgio A, Valeri CR. Monoclonal antibody testing of lymphocytes after overnight storage. J Immunol Methods 70:179-183, 1984.

250. Mackey WC, Connolly RJ, Callow AD, Keough EM, Ramberg-Laskaris K, McCullough JL, O'Donnell TF, Melaragno AJ, Valeri CR, Weiblen BJ. Aspirin decreases platelet uptake on Dacron vascular grafts in baboons. Ann Surg 200:93-99, 1984.

251. Hechtman HB, Valeri CR, Shepro D. Role of humoral mediators in adult respiratory distress syndrome. Chest 86:623-627, 1984.

252. Dunham BM, Hechtman HB, Valeri CR, Shepro D. Antiinflammatory agents inhibit microvascular permeability induced by leukotrienes and by stimulated human neutrophils. Microcirculation, Endothelium, and Lymphatics 1:465-489, 1984.

253. Thompson CB, Galli R, Melaragno AJ, Valeri CR. A method for the separation of erythrocytes on the basis of size using counterflow centrifugation. Am J Hemat 17:177-183, 1984.

254. Bernhard WF, Gernes DG, Clay WC, Schoen FJ, Burgeson R, Valeri CR, Melaragno AJ, Poirier VL. Investigations with an implantable, electrically actuated ventricular assist device. J Thorac Cardiovasc Surg 88:11-21, 1984.
255. Jakubowski JA, Adler B, Thompson CB, Valeri CR, Deykin D. Influence of platelet volume on the ability of prostacyclin to inhibit platelet aggregation and the release reaction. J Lab Clin Med 105:271-276, 1985.
256. Apstein CS, Dennis RC, Briggs L, Vogel WM, Frazer J, Valeri CR. Effect of erythrocyte storage and oxyhemoglobin affinity changes on cardiac function. Am J Physiol 248 (Heart Circ Physiol 17): H508-H515, 1985.
257. Carciero R, Valeri CR. Isolation of mononuclear leukocytes in a plastic bag system using Ficoll-Hypaque. Vox Sang 49:373-380, 1985.
258. Valeri CR, Landrock RD, Pivacek LE, Gray AD, Fink JG, Szymanski IO. Quantitative differential agglutination method using the Coulter Counter to measure survival of compatible but identifiable red blood cells. Vox Sang 49:195-205, 1985.
259. Melaragno AJ, Carciero R, Feingold H, Talarico L, Weintraub L, Valeri CR. Cryopreservation of human platelets using 6% dimethylsulfoxide and storage at -80C. Effects of 2 years of frozen storage at -80C and transportation in dry ice. Vox Sang 49:245-258, 1985.
260. Valeri CR. Measurement of viable ADSOL-preserved human red blood cells. NEJM 312:377-378, 1985.
261. Szymanski IO, Odgren PR, Valeri CR. Relationship between the third component of human complement (C3) bound to stored preserved erythrocytes and their viability in vivo. Vox Sang 49:34-41, 1985.
262. Binder R, Stone PJ, Calore JD, Dunn DM, Snider GL, Franzblau C, Valeri CR. Serum antielastase and neutrophil elastase levels in PiM phenotpye cigarette smokers with airflow obstruction. Respiration 47:267-277, 1985.
263. Lelcuk S, Alexander F, Kobzik L, Valeri CR, Shepro D, Hechtman HB. Prostacyclin and thromboxane A2 moderate post-ischemic renal failure. Surgery 98:207-212, 1985.
264. Leluck S, Alexander F, Valeri CR, Shepro D, Hechtman HB.

Thromboxane A2 moderates permeability after limb ischemia. Ann Surg 202:642-646, 1985.

265. Valeri CR, Ellis A, Donahue K, Curran T, Pivacek L. The viability of young and old baboon red cells stored in the liquid state at 4 C. Prog Clin Biol Res 195:429-441, 1985.

266. Valeri CR. Cryopreservation of human platelets and bone marrow and peripheral blood totipotential mononuclear stem cells. Ann N Y Acad Sci 459:353-66, 1985.

267. Smith JM, Blue B, Clancy E, Valeri CR, Cohen RJ. Subtle altering electrocardiographic morphology as an indicator of decreased cardiac electrical stability. Comput Cardiol 12:109-12, 1985.

268. Valeri CR, Donahue K, Feingold HM, Cassidy GP, Altschule MD. Increase in plasma volume after the transfusion of washed erythrocytes. Surg Gynec Obstet 162:30-36, 1986.

269. Vogel WM, Dennis RC, Cassidy GP, Apstein CS, Valeri CR. Coronary constrictor effect of stroma-free hemoglobin solutions. Am J Physiol (Heart Circ Physiol 20) 251:H413-H420, 1986.

270. Eldrup-Jorgensen J, Connolly RJ, Mackey WC, Ramberg K, O'Donnell TF, Valeri CR, Callow AD. Antiplatelet therapy and vascular grafts: Studies in a baboon ex vivo shunt. Arch Surg 121:778-781, 1986.

271. Manny J, Manny N, Lelcuk S, Alexander F, Feingold H, Kobzik L, Valeri CR, Shepro D, Hechtman HB. Pulmonary and systemic consequences of localized acid aspiration. Surg Gynec Obstet 162:259-267, 1986.

272. Valeri CR, Feingold H, Melaragno AJ, Vecchione JJ. Cryopreservation of dog platelets with dimethylsulfoxide: Therapeutic effectiveness of cryopreserved platelets in the treatment of thrombocytopenic dogs, and the effect of platelet storage at -80 C. Cryobiology 23:387-394, 1986.

273. Feingold HM, Pivacek LE, Melaragno AJ, Valeri CR. Coagulation assays and platelet aggregation patterns in human, baboon and canine blood. Am J Vet Res 47:2197-2199, 1986.

274. Black HM, Shoenfeld NA, O'Donnell TF Jr, Valeri CR, Connolly R, Callow AD. Urinary thromboxane B2 levels in the diagnosis of mesenteric ischemia. Surg Forum 37:146-148, 1986.

275. Lelcuk S, Threlfall L, Valeri CR, Shepro D, Hechtman HB.

Nicotine stimulates pulmonary parenchymal thromboxane synthesis. Surgery 100:836-840, 1986.

276. Shoenfeld NA, Eldrup-Jorgensen J, Connolly R, Callow AD, Valeri CR, Ramberg K, Mackey WC, O'Donnell TF. The effect of low molecular weight dextran on platelet deposition onto prosthetic materials. J Vasc Surg 5:76-82, 1987.

277. Lieberthal W, Vasilevsky ML, Valeri CR, Levinsky NG. Interactions between ADH and prostaglandins in isolated erythrocyte-perfused rat kidney. Am J Physiol 252:F331-F337, 1987.

278. Kaufman RP Jr, Anner H, Kobzik L, Valeri CR, Shepro D, Hechtman HB. Vasodilator prostaglandins (PG) prevent renal damage after ischemia. Ann Surg 205:195-198, 1987.

279. Wilentz JR, Sanborn TA, Haudenschild CC, Valeri CR, Ryan TJ, Faxon DP. Platelet accumulation in experimental angioplasty: time course and relation to vascular injury. Circulation 75:636-642, 1987.

280. Crowley JP, Dennis RC, Pivacek L, Metzger J, Carvalho A, Valeri CR. Effects of granulocytopenia on the hemodynamic responses of dogs during E. coli bacteremia. Circ Shock 22:91-104, 1987.

281. Lieberthal W, Wolf EF, Merrill EW, Levinsky NG, Valeri CR. Hemodynamic effects of different preparations of stroma free hemolysates in the isolated perfused rat kidney. Life Science 41:2525-2533, 1987.

282. Valeri CR, Feingold H, Cassidy G, Ragno G, Khuri S, Altschule MD. Hypothermia-induced reversible platelet dysfunction. Ann Surg 205:175-181, 1987.

283. Lieberthal W, Stephens GW, Wolf EF, Rennke HG, Vasilevsky ML, Valeri CR, Levinsky NG. Effect of erythrocytes on the function and morphology of the isolated perfused rat kidney. Renal Physiol 10:14-24, 1987.

284. Vogel WM, Hsia JC, Briggs LL, Er SS, Cassidy G, Apstein CS, Valeri CR. Reduced coronary vasoconstrictor activity of hemoglobin solutions purified by ATP-agarose affinity chromatography. Life Sciences 41:89-93, 1987.

285. Muza SR, Sawka MN, Young AJ, Dennis RC, Gonzalez RR, Martin JW, Pandolf KB, Valeri CR. Elite special forces: Physiological description and ergogenic influence of blood reinfusion. Aviat Space Environ Med 58:1001-1004, 1987.

286. Sawka MN, Dennis RC, Gonzalez RR, Young AJ, Muza SR, Martin JW, Wenger CB, Francesconi RP, Pandolf KB, Valeri CR. Influence of polycythemia on blood volume and thermoregulation during exercise-heat stress. J Appl Physiol 62:912-918, 1987.

287. Anner H, Kaufman RP Jr, Kobzik L, Valeri CR, Shepro D, Hechtman HB. Pulmonary hypertension and leukosequestration after lower torso ischemia. Ann Surg 206:642-648, 1987.

288. Anner H, Kaufman RP Jr, Kobzik L, Valeri CR, Shepro D, Hechtman HB. Pulmonary leukosequestration induced by hind limb ischemia. Ann Surg 206:162-167, 1987.

289. Klausner JM, Schoof DD, Kobzik LM, Morel N, Highum D, Valeri CR, Eberlein TJ, Shepro D, Hechtman HB. Interleukin-2 leads to lung lymphosequestration, thromboxane generation, and rapid increase in permeability. Surgical Forum 38:428-430, 1987.

290. Klausner JM, Kobzik LM, Valeri CR, Shepro D, Hechtman HB. Selective lung leukosequestration following complement activation. Surgical Forum 38:304-307, 1987.

291. Kaufman RP Jr, Anner H, Kobzik L, Valeri CR, Shepro D, Hechtman HB. A high plasma prostaglandin to thromboxane ratio protects against renal ischemia. Surg Gynec Obstet 165:404-409, 1987.

292. Pitt AM, O'Connor JL, Valeri CR, Castino F. Assessment of blood interactions with the Therapore Apheresis System. Life Support Syst 5(4):347-52, 1987.

293. Crowley JP, Ragosta A, Homans AC, Valeri CR. A solid phase urease-linked cellular immunosorbent assay for circulating polymorphonuclear binding immunoglobulin. Ann Clin Lab Sci 17(5):306-11, 1987.

294. Francesconi RP, Sawka MN, Dennis RC, Gonzalez RR, Young AJ, Valeri CR. Autologous red blood cell reinfusion: Effects on stress and fluid regulatory hormones during exercise in the heat. Aviat Space Environ Med 59:133-137, 1988.

295. O'Murchadha ET, Horland A, Neiman RS, Valeri CR. Morphology of cells grown in the CFU-GEMM tissue culture assay from mononuclear cells obtained from peripheral blood and bone marrow of normal volunteers. Exp Hematol 16:235-239, 1988.

296. Kaufman RP Jr, Klausner JM, Anner H, Feingold H, Kobzik L, Valeri CR, Shepro D, Hechtman HB. Inhibition of thromboxane (Tx) synthesis by free radical scavengers. J Trauma 28:458-464, 1988.

297. Shoenfeld NA, Connolly R, Ramberg K, Valeri CR, Eldrup-Jorgensen J, Callow AD. The systemic activation of platelets by Dacron® grafts. Surg Gynec Obstet 166:454-457, 1988.

298. Klausner JM, Paterson IS, Valeri CR, Shepro D, Hechtman HB. Limb ischemia-induced increase in permeability is mediated by leukocytes and leukotrienes. Ann Surg 208:775-760, 1988.

299. Klausner JM, Anner H, Patterson IS, Kobzik L, Valeri CR, Shepro D, Hechtman HB. Lower torso ischemia-induced lung injury is leukocyte dependent. Ann Surg 208:761-767, 1988.

300. Blumenstock FA, Valeri CR, Saba TM, Cho E, Melaragno A, Gray A, Lewis M. Progressive loss of fibronectin-mediated opsonic activity in plasma cryoprecipitate with storage: Role of fibronectin fragmentation. Vox Sang 54:129-137, 1988.

301. Valeri CR, Pivacek LE, Palter M, Dennis RC, Yeston N, Emerson CP, Altschule MD. A clinical experience with ADSOL® preserved erythrocytes. Surg Gynec Obstet 166:33-46, 1988.

302. Moore FD Jr, Warner KG, Assousa S, Valeri CR, Khuri SF. The effects of complement activation during cardiopulmonary bypass. Attenuation by hypothermia, heparin, and hemodilution. Ann Surg 208:95-103, 1988.

303. Wright JG, Kerr JC, Valeri CR, Hobson RW II. Endothelial permeability to iodine-125-labeled albumin predicts skeletal muscle injury after ischemia reperfusion. Current Surgery 45:25-27, 1988.

304. Wright JG, Kerr JC, Valeri CR, Hobson RW II. Heparin decreases ischemia-reperfusion injury in isolated canine gracilis model. Arch Surg 123:470-472, 1988.

305. Smith JM, Clancy EA, Valeri CR, Ruskin JN, Cohen RJ. Electrical alternans and cardiac electrical instability. Circulation 77:110-121, 1988.

306. Klausner JM, Valeri CR, Shepro D, Hechtman HB. Peripheral ischemia: Part 1. Local consequences. Hospital Therapy 13:73-89, 1988.

307. Klausner JM, Kobzik L, Valeri CR, Shepro D, Hechtman HB.

Selective lung leukosequestration after complement activation. J Appl Physiol 65:80-88, 1988.

308. Crowley JP, Metzger J, Pivacek L, Dennis RC, Valeri CR. Effects of plasma administration on gram negative shock in granulocytopenic dogs. Circulatory Shock 26:287-295, 1988.

309. Shoenfeld NA, Yeager A, Connolly R, Ramberg K, Forgione L, Giorgio A, Valeri CR, Callow AD. A new primate model for the study of intravenous thrombotic potential and its modification. J Vasc Surg 8:49-54, 1988.

310. Lieberthal W, Rennke HG, Sandock KM, Valeri CR, Levinsky NG. Ischemia in the isolated erythrocyte-perfused rat kidney. Protective effect of hypothermia. Renal Physiol Biochem 11:60-69, 1988.

311. Sawka MN, Gonzalez RR, Young AJ, Muza SR, Pandolf KB, Latzka WA, Dennis RC, Valeri CR. Polycythemia and hydration: effects on thermoregulation and blood volume during exercise-heat stress. Am J Physiol 255 (Regulatory Integrative Comp Physiol 24):R456-463, 1988.

312. Wright JG, Kerr JC, Valeri CR, Hobson RW II. Regional hypothermia protects against ischemia-reperfusion injury in isolated canine gracilis muscle. J Trauma 28:1026-1031, 1988.

313. Wright JG, Fox D, Kerr JC, Valeri CR, Hobson RW II. Rate of reperfusion blood flow modulates reperfusion injury in skeletal muscle. J Surg Res 44:754-763, 1988.

314. Stephens GW, Lieberthal W, Oza NB, Valeri CR, Levinsky NG. Vasopressin stimulates urinary kallikrein excretion in the isolated erythrocyte-perfused rat kidney. Renal Physiol Biochem 11:50-59, 1988.

315. Franco RS, Weiner M, Wagner K, Martelo OJ, Ragno G, Pivacek LE, Valeri CR. The 24-hour posttransfusion survival and lifespan of autologous baboon red cells treated with inositol hexaphosphate-polyethylene glycol or inositol hexaphosphate-adenosine triphosphate-polyethylene glycol to decrease oxygen affinity. Vox Sang 55:90-96, 1988.

316. Huval WV, Lelcuk S, Feingold H, Valeri CR, Shepro D, Hechtman HB. Effects of nitroprusside and ketanserin upon pulmonary edema after acid injury. Surg Gynec Obstet 166:527-534, 1988.

317. Anner H, Kaufman RP Jr, Valeri CR, Shepro D, Hechtman HB. Reperfusion of ischemic lower limbs increases pulmonary microvascular permeability. J Trauma 28:607-610, 1988.
318. Katzir A, Bowman F, Asfour Y, Zur A, Valeri CR. Infrared fiber optics for thermometry in microwave. Appl Phys Lett 53:1877-1879, 1988.
319. Katzir A, Bowman F Asfour Y, Zur A, Valeri CR. Infrared fiber radiometer for thermometry in electromagnetic induced therapeutic healing. SPIE, Optical Fibers in Medicine III, Vol 906, pps 120-123, 1988.
320. Vogel WM, Lieberthal W, Apstein CS, Levinsky N, Valeri CR. Effects of stroma-free hemoglobin solutions on isolated perfused rabbit hearts and isolated perfused rat kidneys. Biomater Artif Cells Artif Organs 16(1-3):227-35, 1988.
321. Lieberthal W, Wolf EF, Rennke HG, Valeri CR, Levinsky NG. Renal ischemia and reperfusion impair endothelium-dependent vascular relaxation. Am J Physiol 256 (Renal Fluid Electrolyte Physiol 25):F894-F900, 1989.
322. Sawka MN, Young AJ, Dennis RC, Gonzalez RR, Pandolf KB, Valeri CR. Human intravascular immunoglobulin responses to exercise-heat and hypohydration. Aviat Space Environ Med 60:634-638, 1989.
323. Klausner JM, Morel N, Paterson IS, Kobzik L, Valeri CR, Eberlein TJ, Shepro D, Hechtman HB. The rapid induction by interleukin-2 of pulmonary microvascular permeability. Ann Surg 209:119-128, 1989.
324. Valeri CR, Pivacek LE, Gray AD, Cassidy GP, Leavy ME, Dennis RC, Melaragno AJ, Niehoff J, Yeston N, Emerson CP, Altschule MD. The safety and therapeutic effectiveness of human red cells stored at -80C for as long as 21 years. Transfusion 29:429-437, 1989.
325. Klausner JM, Paterson IS, Morel NML, Goldman G, Gray AD, Valeri CR, Eberlein TJ, Shepro D, Hechtman HB. Role of thromboxane in interleukin 2-induced lung injury in sheep. Cancer Research 49:3542-3549, 1989.
326. Katzir A, Bowman HF, Asfour Y, Zur A, Valeri CR. Infrared fibers for radiometer thermometry in hypothermia and hyperthermia treatment. IEEE Trans Biomed Eng 36:634-637, 1989.

327. Klausner JM, Paterson IS, Kobzik L, Valeri CR, Shepro D, Hechtman HB. Oxygen free radicals mediate ischemia-induced lung injury. Surgery 105:192-199, 1989.
328. Klausner JM, Paterson IS, Kobzik L, Rodzen C, Valeri CR, Shepro D, Hechtman HB. Vasodilating prostaglandins attenuate ischemic renal injury only if thromboxane is inhibited. Ann Surg 209:219-224, 1989.
329. Valeri CR. Effects of preservation on the quality of baboon red blood cells, platelets, and plasma proteins. J Invest Surg 2:223-226, 1989.
330. Sawka MN, Gonzalez RR, Young AJ, Dennis RC, Valeri CR, Pandolf KB. Control of thermoregulatory sweating during exercise in the heat. Am J Physiol 257: (Regulatory Integrative Comp Physiol 26): R311-R316, 1989.
331. Paterson IS, Klausner JM, Goldman G, Pugatch R, Feingold H, Allen P, Mannick JA, Valeri CR, Shepro D, Hechtman HB. Pulmonary edema after aneurysm surgery is modified by mannitol. Ann Surg 210:796-801, 1989.
332. Klausner JM, Paterson IS, Goldman G, Kobzik L, Rodzen C, Lawrence R, Valeri CR, Shepro D, Hechtman HB. Postischemic renal injury is mediated by neutrophils and leukotrienes. Am J Physiol 256:F794-F802, 1989.
333. Zur A, Yekuel A, Drizlikh S, Bowman HF, Katzir A, Valeri CR. Improved infrared fiberoptic radiometer for thermometry in electromagnetic induced heating. SPIE, Optical Fibers in Medicine, 1067:75-82, 1989.
334. Klausner JM, Paterson IS, Kobzik L, Valeri CR, Shepro D, Hechtman HB. Leukotrienes but not complement mediate limb ischemia-induced lung injury. Ann Surg 209:462-470, 1989.
335. Klausner JM, Paterson IS, Goldman G, Kobzik L, Valeri CR, Shepro D, Hechtman HB. Thromboxane A_2 mediates increased pulmonary microvascular permeability following limb ischemia. Circ Res 64:1178-1189, 1989.
336. Klausner JM, Paterson IS, Mannick JA, Valeri CR, Shepro D, Hechtman HB. Reperfusion pulmonary edema. JAMA 261:1030-1035, 1989.
337. Paterson IS, Klausner JM, Goldman G, Kobzik L, Welbourn R, Valeri CR, Shepro D, Hechtman HB. Thromboxane mediates the

ischemia-induced neutrophil oxidative burst. Surgery 106:224-229, 1989.

338. Lieberthal W, Vogel WM, Apstein CS, Valeri CR: Studies of the mechanism of the vasoconstrictor activity of stroma-free hemoglobin in the isolated perfused rat kidney and rabbit heart. Prog Clin Biol Res 319:407-420, 1989.

339. Khabbaz KR, Krisanda J, Wolfe JA, Axford TC, Dearani JA, Marquardt C, Neuringer L, Valeri CR, Khuri SF. Simultaneous in vivo measurements of intracellular and extracellular myocardial pH during repeated episodes of ischemia. Current Surgery 46:399-400, 1989.

340. Belkin M, Valeri CR, Hobson RW II. Intraarterial urokinase increases skeletal muscle viability after acute ischemia. J Vasc Surg 9:161-168, 1989.

341. Goldman G, Welbourn R, Klausner JM, Paterson IS, Kobzik L, Valeri CR, Shepro D, Hechtman HB. Ischemia activates neutrophils but inhibits their local and remote diapedesis. Ann Surg 211:196-201, 1990.

342. Lieberthal W, Sheridan A, Valeri CR. Protective effect of atrial natriuretic factor and mannitol following renal ischemia. Am J Physiol (Renal Fluid and Electrolyte Physiol): 258:R1266-R1272, 1990.

343. McKenney J, Valeri CR, Mohandas N, Fortier N, Giorgio A, Snyder LM. Decreased in vivo survival of hydrogen peroxide-damaged baboon red blood cells. Blood 76:206-211, 1990.

344. Welbourn R, Goldman G, Kobzik L, Valeri CR, Shepro D, Hechtman HB. Involvement of thromboxane and neutrophils in multiple-system organ edema with interleukin-2. Ann Surg 212:728-733, 1990.

345. Welbourn R, Goldman G, Kobzik L, Paterson I, Valeri CR, Shepro D, Hechtman HB. Neutrophil adherence receptors (CD 18) in ischemia. Dissociation between quantitative cell surface expression and diapedesis mediated by leukotriene B_4. J Immunol 145:1906-1911, 1990.

346. Goldman G, Welbourn R, Klausner JM, Alexander S, Kobzik L, Valeri CR, Shepro D, Hechtman HB. Attenuation of acid aspiration edema with phalloidin. Am J Physiol 259:L378-L383, 1990.

347. Jemmott JB 3rd, Hellman C, McClelland DC, Locke SE, Kraus L, Williams RM, Valeri CR. Motivational syndromes associated with natural killer cell activity. J Behav Med 13:53-73, 1990.
348. Crowley JP, Valeri CR, Metzger JB, Pono L, Chazan JA. Lymphocyte subpopulations in long-term dialysis patients: a case-controlled study of the effects of blood transfusion. Transfusion 30:644-647, 1990.
349. Goldman G, Welbourn R, Paterson IS, Klausner JM, Kobzik L, Valeri CR, Shepro D, Hechtman HB. Ischemia-induced neutrophil activation and diapedesis is lipoxygenase dependent. Surgery 107:428-433, 1990.
350. Goldman G, Welbourn R, Kobzik L, Valeri CR, Shepro D, Hechtman HB. Tumor necrosis factor- mediates acid aspiration-induced systemic organ injury. Ann Surg 212:513-520, 1990.
351. Gray AD, Akin MT, McLean R, Valeri CR. Evaluation of the Quantitative Buffy Coat (QBC) method to detect malaria-infected red blood cells. Military Medicine 156:241-245, 1991.
352. Welbourn R, Goldman G, Kobzik L, Valeri CR, Hechtman HB, Shepro D. Attenuation of IL-2-induced multisystem organ edema by phalloidin and antamanide. J Appl Physiol 70:1364-1368, 1991.
353. Goldman G, Welbourn R, Alexander S, Klausner JM, Wiles M, Valeri CR, Shepro D, Hechtman HB. Modulation of pulmonary permeability in vivo with agents that affect the cytoskeleton. Surgery 109:533-538, 1991.
354. Goldman G, Welbourn R, Klausner JM, Valeri CR, Shepro D, Hechtman HB. Thromboxane mediates diapedesis after ischemia by activation of neutrophil adhesion receptors interacting with basally expressed intercellular adhesion molecule-1. Circ Res 68:1013-1019, 1991.
355. Klausner JM, Paterson IS, Goldman G, Kobzik L, Lelcuk S, Skornick Y, Eberlein T, Valeri CR, Shepro D, Hechtman HB. Interleukin-2-induced lung injury is mediated by oxygen free radicals. Surgery 109:169-175, 1991.
356. Goldman G, Welbourn R, Klausner JM, Kobzik L, Valeri CR, Shepro D, Hechtman HB. Intravascular chemoattractants inhibit diapedesis by selective receptor occupancy. Am J Physiol 260:H465-H472, 1991.

357. Valeri CR, MacGregor H, Pompei F, Khuri SF. Acquired abnormalities of platelet function. Letter to the Editor. NEJM 324:1670, 1991.
358. Goldman G, Welbourn R, Valeri CR, Shepro D, Hechtman HB. Thromboxane A_2 induces leukotriene B_4 synthesis that in turn mediates neutrophil diapedesis via CD 18 activation. Microvasc Res 41:367-375, 1991.
359. Welbourn CRB, Goldman G, Paterson IS, Valeri CR, Shepro D, Hechtman HB. Pathophysiology of ischaemia reperfusion injury: central role of the neutrophil. Br J Surg 78:651-655, 1991.
360. Fisher JB, Dennis RC, Valeri CR, Woodson J, Doyle JE, Walsh LM, Pivacek L, Giorgio A, LaMorte WW, Menzoian JO. Effect of graft material on loss of erythrocytes after aortic operations. Surg Gynec Obstet 173:131-136, 1991.
361. Goldman G, Welbourn R, Klausner JM, Kobzik L, Valeri CR, Shepro D, Hechtman HB. Neutrophil accumulations due to pulmonary thromboxane synthesis mediate acid aspiration injury. J Appl Physiol 70:1511-1517, 1991.
362. Welbourn R, Goldman G, O'Riordain M, Lindsay TF, Paterson IS, Kobzik L, Valeri CR, Shepro D, Hechtman HB. Role for tumor necrosis factor as mediator of lung injury following lower torso ischemia. J Appl Physiol 70:2645-2649, 1991.
363. Lieberthal W, McGarry AE, Sheils J, Valeri CR. Nitric oxide inhibition in rats improves blood pressure and renal function during hypovolemic shock. Am J Physiol 261:F868-F872, 1991.
364. Crowley JP, Valeri CR, Metzger J, Gray A, Schooneman F, Man NK, Merrill E: The estimation of whole blood viscosity by a porous bed method. Am J Clin Pathol 96:729-737, 1991.
365. Faris PM, Ritter MA, Keating EM, Valeri CR. Unwashed filtered shed blood collected after knee and hip arthroplasties: A source of autologous red blood cells. J Bone & Joint Surg 73A:1169-1178, 1991.
366. Welbourn CRB, Goldman G, Paterson IS, Valeri CR, Shepro D, Hechtman HB. Neutrophil elastase and oxygen radicals: synergism in lung injury after hindlimb ischemia. Am J Physiol (Heart Circ Physiol 29)260H:1851-1856, 1991.
367. Ichikura T, Tamakuma S, Ito H, Tomimatsu S, Valeri CR. [Effects of syngeneic preserved blood cells on metastatic growth of the

Lewis lung carcinoma.] Nippon Geka Gakkai Zasshi 92(6):734-9, 1991 (Japanese).

368. Crowley JP, Valeri CR, Metzger JB, Pono L, Chazan J. Hemoglobin and 2,3-diphosphoglycerate levels in transfused dialysis patients with myocardial infarction. Ann Clin Lab Science 23:11-17, 1992.

369. Crowley JP, Metzger J, Pivacek, L, Valeri CR. The effect of viable and nonviable autologous red blood cell transfusions on experimental bacteremia. Circ Shock 36:31-37, 1992.

370. Horowitz B, Rywkin S, Margolis-Nunno H, Williams, B, Geacintov N, Prince AM, Pascual D, Ragno G, Valeri CR, Huima-Byron T. Inactivation of viruses in red cell and platelet concentrates with aluminum phthalocyanine (AlPc) sulfonates. Blood Cells 18:141-150, 1992.

371. Sawka MN, Young AJ, Pandolf KB, Dennis RC, Valeri CR. Erythrocyte, plasma, and blood volume of healthy young men. Med Sci Sports Exerc 24:447-453, 1992.

372. Stamler JS, Jaraki O, Osborne J, Simon DI, Keaney J, Vita J, Singel D, Valeri CR, Loscalzo J. Nitric oxide circulates in mammalian plasma primarily as an S-nitroso adduct of serum albumin. Proc Natl Acad Sci USA 89:7674-7677, 1992.

373. Crowley JP, Metzger JB, Merrill EW, Valeri CR. Whole blood viscosity in beta thalassemia minor. Ann Clin Lab Sci 22:229-235, 1992.

374. Axford TC, Dearani JA, Khait I, Park WM, Patel MA, Doursounian M, Neuringer L, Valeri CR, Khuri SF. Electrode-derived myocardial pH measurements reflect intracellular myocardial metabolism assessed by phosphorus 31-nuclear magnetic resonance spectroscopy during normothermic ischemia. J Thorac Cardiovasc Surg 103:902-907, 1992.

375. Khuri SF, Wolfe JA, Josa M, Axford TC, Szymanski I, Assousa S, Ragno G, Patel M, Silverman A, Park M, Valeri CR. Hematologic changes during and after cardiopulmonary bypass and their relationship to the bleeding time and nonsurgical blood loss. J Thorac Cardiovasc Surg 104:94-107, 1992.

376. Valeri CR, Khabbaz K, Khuri SF, Marquardt C, Ragno G, Feingold H, Gray AD, Axford T. Effect of skin temperature on

platelet function in patients undergoing extracorporeal bypass. J Thorac Cardiovasc Surgery 104:108-116, 1992.

377. Cordts PR, LaMorte WW, Fisher JB, DelGuercio C, Niehoff J, Pivacek LE, Dennis RC, Siebens H, Giorgio A, Valeri CR, Menzoian JO. Poor predictive value of hematocrit and hemodynamic parameters for erythrocyte deficits after extensive elective vascular operations. Surg Gynec Obstet 175:243-248, 1992.

378. Goldman G, Welbourn R, Rothlein R, Wiles M, Kobzik L, Valeri CR, Shepro D, Hechtman HB. Adherent neutrophils mediate permeability after atelectasis. Ann Surg 216:372-380, 1992.

379. Welbourn R, Goldman G, Kobzik L, Paterson IS, Valeri CR, Shepro D, Hechtman HB. Role of neutrophil adherence receptors (CD 18) in lung permeability following lower torso ischemia. Circ Res 71:82-86, 1992.

380. Goldman G, Welbourn R, Kobzik L, Valeri CR, Shepro D, Hechtman HB. Reactive oxygen species and elastase mediate lung permeability after acid aspiration. J Appl Physiol 73:571-575, 1992.

381. Goldman G, Welbourn R, Klausner JM, Kobzik L, Valeri CR, Shepro D, Hechtman HB. Mast cells and leukotrienes mediate neutrophil sequestration and lung edema after remote ischemia in rodents. Surgery 112:578-586, 1992.

382. Goldman G, Welbourn R, Kobzik L, Valeri CR, Shepro D, Hechtman HB. Synergism between leukotriene B_4 and thromboxane A_2 in mediating acid-aspiration injury. Surgery 111:55-61, 1992.

383. Goldman G, Welbourn R, Klausner JM, Valeri CR, Shepro D, Hechtman HB. Oxygen free radicals are required for ischemia-induced leukotriene B_4 synthesis and diapedesis. Surgery 111:287-293, 1992.

384. Hill J, Lindsay T, Rusche J, Valeri CR, Shepro D, Hechtman HB. A MAC-1 antibody reduces liver and lung injury but not neutrophil sequestration after intestinal ischemia-reperfusion. Surgery 112:166-172, 1992.

385. Lindsay TF, Hill J, Ortiz F, Rudolph A, Valeri CR, Hechtman HB, Moore Jr FD. Blockade of complement activation prevents

local and pulmonary albumin leak after lower torso ischemia-reperfusion. Ann Surg 216:677-683, 1992.

386. Hill J, Lindsay TF, Simpson R, Valeri CR, Shepro D, Hechtman HB. Disseminated intravascular coagulation but not fibrinolysis follows intestinal ischemia. Surgical Forum 43:42-43, 1992.
387. Dittmer J, Ichikura T, Pivacek LE, Giorgio A, Prusty S, Valeri CR. Intravascular circulation and distribution of human ^{51}Cr-DBBF stroma-free hemoglobin. Biomater Artif Cells Immobilization Biotechnol 20(2-4):751-5, 1992.
388. Vogel WM, Cassidy G, Valeri CR. Effects of o-raffinose-polymerized human hemoglobin on coronary tone and cardiac function in isolated hearts. Biomater Artif Cells Immobilization Biotechnol 20(2-4):673-7, 1992.
389. Lieberthal W, La Raia J, Valeri CR. Role of thromboxane in mediating the intrarenal vasoconstriction induced by unmodified stroma free hemoglobin in the isolated perfused rat kidney. Biomater Artif Cells Immobilization Biotechnol 20(2-4)663-7, 1992.
390. Blevins FT, Shaw B, Valeri CR, Kasser J, Hall J. Reinfusion of shed blood after orthopaedic procedures in children and adolescents. J Bone & Joint Surg 75-A:363-371, 1993.
391. Crowley JP, Chazan JA, Metzger JB, Pono L, Valeri CR. Blood rheology and 2,3-diphosphoglycerate levels after erythropoietin treatment. Ann Clin Lab Sci 23:24-32, 1993.
392. Kestin AS, Valeri CR, Khuri SF, Loscalzo J, Ellis PA, MacGregor H, Birjiniuk V, Ouimet H, Pasche B, Nelson MJ, Benoit SE, Rodino LJ, Barnard MR, Michelson AD. The platelet function defect of cardiopulmonary bypass. Blood 82:107-117, 1993.
393. Crowley JP, Metzger J, Gray A, Pivacek LE, Cassidy G, Valeri CR. Infusion of stroma-free cross-linked hemoglobin during acute gram-negative bacteremia. Circ Shock 41:144-149, 1993.
394. Young AJ, Sawka MN, Quigley MD, Cadarette BS, Neufer PD, Dennis RC, Valeri CR. Role of thermal factors on aerobic capacity improvements with endurance training. J Appl Physiol 75:49-54, 1993.
395. Merrill EW, Crowley JP, Valeri CR. Rapid and simple measurement of apparent whole blood viscosity. Lab Medica Int 10:19-24, 1993.

396. Healy WL, Wasilewski SA, Pfeifer BA, Kurtz SR, Hallack GN, Valerio M, Valeri CR. Methylmethacrylate monomer and fat content in shed blood after total joint arthroplasty. Clin Ortho Rel Res 286:15-17, 1993.

397. Simpson R, Alon R, Kobzik L, Valeri CR, Shepro D, Hechtman HB. Neutrophil and nonneutrophil-mediated injury in intestinal ischemia-reperfusion. Ann Surg 218:444-453, 1993.

398. Hill J, Lindsay T, Valeri CR, Shepro D, Hechtman HB. A CD18 antibody prevents lung injury but not hypotension after intestinal ischemia-reperfusion. J Appl Physiol 74:659-664, 1993.

399. Goldman G, Welbourn R, Klausner JM, Kobzik L, Valeri CR, Shepro D, Hechtman HB. Leukocytes mediate acid aspiration-induced multiorgan edema. Surgery 114:13-20, 1993.

400. Goldman G, Welbourn R, Kobzik L, Valeri CR, Shepro D, Hechtman HB. Lavage with leukotriene B_4 induces lung generation of tumor necrosis factor- that in turn mediates neutrophil diapedesis. Surgery 113:297-303, 1993.

401. Simpson R, Alon R, Valeri CR, Shepro D, Hechtman HB. Integrin dependent neutrophil adhesion following gut ischemia and reperfusion. Behring Inst Mitt 92:210-217, 1993.

402. Valeri CR. Transfusion medicine and surgical practice. Bull Am Coll Surg 78(9):19-24, 1993.

403. Healy WL, Pfeifer BA, Kurtz SR, Johnson C, Johnson W, Johnston R, Sanders D, Karpman R, Hallack GN, Valeri CR. Evaluation of autologous shed blood for autotransfusion after orthopaedic surgery. Clin Ortho Rel Res Number 99, pps 53-59, February 1994.

404. Axford TC, Dearani JA, Ragno G, MacGregor H, Patel MA, Valeri CR, Khuri SF. Safety and therapeutic effectiveness of reinfused shed blood after open heart surgery. Ann Thorac Surg 57:615-622, 1994.

405. Crowley JP, Metzger J, Assaf A, Carleton RC, Merrill E, Valeri CR. Low density lipoprotein cholesterol and whole blood viscosity. Ann Clin Lab Sci 24:533-541, 1994.

406. Michelson AD, MacGregor H, Barnard MR, Kestin AS, Rohrer MJ, Valeri CR. Reversible inhibition of human platelet activation by hypothermia in vivo and in vitro. J Thrombosis and Haemostasis 5:633-640, 1994.

407. Thompson A, McGarry AE, Valeri CR, Lieberthal W. Stroma free hemoglobin increases blood pressure and GFR in the hypotensive rat: role of nitric oxide. J Appl Physiol 77:2348-2354, 1994.

408. Rosenblatt MS, Hirsch EF, Valeri CR. Frozen red blood cells in combat casualty care: Clinical and logistical considerations. Milit Med 159:392-397, 1994

409. Weiser MR, Gibbs SAL, Kobzik L, Valeri CR, Shepro D, Hechtman HB. P-selectin mediates local reperfusion injury after lower torso ischemia. Surgical Forum 45:389-391, 1994.

410. Mitchell JD, Lee R, Hodakowski GT, Neya K, Harringer W, Valeri CR, Vlahakes GJ. Prevention of postoperative pericardial adhesions with a hyaluronic acid coating solution: Experimental safety and efficacy studies. J Thorac Cardiovasc Surg 107:1481-1488, 1994.

411. Goldman G, Welbourn R, Kobzik L, Valeri CR, Shepro D, Hechtman HB. Neutrophil adhesion receptor CD18 mediates remote but not localized acid aspiration injury. Surgery 17:83-89, 1995.

412. Rodriguez AA, Gardner GP, LaMorte WW, Obi-Tabot ET, Valeri CR, Hirsch EF. Comparison of skeletal muscle laser Doppler flowmetry to changes in central hemodynamics in detecting the physiological response to moderate hemorrhage. J Surg Res 58:189-192, 1995.

413. Kiefer CR, Trainor JF, McKenney JB, Valeri CR, Snyder LM. Hemoglobin-spectrin complexes: Interference with spectrin tetramer assembly as a mechanism for compartmentalization of Band 1 and Band 2 complexes. Blood 86:366-371, 1995.

414. Valeri CR, MacGregor H, Cassidy G, Tinney R, Pompei F. Effects of temperature on bleeding time and clotting time in normal male and female volunteers. Crit Care Med 23:698-704, 1995.

415. Martin D, Garcia J, Valeri CR, Khuri SF. The effects of normothermic and hypothermic cardiopulmonary bypass on defibrillation energy requirements and transmyocardial impedance. Implications for implantable cardioverter-defibrillator implantation. J Thorac Cardiovasc Surg 109:981-988, 1995.

416. Freund BJ, Montain SJ, Young AJ, Sawka MN, DeLuca JP, Pandolf KB, Valeri CR. Glycerol hyperhydration: Hormonal, renal and vascular fluid responses. J Appl Physiol 79:2069-2077, 1995.

417. Thompson A, Valeri CR, Lieberthal W. Endothelin receptor A blockade alters hemodynamic response to nitric oxide inhibition in rats. Am J Physiol 269 (Heart and Circ Physiol 38):H-743-H748, 1995.

418. Fast LD, Valeri CR, Crowley JP. Immune responses to major histocompatibility complex homozygous lymphoid cells in murine F_1 hybrid recipients: Implications for transfusion-associated graft-versus-host disease. Blood 86:3090-3096, 1995.

419. Khuri SF, Valeri CR, Loscalzo J, Weinstein MJ, Birjiniuk V, Healey NA, MacGregor H, Doursounian M, Zolkewitz MA. Heparin causes platelet dysfunction and induces fibrinolysis before cardiopulmonary bypass. Ann Thorac Surg 60:1008-14, 1995.

420. Freedman JE, Loscalzo J, Benoit SE, Valeri CR, Barnard MR, Michelson AD. Decreased platelet inhibition by nitric oxide in two brothers with a history of arterial thrombosis. J Clin Invest 97:979-987, 1996.

421. Valeri CR, Pivacek LE. Effects of the temperature, the duration of frozen storage, and the freezing container on in vitro measurements in human peripheral blood mononuclear cells. Transfusion 36:303-308, 1996.

422. Young AJ, Sawka MN, Muza SR, Boushel R, Lyons T, Rock PB, Freund BJ, Waters R, Cymerman A, Pandolf KB, Valeri CR. Effects of erythrocyte infusion on VO_2max at high altitude. J Appl Physiol 81:252-259, 1996.

423. Gibbs SAL, Weiser MR, Kobzik L, Valeri CR, Shepro D, Hechtman HB. P-selectin mediates intestinal ischemic injury by enhancing complement deposition. Surgery 119:652-656, 1996.

424. Weiser MR, Gibbs SAL, Valeri CR, Shepro D, Hechtman HB. Anti-selectin therapy modifies skeletal muscle ischemia and reperfusion injury. Shock 5:402-407, 1996.

425. Upchurch GR, Valeri CR, Khuri SF, Rohrer MJ, Welch GN, MacGregor H, Ragno G, Francis S, Rodino LJ, Michelson AD, Loscalzo J. Effect of heparin on fibrinolytic activity and platelet function in vivo. Am J Physiol 271 (Heart Circ Physiol 40): H528-H534. 1996.

426. Vander Salm TJ, Kaur S, Lancey RA, Okike ON, Pezzella AT, Stahl RF, Leone L, Li J-M, Valeri CR, Michelson AD. Reduction of bleeding after heart operations through the prophylactic use of

epsilon-aminocaproic acid. J Thorac Cardiovasc Surg 112:1098-1107, 1996.

427. Lieberthal W, Thompson A, Valeri CR. Effects of nitric oxide inhibition on systemic and renal hemodynamics in the hemorrhaged rat. Kidney Blood Res 19:340-346, 1996.

428. Michelson AD, Barnard MR, Hechtman HB, MacGregor H, Connolly RJ, Loscalzo J, Valeri CR. In vivo tracking of platelets: circulating degranulated platelets rapidly lose surface P-selectin but continue to circulate and function. Proc Natl Acad Sci USA 93:11877-11882, 1996.

429. Sawka MN, Young AJ, Rock PB, Lyons TP, Boushel R, Freund BJ, Muza SR, Cymerman A, Dennis RC, Pandolf KB, Valeri CR. Altitude acclimatization and blood volume: Effects of exogenous erythrocyte volume expansion. J Appl Physiol 81:636-642, 1996.

430. Valeri CR, Ragno G, MacGregor H, Pivacek LE. The effect of disinfection on viability and function of baboon red blood cells. Photochem Photobiol 65:446-450, 1997.

431. Waugh RE, McKenney JB, Bauserman RG, Brooks DM, Valeri CR, Snyder LM. Surface area and volume changes during maturation of reticulocytes in the circulation of the baboon. J Lab Clin Med 129:527-535, 1997.

432. Crowley JP, Metzger JB, Valeri CR. The volume of blood shed during the bleeding time correlates with the peripheral venous hematocrit. Am J Clin Pathol 108:579-584, 1997.

433. Montain SJ, Sawka MN, Latzka WA, Valeri CR. Thermal and cardiovascular strain from hypohydration: influence of exercise intensity. Int J Sports Med 19:87-91, 1998.

434. Furman MI, Benoit SE, Barnard MR, Valeri CR, Borbone ML, Becker RC, Hechtman HB, Michelson AD. Increased platelet reactivity and circulating monocyte-platelet aggregates in patients with stable coronary artery disease. J Am Coll Cardiol 31:352-358, 1998.

435. Valeri CR, Crowley JP, Loscalzo. The red cell transfusion trigger: has a sin of commission now become a sin of omission? Transfusion 38:602-610, 1998.

436. Kiefer CR, McKenney JB, Trainor JF, Lambrecht RW, Bonkovsky HL, Lifshitz LM, Valeri CR and Snyder LM. Porphyrin loading

of lipofuscin granules in inflamed striated muscle. Am J Path 153:703-708, 1998.

437. Peyton BD, Rohrer MJ, Furman MI, Barnard MR, Rodino LJ, Benoit SE, Hechtman HB, Valeri CR, Michelson AD. Patients with venous stasis ulceration have increased monocyte-platelet aggregation. J Vasc Surg 27:1109-1116, 1998.

438. Pandolf KB, Young AJ, Sawka MN, Kenney JL, Sharp MW, Cote RR, Freund BJ, Valeri CR. Does erythrocyte infusion improve 3.2-km run performance at high altitude? Europ J Appl Physiol Occup Physiol 79:1-6, 1998.

439. Michelson AD, Barnard MR, Khuri SF, Rohrer MJ, MacGregor H, Valeri CR. The effects of aspirin and hypothermia on platelet function in vivo. Br J Haematol 104:64-68, 1999.

440. Khuri SF, Healey N, MacGregor H, Barnard MR, Szymanski IO, Birjiniuk V, Michelson AD, Gagnon DR, Valeri CR. Comparison of the effects of transfusions of cryopreserved and liquid-preserved platelets on hemostasis and blood loss after cardiopulmonary bypass. J Thorac Cardiovasc Surg. 117:172-184, 1999.

441. O'Neill EM, Rowley J, Hansson-Wicher M, McCarter S, Ragno G, Valeri CR. Effect of 24-hour whole-blood storage on plasma clotting factors. Transfusion 39:488-491, 1999.

442. Lieberthal W, Fuhro R, Freedman JE, Toolan G, Loscalzo J, Valeri CR. O-raffinose cross-linking markedly reduces systemic and renal vasoconstrictor effects of unmodified human hemoglobin. J Pharmac Exp Therap 288:1278-1287, 1999.

443. Barnard MR, MacGregor H, Mercier R, Ragno G, Pivacek LE, Hechtman HB, Michelson AD, Valeri CR. Platelet surface p-selectin, platelet-granulocyte heterotypic aggregates, and plasma soluble p-selectin during plateletpheresis. Transfusion 39:735-741, 1999.

444. Barnard MR, MacGregor H, Ragno G, Pivacek LE, Khuri SF, Michelson AD, Valeri CR. Fresh, liquid-preserved, and cryopreserved platelets: adhesive surface receptors and membrane procoagulant activity. Transfusion 39:880-888, 1999.

445. Valeri CR, Pivacek LE, Crowley JP. Transfusion medicine (letter, comment). NEJM 341:124-127, 1999.

446. Valeri CR, Pivacek LE, Cassidy GP, Ragno G. Posttransfusion survival (24-hour) and hemolysis of previously frozen,

deglycerolized RBCs after storage at 4 C for up to 14 days in sodium chloride alone or sodium chloride supplemented with additive solutions. Transfusion 40:1337-1340, 2000.

447. Valeri CR, Pivacek LE, Cassidy GP, Ragno G. The survival, function, and hemolysis of human RBCs stored at 4 C in additive solution (AS-1, AS-3, or AS-5) for 42 days and then biochemically modified, frozen, thawed, washed, and stored at 4 C in sodium chloride and glucose solution for 24 hours. Transfusion 40:1341-1345, 2000.

448. Valeri CR, Ragno G, Pivacek LE, Cassidy GP, Srey R, Hansson-Wicher M, Leavy ME. An experiment with glycerol-frozen red blood cells stored at −80 C for up to 37 years. Vox Sang 79:168-174, 2000.

449. Ben-Hur E, Chan WS, Yim Z, Zuk MM, Dayal V, Roth N,Heldman E, Lazo A, Valeri CR, Horowitz B. Photochemical decontamination of red blood cell concentrates with the silicon phthalocyanine PC 4 and red light. Dev Biol 102:149-155, 2000.

450. Lieberthal W, Fuhro R, Andry C, Valeri CR. Effects of hemoglobin-based oxygen-carrying solutions in anesthetized rats with acute ischemic renal failure. J Lab Clin Med 135:73-81, 2000.

451. Woodcock SA, Kyriakides C, Wang Y, Austen WG Jr, Moore FD Jr, Valeri CR, Hartwell D, Hechtman HB. Soluble p-selectin moderates complement dependent injury. Shock 14:610-615, 2000.

452. Valeri CR, Ichikura T, Pivacek LE, Giorgio A, Prusty S, Dittmer J. Intravascular circulation and distribution of human 51Cr-DBBF stroma-free hemoglobin, 51Cr-plasma, 51Cr-saline, 59Fe-plasma, and 125I-albumin in the mouse. Artificial Cells, Blood Substitutes, & Immobilization Biotechnology 28:451-475, 2000.

453. Kyriakides C, Woodcock SA, Wang Y, Favuzza J, Austen Jr WG, Kobzik L, Moore FD, Valeri CR, Shepro D, Hechtman HB. Soluble p-selectin moderates complement-dependent reperfusion injury of ischemic skeletal muscle. Am J Physiol Cell Physiol 279:C520-C528, 2000.

454. Valeri CR, Pivacek LE, Cassidy GP, Ragno G. In vitro and in vivo measurements of human RBCs frozen with glycerol and subjected to various storage temperatures before

deglycerolization and storage at 4 C for 3 days. Transfusion 41:401-405, 2001.

455. Valeri CR, Pivacek LE, Cassidy GP, Ragno G. In vitro and in vivo measurements of gamma-radiated, frozen, glycerolized RBCs. Transfusion 41:545-549, 2001.

456. Valeri CR, Ragno G, Pivacek LE, Srey R, Hess JR, Lippert LE, Mettille F, Fahie R, O'Neill EM, Szymanski IO. A multicenter study of in vitro and in vivo values in human RBCs frozen with 40-percent (wt/vol) glycerol and stored after deglycerolization for 15 days at 4 C in AS-3: assessment of RBC processing in the ACP 215. Transfusion 41:933-939, 2001.

457. Valeri CR, Ragno G, Pivacek LE, O'Neill EM. In vivo survival of apheresis RBCs, frozen with 40-percent (wt/vol) glycerol, deglycerolized in the ACP 215, and stored at 4 C in AS-3 for up to 21 days. Transfusion 41:928-932, 2001.

458. Valeri CR, Cassidy G, Pivacek LE, Ragno G, Lieberthal W, Crowley JP, Khuri SF, Loscalzo J. Anemia-induced increase in the bleeding time: implications for treatment of nonsurgical blood loss. Transfusion 41:977-983, 2001.

459. Valeri CR, Pivacek LE, Cassidy GP, Ragno G. 24-hour ^{51}Cr post-transfusion survival, ^{51}Cr life span and haemolysis of red blood cells stored at 4 C for 56 days in AS-3. Vox Sang 80:48-50, 2001.

460. O'Neill EM, Zalewski WM, Eaton LJ, Popovsky MA, Pivacek LE, Ragno G, Valeri CR. Autologous platelet-rich plasma isolated using the Haemonetics Cell Saver 5 and Haemonetics MCS+ for the preparation of platelet gel. Vox Sang 81:172-175, 2001.

461. Valeri CR, Pivacek LE, Ragno G. The quality of RBC stored in CPD/ADSOL (letter). Transfusion 41:1072-73, 2001.

462. Michelson AD, Barnard MR, Krueger LA, Valeri CR, Furman I. Circulating monocyte-platelet aggregates are a more sensitive marker of in vivo platelet activation than platelet surface P-selectin: Studies in baboons, human coronary intervention, and human acute myocardial infarction. Circulation 104(13):1533-7, 2001.

463. Valeri CR, Ragno G, Pivacek LE, Dennis RC, Hechtman HB, Khuri SF. Survival and function of baboon RBCs released from clotted blood and washed before autologous transfusion. Transfusion 41:1384-1389, 2001.

464. Valeri CR, Dennis RC, Ragno G, Pivacek LE, Hechtman HB,

Khuri SF. Survival, function, and hemolysis of shed red blood cells processed as nonwashed blood and washed red blood cells. Ann Thoracic Surg 72:1598-1602, 2001.

465. Battinelli E, Willoughby SR, Foxall T, Valeri CR, Loscalzo J. Induction of platelet formation from megakaryocytoid cells by nitric oxide. Proc Natl Acad Sci USA 98(25):14458-63, 2001.

466. Lieberthal W, Fuhro R, Andry CC, Rennke H, Abernathy VA, Koh JS, Valeri CR, Levine JS. Rapamycin impairs recovery from acute renal failure: role of cell-cycle arrest and apoptosis of tubular cells. Am J Physiol Renal Physiol 281:F693-F706, 2001.

467. Lieberthal W, Fuhro R, Alam H, Rhee P, Szebeni J, Hechtman HB, Favuzza J, Veech RL, Valeri CR. Comparison of the effects of a 50% exchange-transfusion with albumin, hetastarch and modified hemoglobin solutions. Shock 17:61-69, 2002.

468. Purmal A, Valeri CR, Dzik W, Pivacek L, Ragno G, Lazo A, Chapman J. Process for the preparation of pathogen-inactivated RBC concentrates using PEN110 chemistry: Preclinical studies. Transfusion 42:139-145, 2002.

469. Valeri CR, Pivacek LE, Cassidy GP, Ragno G. Volume of RBCs, 24- and 48-hour posttransfusion survivals, and the lifespan of (51)Cr and biotin–X-N-hydroxysuccinimide (NHS)-labeled autologous baboons RBCs: effect of the anticoagulant and blood pH on (51)Cr and biotin-X-NHS elution in vivo. Transfusion 42:343-8, 2002.

470. Valeri CR, MacGregor H, Giorgio A, Ragno G: Circulation and hemostatic function of autologous fresh, liquid preserved, and cryopreserved baboon platelets transfused to correct an aspirin-induced thrombocytopathy. Transfusion 42:1206-16, 2002.

471. Valeri C: Status report on the quality of liquid and frozen red blood cells. Vox Sang 83(1):193-196, 2002.

472. Valeri CR, Giorgio A, MacGregor H, Ragno G. Circulation and distribution of autotransfused fresh, liquid-preserved and cryopreserved baboon platelets. Vox Sang 83:347-351, 2002

473. Valeri CR, Lane J, Srey R, Ragno G. Incidence of breakage of human RBCs frozen with 40 percent wt/vol glycerol using two different methods for storage at -80 C. Transfusion 43 (3):411-414, 2003.

474. Koustova E, Rhee P, Hancock T, Chen H, Inocencio R, Valeri

CR, Alam HB. Ketone and pyruvate Ringer's solutions decrease pulmonary apoptosis in a rat model of severe hemorrhagic shock and resuscitation. Surgery 134(2):267-274, 2003.

475. Valeri CR, Ragno G, Srey R. Restoration of red blood cell volume following 2-unit red blood cell apheresis. Vox Sang 85(2):85-7, 2003.

476. Valeri CR, Srey R, Lane JP, Ragno G. Effect of WBC reduction and storage temperature on PLTs frozen with 6 percent DMSO for as long as 3 years. Transfusion 43(8):1162-7, 2003.

477. Valeri CR, MacGregor H, Giorgio A, Srey R, Ragno G. Comparison of radioisotope methods and a non-radioisotope method to measure the RBC volume and RBC survival in the baboon. Transfusion 43(10): 1366-1373, 2003

478. Valeri CR, Khuri S, Ragno G. Role of Hct in the treatment of thrombocytopenic patients. Letter to the Editor. Transfusion 43:1761-1762, 2003.

479. Valeri CR, Ragno G, Marks PE, Kuter DJ, Rosenberg RD, Stossel TP. Effect of thrombopoietin alone and a combination of cytochalasin B and ethylene and glycol bis (beta-aminoethyl ether) N, N'-tetraacetic acid-AM on the survival and function of autologous baboon platelets stored at 4C for as long as 5 days. Transfusion 44:865-870, 2004.

480. Valeri CR, Srey R, Tilahun D, Ragno G. In vitro effects of polymerized N-acetyl glucosamine (NAG) on the activation of platelets in platelet rich plasma with and without red blood cells. J Trauma (suppl) 57:S22-S25, 2004.

481. Valeri CR, Ragno G, Popovsky MA. Red cell freezing and its impact on the supply chain. Letter to the Editor Transfusion Medicine 14:1-2, 2004.

482. Valeri CR, MacGregor H, Barnard MR, Summaria L, Michelson AD, Ragno G. In vitro testing of fresh and lyophilized reconstituted human and baboon platelets. Transfusion 44:1505-1512, 2004.

483. Valeri CR, Srey R, Tilahun D, Ragno G. In vitro quality of red blood cells frozen with 40% w/v glycerol at -80C for 14 years, deglycerolized with the Haemonetics ACP 215, and stored at 4C in additive solution-1 or additive solution-3 for up to 3-weeks. Transfusion 44:990-995, 2004.

484. Valeri CR, Ragno G: The 24-hour posttransfusion survival of

baboon RBC preserved in CPD/ADSOL (CPD/AS-1) for 49 days. Contemporary Topics in Laboratory Animal Science 44(1):38-40, 2005.

485. Valeri CR, MacGregor H, Barnard MR, Summaria L, Michelson AD, Ragno G. Survival of biotin-X-NHS and 111-in-oxine-labeled autologous fresh and lyophilized reconstituted baboon platelets. Vox Sang 88(2):122-129, 2005.

486. Valeri CR, MacGregor H, Giorgio A, Ragno G: Comparison of radioisotope methods and a non-radioisotope method to measure platelet survival in the baboon. Transfusion and Apheresis Science 32(3):275-81 2005.

487. Valeri CR, MacGregor H, Ragno G. Correlation between in vitro aggregation and thromboxane A2 production in fresh, liquid-preserved, and cryopreserved human platelets: Effects of agonists, pH, and plasma and saline resuspension. Transfusion 45:596-603, 2005.

488. Valeri CR, MacGregor H, Giorgio A, Ragno G. Circulation and distribution of 111in-oxine-labeled autologous baboon platelet aggregates and buffy coat. Transfusion and Apheresis Science 32:139-146, 2005.

489. Valeri CR, Ragno G: Breakage of RBC frozen with 40% W/V glycerol in 800 ml polyvinylchloride (PVC) plastic bags stored in rigid cardboard boxes at -80 C. Letter to the Editor Transfusion 45(5):822-3, 2005.

490. Valeri CR, Ragno G, Khuri S. Freezing human platelets using 6% DMSO with removal of the supernatant solution prior to freezing and storage at -80C without post-thaw processing. Transfusion 45(12):1890-1898, 2005.

491. Valeri CR, Ragno G. Automation of the glycerolization of the RBC using the 275 ml high separation bowl in the Haemonetics ACP215 instrument. Transfusion 45(10):1621-1627, 2005.

492. Valeri CR, Dennis RC, Ragno G, MacGregor H, Menzoian JO, Khuri SF. Limitations of the hematocrit to assess the need for RBC transfusion in hypovolemic anemic patients. Transfusion 46:365-371, 2006.

493. Jaskille A, Koustova E, Rhee P, Britten-Web J, Chen H, Valeri CR, Kirkpatrick JR, Alam HB. Hepatic apoptosis following hemorrhagic shock in rats can be reduced through modifications

of conventional Ringer's solution. J Am Coll Surg 202(1):25-35, 2006.

494. Chan RK, Liu P, Lew DH, Ibrahim SI, Srey R, Valeri CR, Hechtman HB, Orgill DP. Expired liquid preserved platelet releasates retain proliferative activity. J Surg Res Jun 1;126(1):55-58, 2005.

495. Bonegio RGB, Fuhro R, Wang Z, Valeri CR, Andry C, Salant DJ, Lieberthal W. Rapamycin ameliorates proteinuria-associated tubulo-interstitial inflammation and fibrosis in experimental membranous nephropathy. J Am Soc Nephrol 16(7):2063-72, 2005.

496. Valeri CR, Morse DS, Ragno G, Dennis RC. Hemostatic defect in baboons infused non-treated and treated autologous plasma. J Card Surg 21(6):565-571, 2006.

497. Valeri CR, Ragno G. The effect of storage of fresh frozen plasma at -80C for as long as 14 years on plasma clotting proteins. Transfusion 45(11):1829-1830, 2005.

498. Valeri CR, Ragno G, Veech RL. Effects of the resuscitation fluid and the HBOC excipient on the toxicity of the HBOC: Ringer's D, L-Lactate, Ringer's L-Lactate, and Ringer's ketone solutions. Art Cells, Blood Substitutes and Biotechnology 34(6) 601-606, 2006.

499. Bonegio RGB, Fuhro R, Lieberthal W, Ragno G, Valeri CR. A comparison of the acute hemodynamic and delayed effects of 50% exchange transfusion with two different cross-linked hemoglobin based oxygen carrying solutions and pentastarch. Art Cells, Blood Substitutes and Biotechnology 34(2):145-157, 2006.

500. Valeri CR, MacGregor H, Ragno G, Healey N, Fonger J, Khuri SF. Effects of centrifugal and roller pumps on the survival of autologous red blood cells in cardiopulmonary bypass surgery. Perfusion 21:291-6, 2006.

501. Valeri CR, Saleem B, Ragno G. Release of platelet derived growth factors and proliferation of fibroblasts in the releasates from platelets stored in the liquid state at 22C following stimulation with agonists. Transfusion 46:225-229, 2006.

502. Valeri CR, Ragno G. In vitro testing of platelets using the thromboelastogram, platelet function analyzer, and the clot

signature analyzer to predict the bleeding time. Trans Aph Sci 35(1):33-41, 2006.

503. Ayuste EC, Chen H, Koustova E, Rhee P, Ahuja N, Chen Z, Valeri CR, Spaniolas K, Mehrani T, Alam HB. Hepatic and pulmonary apoptosis after hemorrhagic shock in swine can be reduced through modifications of conventional ringer's solution. J Trauma 60(1):52-63, 2006.

504. Valeri CR, Ragno G. Role of RBC and platelet transfusion in the treatment of anemic and thrombocytopenic patients. Letter to the Editor. Transfusion 46:1210-1211, 2006.

505. Valeri CR. "Making sense of the preclinical literature on advanced hemostatic products" Letter to the Editor. J Trauma 61:240-241, 2006.

506. Valeri CR, Ragno G. The survival and function of baboon red blood cells, platelets, and plasma proteins: A review of the experience from 1972 to 2002 at the Naval Blood Research Laboratory, Boston, Massachusetts. Transfusion 46(8): 1-42, 2006.

507. Valeri CR, Ragno G. Cryopreservation of human blood products. Trans Apher Sci 34:271-287, with an editorial on pages 267-269, 2006.

508. Ragno G, Valeri CR: Salvaging of liquid-preserved O-positive and O-negative red blood cells by rejuvenation and freezing. Trans Apher Sci 35:137-143, 2006.

509. Valeri CR, Ragno G. Use of supernatant osmolality and supernatant refraction to assess the glycerol concentration in glycerolized and deglycerolized previously frozen RBC. Trans Aph Sci 36:133-137, 2007.

510. Valeri CR, Ragno G. Platelet radiolabeling procedure. Letter to the Editor Transfusion 47(5):946-7, 2007.

511. Valeri CR, Khuri S, Ragno G. Non-surgical bleeding diathesis in anemic thrombocytopenic patients: Role of temperature, RBC, platelets, and plasma clotting proteins. Transfusion 47:206S-248S, 2007.

512. Valeri CR, Ragno G. The effects of preserved red blood cells on the severe adverse events observed in patients infused with hemoglobin based oxygen carriers. Artificial Cells, Blood Substitutes and Biotechnology 36 (1):3-18, 2008.

513. Valeri CR, Ragno G, Veech RL. Severe adverse events associated

with hemoglobin based oxygen carriers: role of resuscitative fluids and liquid preserved RBC. Trans Apher Sci 39:205-211, 2008.

514. Valeri CR, Ragno G. Prevention of TRALI. Vox Sang 94:81, 2008.
515. Belak M, Valeri CR, Wright DG. Exploring the feasibility of selection of T lymphocyte subsets by whole blood immunoabsorption cytapheresis. Clin Exp Immunol 150(3):477-486, 2007.
516. Valeri CR, Ragno G. Role of nitric oxide in the prevention of severe adverse events associated with blood products. Trans Apher Sci 39:241-245, 2008.
517. Estep T, Bucci E, Farmer M, Greenburg G, Harrington, J, Kim HW, Klein H, Mitchell P, Nemo G, Olsen K, Palmer A, Valeri CR, Winslow R. Basic science focus on blood substitutes: a summary of the NHLBI Division of Blood Diseases and Resources Working Group Workshop, March 1, 2006. Transfusion 48:776-782, 2008.
518. Pietramaggiori G, Yang HJ, Scherer SS, Kaipainen A, Chan RK, Alperovich M, Newalder J, Demcheva M, Vournakis JN, Valeri CR, Hechtman HB, Orgill DP. Effects of poly-N-acetyl glucosamine (pGLcNAc) patch on wound healing in db/db mouse. J Trauma 64(3):803-808, 2008.
519. Fischer TH, Valeri CR, Smith CJ, Scull CM, Merricks EP, Nichols TC, Demcheva M, Vournakis JN. Non-classical processes in surface hemostasis: mechanisms for the poly-N-acetyl glucosamine-induced alteration of red blood cell morphology and surface prothombogenicity. Biomed Mater 3(1):1-9, 2008.
520. Valeri CR, Ragno G. Comments on Lundby et al.'s "testing for recombinant human erythropoietin in urine: problems associated with current anti-doping testing". J Appl Physiol 105:1993, 2008
521. Pietramaggiori G, Scherer SS, Mathews JC, Lancerotto L, Gennaoui A, Ragno G, Valeri CR, Orgill DP. Quiescent platelets stimulate angiogenesis and diabetic wound repair. J Surg Res 160:169-177, 2010.
522. Valeri CR, Ragno G. Massive transfusion in patients with severe traumatic injuries. Letter to the editor. Vox Sang 96:180, 2009.
523. Valeri CR, Ragno G. An approach to prevent the severe adverse events associated with transfusion of FDA-approved blood products. Trans Aph Sci 42:223-233, 2010.

524. Scherer SS, Pietramaggiori G, Matthews J, Perry S, Assmann A, Carothers A, Demcheva M, Muise-Helmericks RC, Seth A, Vournakis JN, Valeri CR, Fischer TH, Hechtman HB, Orgill DP. Poly-N-acetyl glucosamine (pGLcNAc) nano-fibers: a new bioactive material to enhance diabetic wound healing by cell migration and angiogenesis. Ann Surg 250:322-330, 2009.
525. Valeri CR, Ragno G. Therapeutic efficacy of platelet transfusion in patients with acute leukemia. Transfusion 50(11):2504, 2010.
526. Valeri CR, Ragno GM. Prophylactic platelet transfusions. N Engl J Med 362:2141, 2010.
527. Scherer SS, Pietramaggiori G, Matthews JC, Gennaoui A, Demcheva M, Fischer TH, Valeri CR, Orgill DP. Poly-N-acetyl glucosamine fibers induce angiogenesis in ADP inhibitor-treated diabetic mice. J Trauma 71:S183-S186, 2011.
528. Erba P, Adini A, Demcheva M, Valeri CR, Orgill DP. Poly-N-acetyl glucosamine fibers are synergistic with vacuum-assisted closure in augmenting the healing response of diabetic mice. J Trauma 71:S187-S193, 2011.
529. Fischer TH, Hays WE, Valeri CR. Poly-N-acetyl glucosamine fibers accelerate hemostasis in patients treated with antiplatelet drugs. J Trauma 71:S176-S182, 2011.
530. Valeri CR, Vournakis JN. mRDH bandage for surgery and trauma: data summary and comparative review. J Trauma 71:S162-S166, 2011.
531. Valeri CR. The experience of Stinner, et al. Mil Med 176:i, 2011.
532. Valeri CR, Ragno G. The rheologic properties of leukoreduced red blood cells. Transfusion 50:2506, 2011.
533. Veech RL, Valeri CR, VanItallie TB. The mitochondrial permeability transition pore provides a key to the diagnosis and treatment of traumatic brain injury. IUBMB Life 64:203-207, 2012.
534. Valeri CR, Veech RL. The unrecognized effects of the volume and composition of the resuscitation fluid used during the administration of blood products. Trans Aph Sci 46:121-123, 2012.
535. Veech RL, Valeri CR. The hyperglycemia of trauma (submitted for publication).

B. BOOKS

1. Preservation of Red Blood Cells, Ed. H Chaplin, Jr, ER Jaffe, D Lenfant, and CR Valeri, National Academy of Sciences, Washington, DC, 1973.
2. Platelet Preservation and Transfusion, Ed., CR Hogman, HW Krijnen, and CR Valeri, Proc Int Soc Blood Transf Meeting, Amsterdam, 1974.
3. Valeri CR. Blood Banking and the Use of Frozen Blood Products, Chemical Rubber Company, Boca Raton, Florida, 1976.
4. Valeri CR and Altschule MD. Hypovolemic Anemia of Trauma: The Missing Blood Syndrome, Chemical Rubber Company, Boca Raton, Florida, 1981.

C. MONOGRAPHS

1. Valeri CR. Recent advances in techniques for freezing red cells. CRC Crit Rev Clin Lab Sci 1:381-425, 1970.
2. Valeri CR. The viability and function of preserved red cells. Presented at the XXIII John G Gibson, II, Lecture, April 2, 1970.
3. Valeri CR. The current state of platelet and granulocyte cryopreservation. CRC Crit Rev Clin Lab Sci 14:21-74, 1981.
4. Valeri CR. Use of rejuvenation solutions in red blood cell preservation. CRC Crit Rev Clin Lab Sci 17:299-374, 1982.
5. Valeri CR, Ragno G: The survival and function of baboon red blood cells, platelets and plasma proteins: a review of the experience from 1972 to 2002 at the Naval Blood Research Laboratory, Boston, Massachusetts. Transfusion 46(8): 1-42, 2006
6. Valeri CR, Khuri S, Ragno G. Non-surgical bleeding diathesis in anemic thrombocytopenic patients: Role of temperature, RBC, platelets, and plasma clotting proteins. Transfusion 47:206S-248S, 2007.

www.ingramcontent.com/pod-product-compliance
Lightning Source LLC
Chambersburg PA
CBHW031815170526
45157CB00001B/59
9781483402765